# Nuclear Physics

# Nuclear Physics

*A Course Given by* ENRICO FERMI
*at the University of Chicago. Notes Compiled by*
*Jay Orear, A. H. Rosenfeld, and R. A. Schluter*

Revised Edition

THE UNIVERSITY OF CHICAGO PRESS
CHICAGO & LONDON

The University of Chicago Press, Chicago 60637
The University of Chicago Press, Ltd., London


Published 1949. Revised Edition 1950. Paperback edition 1974
Printed in the United States of America

ISBN: 0-226-24365-6

♾ The paper used in this publication meets the minimum requirements of the American National Standard for Information Sciences—Permanence of Paper for Printed Library Materials, ANSI Z39.48-1992.

# P R E F A C E

This material is a reproduction, with some amplification, of our notes on lectures in Physics 262-3: Nuclear Physics, given by Enrico Fermi, Jan.-June 1949. The course covered a large number of topics, both experimental and theoretical.

The lectures presupposed a familiarity with physics generally acquired by a student who has completed one course in quantum mechanics (this to include a discussion of the Pauli spin operators and of perturbation theory, both time-independent and time-dependent). We shall make some use of elementary concepts of such topics as statistical mechanics and electrodynamics, but we give references, and a reader could probably pick up the necessary ideas as he goes along; or he could omit a few sections.

Dr. Fermi has not read this material; he is not responsible for errors. We have made some attempt to confine the classroom presentation to the text proper, putting much of our amplifications in footnotes, appendices, and in the solutions to the problems. Most of the problems were assigned in class, but the solutions are not due to Dr. Fermi.

The literature references in the text apply to the list on page 239. At the end of the book there is also a summary of the notation and a list of pertinent constants, values, and relationships.

We would very much appreciate your calling errors to our attention; we would like to hear any suggestions and comments that you may have.

May we thank warmly all those who have helped us to prepare these notes.

Jay Orear
A.H. Rosenfeld
R.A. Schluter

January, 1950

This second printing of these notes differs from the first in that corrections and minor revisions have been made on approximately 70 pages in the first nine chapters, and major revisions have been made in the chapter on cosmic rays. We are grateful to the many people who have given suggestions and corrections; in particular, we are indebted to Prof. Marcel Schein for his suggestions and generous aid in revision of Chapter X.

JO, AHR, RAS

September 1950

An attempt to bring this second printing of the revised edition up to date has been made by adding new footnotes and two pages (237,238) of recent developments. Corrections and minor revisions have been made on approximately 40 pages.

JO, AHR, RAS

July 1951

Minor revisions have been made, mainly in bringing some of the references up to date.

JO, AHR, RAS

C O N T E N T S

CHAPTER VIII. NUCLEAR REACTIONS

# CHAPTER I

# PROPERTIES OF NUCLEI*

## A. ISOTOPES, CHARTS & TABLES

All nuclei are composed of Z protons + N neutrons. The mass number, A, is given by A = Z + N. Examine an isotope chart, such as at the back of this book, and notice that the stable elements lie along a curve starting out with N/Z = 1, ending with N/Z = 1.6. Nuclei with common Z are called isotopes, those with common A are called isobars, and those with common N, isotones.

Radioactivity

Nuclei richer in N than their stable isobar(s) are $\beta^-$-active, i.e. $_Z(\ )^A \rightarrow {}_{Z+1}(\ )^A + \beta^-$

where $\beta^-$ represents an electron emitted from the nucleus. Nuclei relatively poor in N either emit $\beta^+$ particles or else capture orbital electrons (Chap. IV).

Some nuclei, for example: $_{29}Cu^{64}$, $_{33}As^{74}$, can decay either way ($\beta^+$ or $\beta^-$).

Nuclei near the end of the periodic table also tend to emit particles or to break in two (fission). This explains why elements with Z > 94 are not found in nature. See section C3, page 9.

Radioactive nuclei emit particles according to the statistical law:

$$dn = -\lambda n\, dt$$

which, integrated, gives $n(t) = n(0)\, e^{-\lambda t}$ I.1

where n means number of nuclei remaining after time t, and $\lambda$ is the probability of decay per unit time per atom. $\lambda$ is called the decay constant.

The mean life, $\tau$, is easily shown $= 1/\lambda$. However, in tables, it is customary to give the half life, $T = (\ln 2)\tau = 0.693\tau$. After time $\tau$, the number of nuclei present is 1/e times the original number; after time T, the fraction is 1/2**

---

* Much of this material is covered in Chap. I, Goodman.

** If there are two competing processes, so that a particle, $n_o$, may emit one sort of particle, $p_1$, according to

$$dn_1 = -\lambda_1 n_o\, dt$$

or another type of particle, $p_2$,

$$dn_2 = -\lambda_2 n_o\, dt$$

then

$$dn_o = dn_1 + dn_2 = -(\lambda_1 + \lambda_2) n_o\, dt .$$

Thus mean lives combine as the sum of reciprocals.

Atomic Masses

There are two units of atomic mass. In the chemical system, the "natural" isotopic mixture of oxygen is assigned mass 16.0; on the physical scale, $_8O^{16}$ is given the mass 16.0 amu (atomic mass units).

$$\frac{\text{Mass on Physical Scale}}{\text{Mass on Chemical Scale}} = 1.000272$$

Isotope charts generally use the physical scale. The masses listed are not for nuclei, but for neutral atoms. To get the nuclear mass, subtract Z x m, the mass of the Z atomic electrons (whose binding energy can be neglected - see top footnote, p. 3).

One amu, that is, the mass in grams of a fictitious atom $M_1$, of mass 1.000, is given by the reciprocal of Avogadro's number ($N_o = 6.023 \times 10^{23}$).

| Particle | | Mass | Rest Energy | |
|---|---|---|---|---|
| $M_1$ | | $1.6603 \times 10^{-24}$g | 931 | Mev |
| Electron | ($1/1838\ M_1$) | $= 0.9107 \times 10^{-27}$g | 0.51 | Mev |

Another useful constant: 1 Mev = $1.601 \times 10^{-6}$ erg. A table giving other quantities may be found on page 240.

The following masses are of particular interest:

| | | | |
|---|---|---|---|
| Electron | $_{-1}e^0$ | 0.000548 amu | |
| Proton | $_1P^1$ | 1.00759 | |
| Hydrogen | $_1H^1$ | 1.00813 | |
| Neutron | $_0N^1$ | 1.00898 | $M_N - M_H \approx 0.79$ Mev |
| Deuterium | $_1H^2$ | 2.01471 | |
| Helium | $_2He^4$ | 4.00390 | |

By Einstein's mass-energy relationship, $E = mc^2$, an isolated system appears to decrease in mass when its energy decreases. Thus the total energy and the mass of two attracting particles decrease as they approach one another, losing their excess energy by radiation.

## B. PACKING FRACTION AND BINDING ENERGY

The packing fraction, f, is defined by

$$f = \frac{M(A,Z) - A}{A}$$

where M(A,Z) is the mass, in amu, of the nucleus of mass number A, charge Z.

Experiment shows that f is very small throughout the periodic table (FIG. I.1).

The numerator, M-A, is called the Mass Defect*.

When a nucleus disintegrates, the BINDING ENERGY with which the daughter particles were bound is defined as the sum of the resultant masses ($\Sigma M_f$) minus the initial nuclear mass; i.e.

*Some authors define the mass defect as $-TBE/c^2$, where TBE is the Total Binding Energy defined on p. 3.

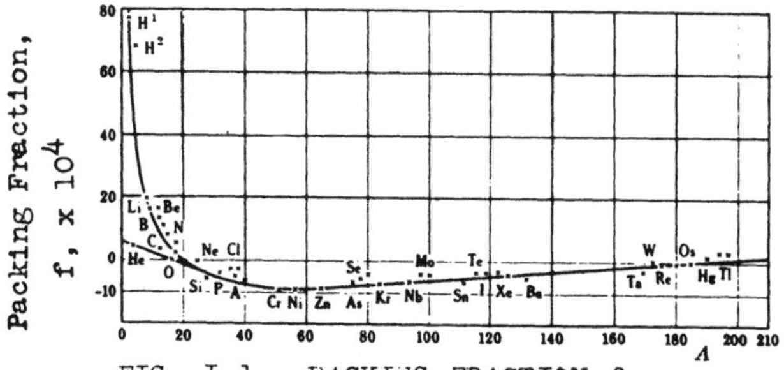


FIG. I.1 PACKING FRACTION f

---

$$BE = \Sigma M_f - M_i$$

Total Binding Energy is defined as that amount of work we would have to supply in order completely to dissociate a nucleus into its component nucleons. We have been expressing mass in terms of energy units and vice versa. This will be done frequently.

Average Binding Energy per Nucleon

As shown in Fig. I.1 above, the mass of any nucleus is very close to **A**. Since nuclei contain about the same number of N and P, the sum of the constituent masses is about 1.0085**A**. Thus there is a total BE of about (0.0085 x 931**A**) Mev, or, dividing by **A**, about 8 Mev per nucleon. The $\alpha$ particle has a BE of about 7 Mev per nucleon**. See Fig. I.2, page 4.

The BE of any process (i.e. of the particles emitted in the process) must be negative for the process to proceed spontaneously.

---

*In calculating BE's from an atomic mass table, it is only necessary to correct for the orbital electrons in one case, namely, $\beta^+$ emission. For the other cases, the correction is automatic. Thus suppose we wished to calculate BE($\beta^-$) for the reaction, $N \rightarrow P + \beta^-$. Following our rule, we would write $BE = M(P) + M(\beta^-) - M(N)$. But, by coincidence, if you wish, this is just*** $BE = M(H \text{ atom}) - M(N)$.

Now, any $\beta^-$ process, $_Z(\ )^A \rightarrow {}_{Z+1}(\ )^A + \beta^-$, can arbitrarily be written as the reaction, $N \rightarrow P + \beta$, considering the inert nucleus (A - 1, Z) as going along unchanged. Therefore $BE(\beta^-) = M(A,Z+1) - M(A,Z)$.

But for $P \rightarrow N + \beta^+$ we must set

$$BE(\beta^+) = M(N) + M(\beta) - M(P)$$
$$= M(N) + M(\beta) - M(_1H^1) + m$$

or

$$BE(\beta^+) = M(A,Z) - M(A,Z+1) + 2m$$

**Nuclear binding energies are much larger than those of orbital electrons. According to the Fermi-Thomas statistical model of the atom, the total BE of all the electrons is 1.55 Z Rydberg.

***Neglecting the 13.5 ev BE of the orbital electron.

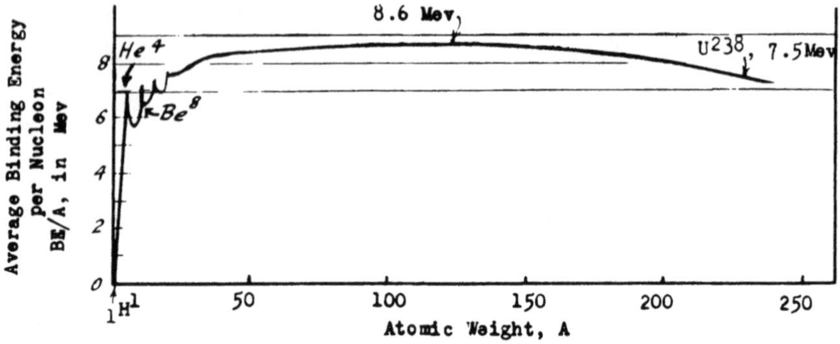


FIG. I.2: AVERAGE BINDING ENERGY PER NUCLEON

Experimental Mass Determinations

The most valuable tool is the mass spectrograph. This will not be discussed. Further data may be obtained from momenta of particles taking part in nuclear reactions. M(neutron) may be obtained from the photodisintegration threshold for deuterium.

Problem. Consider the reaction

$$_0n^1 + {}_5B^{10} \rightarrow {}_3Li^7 + {}_2He^4 + Q$$

Q is defined as the kinetic energy (T) of the resultant, minus the T of the initial, particles. In other words, Q is the exothermic heat of reaction.

Assume that thermal neutrons ($T \rightarrow 0$), react with a fixed B target. Calculate the velocities of the products. In the reverse reaction, where $\alpha$ particles are shot at a fixed Li target, what is the reaction threshold energy of the $\alpha$'s?

Solution. From mass tables, Q = 0.00304 amu = 2.83 Mev. Next show that the velocities involved are non-relativistic.

Relativity. For this problem, and for future reference, we set down some relations from special relativity. For a free particle, moving with velocity v,

$$\beta \equiv \frac{v}{c} \leq 1 \; ; \qquad \gamma \equiv \frac{1}{\sqrt{1-\beta^2}} \geq 1$$

Throughout this text we shall use M as the rest-mass of a general particle, $M\gamma$ for the relativistic mass. m is reserved for the electron.

$$\left.\begin{aligned} &\text{Then, the momentum} && \underline{p} = M\gamma\underline{v} \\ &\text{force} && \underline{F} = \dot{\underline{p}} \\ &\text{energy} && W = Mc^2 + T \quad (T \equiv \text{kinetic energy}) \\ &&& \quad = M\gamma c^2 \\ &&& \quad = \sqrt{M^2 c^4 + p^2 c^2} \end{aligned}\right\} \quad \text{I.2}$$

W does not include interaction energy.

Expand $$T = Mc^2(\gamma - 1) = Mc^2(\tfrac{1}{2}\beta^2 + \tfrac{3}{8}\beta^4 + \ldots) \qquad \text{(A)}$$

$$= \frac{Mv^2}{2}(1 + \tfrac{3}{4}\beta^2 + \ldots) \qquad \text{(B)}$$

At $T = \frac{Mc^2}{10}$, (A) says $\frac{\beta^2}{2} \approx 0.1, \quad \beta \approx 0.45$

(B) says $T \approx T_{classical}\ (1 + 0.15)$

$\underline{p} \approx \underline{p}_{classical}\ (1 + 0.1)$

Thus for kinetic energies below one-tenth rest energy, the relative error in the classical expressions is on the order of the fraction, $T/Mc^2$. Hence the problem at hand may be considered strictly classical.

Reverse Threshold.

In a two-body problem, the only part of the KE that can enter into reactions is the KE of the particles relative to the center of mass. This is easily shown to be

$$\boxed{T_{rel} = \tfrac{1}{2}\mu \dot{r}_{12}^{\,2}}$$

where $\mu \equiv \frac{M_1 M_2}{M_1 + M_2}$ = reduced mass

$\underline{r}_{12}$ = relative position, $\underline{r}_1 - \underline{r}_2$

Thus $T_{rel}$ is just the KE of the "reduced mass particle".

Now, if one particle ("target") is fixed in the lab coordinates, the total KE of the system is the KE of the bombarding particle

$$T_{total} = \tfrac{1}{2} M_1 \dot{r}_1^2 = \tfrac{1}{2} M_1 \dot{r}_{12}^2$$

where $M_1$ stands for the mass of the bombarding particle.

Thus, for fixed target,

$$T_{rel} = \frac{\mu}{M_1} T_{total}$$

i.e.

$$T_{rel} = \frac{M_{target}}{M_{total}} T_{total}$$ where $T_{total}$ is the lab system KE of bombarding particle

For the problem at hand, $T_{rel}$ must equal **Q**.
$T_{threshold} = 11/7\ Q = 4.45$ Mev.

## C. LIQUID DROP MODEL

Experiment (scattering, quadrupole moment, etc.) shows that nuclei are roughly spherical, with volume directly proportional to **A**, so that a nucleus is analogous to a drop of incompressible fluid of very high density ($10^{14}$g cm$^{-3}$). We shall use

$$R = 1.5 \; A^{1/3} \times 10^{-13} \text{ cm}$$

Equation I.4

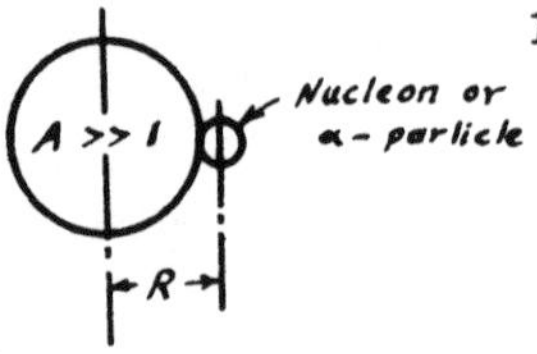


where R is defined by the sketch and is derived from the theory of α mean lives (Ch. III).

## 1. Semi-Empirical Atomic Mass Formula

We shall use this concept, along with other classical ideas (surface tension, electrostatic repulsion) to set up a semi-empirical formula for the mass of any atom, $M(A,Z)$. This formula can then be used to predict the stability of nuclei against particle emission, also the energy release and stability of nuclei for fission (Ch. VIII, Sec. J).

Naturally the first term will be the masses of the constituent particles 1.00813 Z + 1.00898 N,

$$M_0 = 1.00813\, Z + 1.00898\, (A-Z).$$

The first correction term will be the bulk "heat of condensation," due to short-range nuclear forces. On page 3 we mentioned that the BE per nucleon was about 8 Mev. Thus we suspect that this correction will be of order of magnitude 10 Mev = .01amu.

$$M_1 = -a_1 A$$

In assigning the same energy of attraction to all nucleons, we have actually over-corrected, since the surface nucleons will be attracted from only one side. We therefore introduce a surface tension correction to the large correction, $M_1$, which is proportional to the surface of the drop.

$$M_2 = +a_2 A^{2/3}$$

We next notice that stable nuclei tend to form themselves of N-P pairs. We add a positive mass correction for the number of unpaired nucleons.

$$M_3 = a_3 \frac{(A/2 - Z)^2}{A}$$

This form is derived in appendix I.3, p.22. See also Fig. I.3 and discussion, p. 8,9.

Next we add a positive term for electrostatic repulsion. The potential energy of a uniformly charged sphere is

$$U = \frac{3}{5} \frac{(Ze)^2}{R} \text{ erg} = \frac{1}{M_1 c^2} \frac{3}{5} \frac{(Ze)^2}{R} \text{amu}$$

so, inserting $R = 1.5 \times 10^{-13} A^{1/3}$ ,

$$M_4 = 0.000627 \frac{Z^2}{A^{1/3}}$$

The final correction term will be called $\delta$. It depends upon the stability of nuclei with respect to whether the number of Z and N is even or odd*.

---

*Empirical data on the stability of nuclei give the following table, where e = even, o = odd. We therefore construct a correction function $\delta(A,Z)$ as shown.

(footnote continued at bottom of next page)

Let us now group all terms and proceed to fit to experimental values $\delta(A,Z)$ and the three constants $a_i$:

$$M(A,Z) = 1.00898\ A - 0.00085\ Z - a_1\ A + a_2 A^{2/3} + a_3 \frac{(A/2 - Z)^2}{A} + 0.000627\ \frac{Z^2}{A^{1/3}} + \delta(A,Z) \qquad \text{I.5}$$

This formula must fit three criteria:

1. It must give the correct A vs. Z curve for the stable elements. See below - discussion of evaluation of $a_3$.
2. It must give the mass of the odd-A elements on this curve. (For odd-A, $\delta = 0$; $\delta$ can then be adjusted to give the even-A masses and $\beta$-energies (see discussion, pp 8,9)).
3. M(A,Z) plotted against Z is a parabola. The third criterion is that its minimum must fall at the stable elements. Thus we can solve for $a_3$ directly by setting

$$\frac{\partial M}{\partial Z} = 0 = -0.00085 - 2a_3 \frac{(\frac{A}{2} - Z)}{A} + 0.000627\ \frac{2Z}{A^{1/3}} \qquad \text{I.6}$$

Comparison with known stable elements gives

$$a_3 = 0.083$$

Putting $a_3$ back into I.6, we get the A vs. Z curve of stable elements mentioned in criterion 1.

$$Z = \frac{A}{1.98 + 0.015\ A^{2/3}} \qquad \text{I.7}$$

showing that A/Z starts out at 2 and increases because of the term in $A^{2/3}$ which comes from the electrostatic repulsion term, $M_4$.

Equation I.7 predicts correctly the stable elements from 1 to 92.

Putting $a_3$ and I.7 into I.5, above, we get an equation for M(A only) in the two unknowns, $a_1$ and $a_2$. From the mass data for the stable elements (step 2.) we get as a best fit

$$a_1 = 0.01507; \qquad a_2 = 0.014$$

Equation I.5 for the mass of an atom, A, Z, then gives

$$M(A,Z) = 0.99391\ A - 0.00085\ Z + 0.014\ A^{2/3} + 0.083\ \frac{(\frac{A}{2} - Z)^2}{A} + 0.000627\ \frac{Z^2}{A^{1/3}} + \delta(A,Z) \qquad \text{I.8}$$

*Footnote continued from page 6-

| A | Z | N | Comment | $\delta(A,Z)$ |
|---|---|---|---|---|
| e | e | e | Most stable | $\delta(e,e) = -f(A)$ |
| o | e | o | Moderately stable; roughly | $\delta(o,e) = 0$ |
| o | o | e | equal numbers observed | $\delta(o,o) = 0$ |
| e | o | o | Least stable | $\delta(e,o) = +f(A)$ |

The theoretical justification for the behaviour of $\delta$ is as follows. Use the Fermi-gas model of the nucleus (Ch. VIII, Sec. H.). Because the nucleons have spin 1/2 (hence can have spin up or down), each momentum state of a P or an N is, roughly, twofold degenerate. But all our mass terms have assumed that M(A,Z) varies smoothly every time N or Z changes. $\delta$ corrects for this. See Bethe A, para. 10.

and, by step 3.

$$\begin{aligned} \delta(\text{A-even, Z-even}) &= -0.036\ A^{-3/4} \\ \delta(\text{A-odd, Z-anything}) &= 0 \\ \delta(\text{A-even, Z-odd}) &= +0.036\ A^{-3/4} \end{aligned} \qquad \text{I.9}$$

Equation I.8 describes quite well the atomic masses for $A > 15$. It can be used to predict the most stable Z. Its numerical value for all atoms from $_3(\ )^{15}$, to $_{97}(\ )^{249}$ has been published by N. Metropolis, Inst. for Nuclear Studies, U. of Chgo., 1948.

The accuracy of I.8 is indicated by the following data:

| Nucleus | $_8O^{16}$ | $_{24}Cr^{52}$ | $_{42}Mo^{98}$ | $_{79}Au^{197}$ | $_{92}U^{238}$ |
|---|---|---|---|---|---|
| Experimental Mass | 16.00000 | 51.956 | 97.943 | 197.04 | 238.12 |
| Mass Formula | 15.99615 | 51.959 | 97.946 | 197.04 | 238.12 |

For an empirical correction term for the heavy isotopes, based on decay measurements, see Stern, Rev. Mod. Phys. 21, 316 (1949).

For a review article, see Feenberg, Rev. Mod. Phys. 19, 239 (1947)

## 2. Isobaric Behavior

For some particular odd-A, I.8 is given in Fig. 3, below. Note that the form, a parabola, is governed by the "unpaired" term, $M_3$, but that Z-minimum need not be an integer.

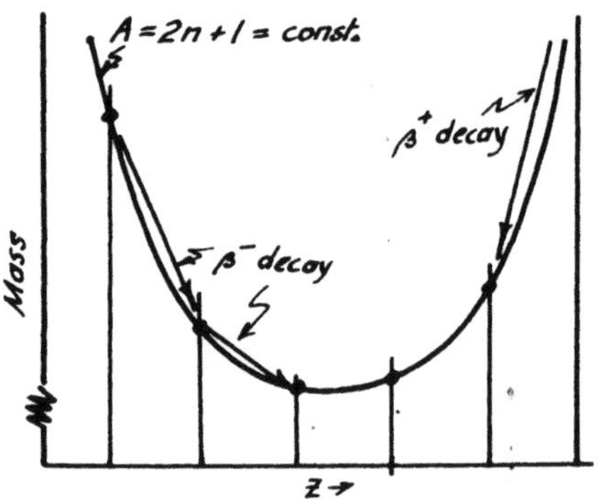


FIG. I.3, Isobars, ODD-A

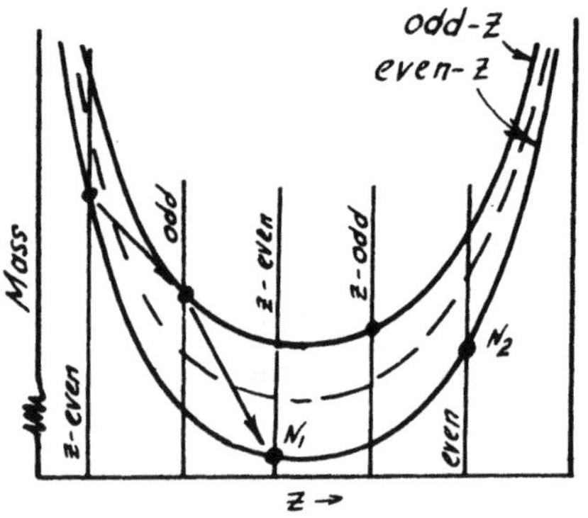


FIG. I.4, EVEN-A

Arrows are possible decay chains, mentioned below.

In Fig. I.4, the dashed line represents all the terms of I.8 except $\delta$. The solid lines (representing some series of even-A isobars) are raised or lowered by $\delta$. This effect has been exaggerated. The dots are possible nuclei. Note that, in general, there is only one stable odd-A nucleus, but that, in the case of even-A, there may be two, or even three. Thus in Fig. I.4, $N_1$ and $N_2$ would both be stable.

Fig. I.4 shows that even-A, odd-Z nuclei are unstable, and should not be found in nature. Actually, there are a few, but none heavier than $_7N^{14}$, except for $_{19}K^{40}$ (which is radioactive) and $_{23}V^{50}$.

We shall show in Ch. IV that $\beta$ lifetimes increase as $\beta$ energies decrease. Using this information, we see that the theory is corroborated by the following chains. Thus, the odd-**A** (139) $\beta^-$-active chain produced by the fission of $_{92}U^{239}$

$$\text{Xe} \xrightarrow{\text{41 S}} \text{Cs} \xrightarrow{\text{7 M}} \text{Ba} \xrightarrow{\text{85 M}} \text{La (stable)}$$

shows steadily increasing lifetimes. The following even-**A** chain (140) shows, in addition, an alternation of lifetimes.

$$\underset{\delta^-}{\text{Xe}} \xrightarrow{\text{16 S}} \underset{\delta^+}{\text{Cs}} \xrightarrow{\text{40 S}} \underset{\delta^-}{\text{Ba}} \xrightarrow{\text{12.8 d}} \underset{\delta^+}{\text{La}} \xrightarrow{\text{40 h}} \underset{\delta^-}{\text{Ce}} \text{ (stable)}$$

Every second transition is from a $\delta_+$ curve to a $\delta_-$ curve, with a short life.

## 3. Alpha Emission

We can use I.8, p. 7, to compute the BE of neutrons in a nucleus:

$$\text{BE(N)} = \text{M}(\mathbf{A}-1,\text{Z}) + 1.00898 - \text{M}(\mathbf{A},\text{Z})$$

This always turns out positive for the stable elements, showing that these do not tend to emit neutrons spontaneously. However,

$$\text{BE}(\alpha) = \text{M}(\mathbf{A}-4,\text{Z}-2) + 4.00390 - \text{M}(\mathbf{A},\text{Z}) \qquad \text{I.10}$$

goes negative in the middle of the periodic table, long before the "natural $\alpha$-emitters" are reached. The intervening elements are stable against $\alpha$-decay only because the $\alpha$-energies are so small that the lifetimes are prohibitively long. (See problem on p. 59, Ch. III.)

The periodic table ends in the region Z = 90-100, because of the increasingly negative value of the BE for $\alpha$-emission and fission.

## 4. Periodic Shell Structure

In addition to the general behavior predicted by the liquid drop idea, nuclei are observed to have periodic variations in BE which are not predicted by I.8, p. 7. These variations have nothing to do with the periodic atomic (chemical) properties.

The lighter nuclei have total BE maxima at **A** = 4, 8, 12, 16, etc., showing the saturated character of sub-units containing 2N + 2P.

A larger shell structure is also observed. See Ch. VIII, Sec. K, p. 167.

## D. SPIN AND MAGNETIC MOMENT*

The total angular momentum of the <u>nucleus</u> is denoted by <u>I</u> instead of <u>J</u>, the symbol usually applied to atoms.

---

* For more complete discussion, see Bethe, **A**, para. 4 and 5, and also Ch. VIII; Bethe, D, Ch. V. For experimental references, see Cork, "Radioactivity and Nuclear Physics", Van Nostrand, 1946, p. 131.

$$\underline{I} = \underline{L} + \underline{S}$$

In the literature, $\underline{I}$ is often referred to as the "spin" of the nucleus. In this chapter we shall use "spin" for the inherent (non-orbital) part of the angular momentum.

As a general consequence of quantum mechanics, an isolated system (nucleus) has its total angular momentum, $\underline{I}$, quantized to integral and half-integral values, in units of $\hbar$. The half-integral values arise from the inherent spin of the nucleons.

For an even number of nucleons, we expect $I = 0, 1, \ldots$, whereas for A-odd, we expect $I = 1/2,\ 1\ 1/2, \ldots$. No exceptions to this rule have been found.

In general, nuclei have a magnetic moment, $\underline{\mu}_n$, associated with $\underline{I}$.

We shall discuss, briefly, four methods of determining $\underline{I}$, $\underline{\mu}_n$.

1. Hyperfine structure of optical spectra

This arises either from nuclear magnetic moments or (and we ignore these) nuclear quadupole moments or isotopic mass differences.

Let us calculate the interaction energy, U, between the magnetic moment, $\mu_n$, of the nucleus and the magnetic field, $\underline{\mathcal{H}}_e$, of the atomic electrons.

This is easily done by setting

$$\overline{U} = -\underline{\mu}_n \cdot \overline{\underline{\mathcal{H}}_e} \qquad \text{I.11}$$

where $\overline{\underline{\mathcal{H}}_e}$ is evaluated at the nucleus and we consider time averages.

We must now express $\underline{\mu}_n$ and $\underline{\mathcal{H}}_e$ in terms of the nuclear and electronic angular momenta, $\underline{I}$ and $\underline{J}$. The nucleus and electrons are assumed weakly coupled, so that $\underline{I}$ and $\underline{J}$ each stay fixed in space (Fig. I.6).

Next we assume that $\underline{\mu}_n = \mu_n \dfrac{\underline{I}}{I}$, I.12

where we shall try to solve for $\mu_n$, and that

$$\overline{\underline{\mathcal{H}}_e} = -A' \frac{\underline{J}}{J} \qquad \text{I.13}$$

where A′ is a constant that must be calculated for each atomic configuration (Mattauch, p. 31). A plausibility argument for the assumption that $\underline{\mathcal{H}}_e$ is parallel to $\underline{J}$ is given in Fig. I.5. Use the vector model, and assume L-S coupling. $\underline{\mu}_e$ rotates rapidly around $\underline{J}$, so that its component perp. to $\underline{J}$, $\mu_\perp$, is, on time average, zero. Thus, $\overline{\underline{\mu}_e} = \overline{\underline{\mu}_\parallel}$. Rather naturally, $\underline{\mathcal{H}}_e$ is parallel to $\underline{\mu}_e$, in all cases: $\underline{J} = \underline{L}$, $\underline{J} = \underline{S}$, $\underline{J} = \underline{L} + \underline{S}$.

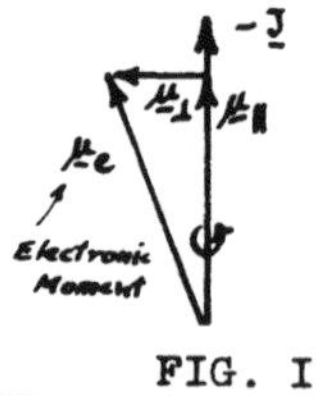

FIG. I.5
Electronic Vectors only.

Using I.12 and I.13, I.11 becomes

$$\overline{U} = +\mu_n A' \frac{\underline{I} \cdot \underline{J}}{IJ} \qquad \text{I.14}$$

Next, see Fig. I.6, we define the total atomic angular momentum, $\underline{F}$, and express $\underline{I}\cdot\underline{J}$ in terms of the quantization condition,

$$|\underline{F}| = |\underline{I} + \underline{J}| = I+J,\ I+J-1,\ I+J-2\ \ldots\ |I-J| \qquad \text{I.15}$$

classically, the law of cosines gives

$$F^2 = I^2 + J^2 + 2\underline{I}\cdot\underline{J}$$

or

$$\underline{I}\cdot\underline{J} = \tfrac{1}{2}(F^2 - I^2 - J^2)$$

To get quantitative, quantum-mechanical results from the vector model, we must always replace the squares of angular momenta, $J^2$, $L^2$, $S^2$, by $J(J+1)$, etc.* Hence I.14 becomes

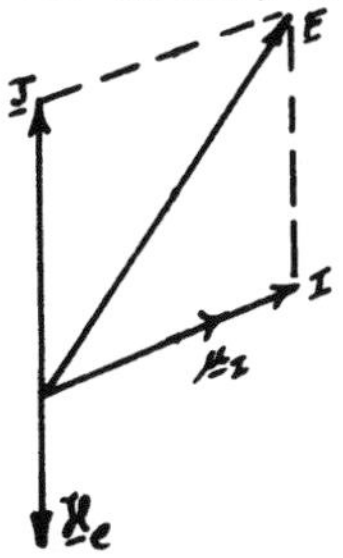


FIG. I.6
Whole Atom

$$\overline{U} = \mu_n A' \frac{F(F+1) - I(I+1) - J(J+1)}{2\sqrt{I(I+1)\,J(J+1)}} \qquad \text{I.16}$$

where, as indicated in I.15, F can take on a multiplicity, M, of values

$$M \text{ smaller of} \begin{cases} 2I + 1 \\ 2J + 1 \end{cases}$$

Different values of F give the interval rule for hyperfine multiplet structure.

Notice that merely by counting the <u>maximum</u> M in the spectrum of an isotope, I can be determined. There will be, of course, some lines with an M limited by a low J, but there will always be some lines given by transitions to or from a J greater than I.

<u>Experiment</u> shows that for
**Even-A** nuclei, I is integral ( 0,1,2, ...)
**Odd -A** nuclei, I is half integral ( $\frac{1}{2}, \frac{3}{2}, \frac{5}{2}$, ...)

as predicted on page 10. Note that we would not have this result if the nucleus were composed of protons and electrons rather than P + N.

Apart from Hydrogen, unfortunately, we do not know enough about atomic wave functions to calculate **A**' to within better than 10%, even with the help of empirical fine-structure data.*** The values of $\mu_n$ from hyperfine structure are thus uncertain to 10%.

<u>2. Alternation of Intensity of Band Spectra</u> ...**

... of diatomic molecules containing identical nuclei. This will give I, but not $\mu_n$.

*This does not apply to other vectors, like $\underline{\mu}$. See appendix, p. 19ff

** For more discussion, see Mayer and Mayer, "Statistical Mechanics", p 172ff.

***Do not confuse this with <u>hyper</u>-fine-structure data, which is what we use to obtain differences of U (I.16).

The alternation depends upon the fact that the state function for the molecule with identical nuclei,

$$\Psi(x_1, x_2)$$

must be either symmetric or antisymmetric with respect to exchange of $x_1$ and $x_2$. x stands for all the coordinates, including the spin, of a nucleus.

Experiment shows that all even-**A** nuclei obey Bose-Einstein statistics, and all odd-**A**, Fermi-Dirac statistics, as is to be expected if protons and neutrons individually obey Fermi statistics.

For even-**A** $\quad \Psi(x_1,x_2) = +\Psi(x_2,x_1)$

For odd -**A** $\quad \Psi(x_1,x_2) = -\Psi(x_2,x_1)$

Now, $\Psi$ may be written *

$$\Psi = \Psi_{elect}\ \phi_v(r_{12})_{vibr}\ \rho_J(\theta,\phi)_{rot}\ \sigma_{I\ nuclear}$$

$\Psi_{el}$ defines the position of the electrons. For identical nuclei it is usually symmetric.
$\phi_v$ is symm. because it depends only on a separation distance.
$\rho_J$ is symm. for even J, anti-symm. for odd.
$\sigma_I$ may be either symm. or anti-symm. For I = 0, $\sigma$ is symm. Moreover (Bethe, D, p. 18), the statistical wt. of symm. relative to anti-symm. $\sigma$ states is (I+1)/I. The important thing is that the symmetry of $\sigma_I$ determines the symmetry of $\rho_J$, i.e. allows only certain values of J. The energy of a level depends upon J.

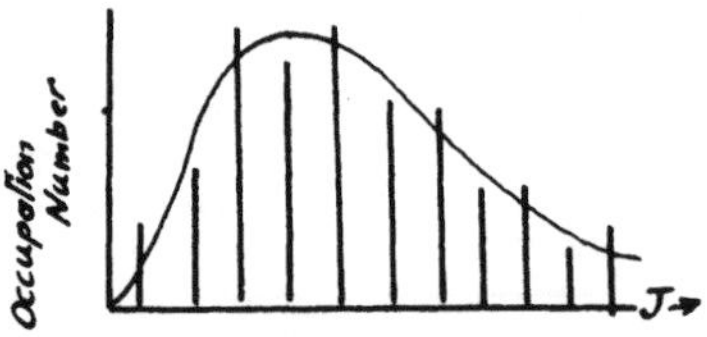


Alternation of Intensity in Band Spectra.

For example, consider $H_2$, which obeys Fermi statistics, has $\Psi$ odd. Each odd $\rho_J$ must be combined with an even $\sigma_I$, and vice-versa. Because of the statistical weights attached to the $\sigma_I$, the intensity of the $\rho_{odd}\ \sigma_{even}$ lines will be three times as great as that of the $\rho_{even}\ \sigma_{odd}$ lines.

## 3. Atomic and Molecular Beams

The $\mu$ of the nucleus is determined by a Stern-Gerlach type experiment, splitting a beam of atoms in an inhomogeneous $\mathcal{H}$.

Nuclear moments are, however, hard to observe, since they tend to be masked by electronic moments on the order of 1000 times as great, unless J = 0.**

## 4. Magnetic Resonance (Nuclear Induction)

Rabi has passed neutral atoms through a strong $\mathcal{H}$. I and J are decoupled and precess independently about $\mathcal{H}$ at a frequency

*approximately -- see Pauling and Wilson, "Introduction to Quantum Mechanics, page 260

**Esterman, Molecular Beam Technique, Rev. Mod. Phys. 18, 300 ('46)

$g\frac{e\mathcal{H}}{2Mc}$, where g is the gyromagnetic ratio involved. For I this is the radiofrequency $\omega_I = \frac{\mu_n \mathcal{H}}{\hbar I}$. If a small secondary $\mathcal{H}'$, perpendicular to the main $\mathcal{H}$, oscillates at $n\omega_I$ ($1 < n < 2I+1$), it will induce a change in the orientation of $\underline{I}$. This flip may be detected because of its effect upon the atom's trajectory.

This technique of resonance-inducing a flip has recently been applied to stationary nuclei, for example simply to a drop of water. The transition is observed by its emission or absorption of radiation. The relative $\mu$'s of different nuclei can thus be determined to one part in $10^6$. Also, now that we know $\mu_{Proton}$ very accurately, the experiment can be run "backward" to furnish the most precise (and a convenient) method for determining an unknown $\mathcal{H}$.*

5. Nuclear Reactions, Gamma and Beta Decay If the spin of an initial nucleus is known, and then there is $\beta$- or $\gamma$-decay for example, we can then tell something about the spin of the end-product, or vice-versa. This is discussed in later chapters.

## Experimental Results

In the same way that electronic moments are measured in Bohr magnetons,

$$\mu_B = \frac{e\hbar}{2mc} = 0.9273 \times 10^{-20} \text{ erg gauss}^{-1},$$

so nuclear moments are measured in nuclear magnetons,

$$\mu_n = \frac{e\hbar}{2M_pc} = \frac{\mu_B}{1837} = 5.04 \times 10^{-24} \text{ erg gauss}^{-1}$$

Some observed spins are given below:

| Particle | Spin | Moment |
|---|---|---|
| Electron | 1/2 | -1.002 Bohr magneton |
| Proton | 1/2 | +2.7896 Nuclear " |
| Neutron | 1/2 | -1.9103 " " |
| Deuteron | 1 | +0.85647 " " |

$\underline{\mu}$ is considered + when directed along $\underline{s}$. The electron moment of one is predicted by the Dirac theory of the electron, but we have no well accepted theory for the nuclear moment. The answer may lie in meson theory.

It is rather interesting to note that

$$|\mu_p| - |\mu_n| \approx 1$$

even though the deviation from one is well outside the limits of experimental error.

----

This integral difference would be explained by a naive little fantasy. Suppose that what we call a "neutron" is really a mixture of states, i.e., that most of the time a "neutron" is really an "ideal" neutron, with $\mu = 0$, but that, for a fraction of each second, t,

$$N \rightarrow P + \pi^-$$

where P is an "ideal" proton, with $\mu = 1$, and $\pi^-$ is a negative meson, with mass much less than that of a P, moving orbitally around the P, so that its orbital magnetic moment, $\mu_\pi$, is greater than, and antiparallel to, the unit moment of the ideal proton.

*Kellogg and Millman, "Molecular Beam Magnetic Resonance Method" Rev. Mod. Phys. 18, 323 ('46).
Hopkins, N.J., Rev. Sci. Inst. 20, 401 ('49).

Write $\mu_{\pi^-} = -\mu_\pi$ ; and $\mu_{\pi^+} = \mu_\pi$

The observed $\mu_N$ would then be given by

$$\mu_N = (1 - \mu_\pi)t \qquad \text{I.16}$$

Similarly, assume that a "proton" spends most of its time in an ideal proton state, $\mu = 1$, but that, for the same fraction of each second, t,

$$P \rightarrow N + \pi^+$$

The observed $\mu_P$ would be given by

$$\mu_P = (1 - t) + t\mu_\pi = 1 + t(\mu_\pi - 1)$$

But from I.16, $(\mu_\pi - 1) = \dfrac{-\mu_N}{t}$

So $\mu_P = 1 - \mu_N = 1 + 1.91 = 2.91$

In the above model, angular momentum is not conserved.

----

A less fantastic relation in the table on the previous page is the correlation between $\mu_P$, $\mu_N$, and $\mu_D$. If we assume that deuterium is mainly in a state where the N and P are alined spin-parallel (moment-antiparallel), we would expect $\mu = 2.7896 - 1.9103 = 0.8793$, fairly close to the observed value. We may assume that the discrepancy is because deuterium is not in a pure S-state, but that there is some small orbital contribution to the moment. Other evidence bears this out. See Ch. VI, p. 114.

Even-Z, Even-N nuclei, and the effect of adding one nucleon.

It was mentioned before that these nuclei are very stable. It is also observed that they all have spin and moment zero. They seem analogous to the closed shell atoms of chemistry.

We can test our ideas on nuclei, and the application of atomic quantum mechanics thereto, by predicting the moment of closed-shell-plus-one nuclei. The odd nucleon may be either P or N.

We then assume that the closed-shell part of the nucleus merely acts as an inert center of mass (central force field) around which the odd nucleon revolves like an atomic electron, complete with spin.

As shown in appendix I.1, p.199ff, the maximum component of $\underline{\mu}$ along $\underline{\mathcal{H}}$ is given by

Schmidt Equations*

$$\mu_Z = I + 2.29 \qquad \text{odd-P, } I = \ell + 1/2 \qquad \text{I.17}$$

$$\mu_Z = \frac{I^2 - 1.29\, I}{I + 1} \qquad \text{odd-P, } I = \ell - 1/2 \qquad \text{I.18}$$

$$\mu_Z = -1.91 \qquad \text{odd-N, } I = \ell + 1/2$$

$$\mu_Z = \frac{1.91\, I}{I + 1} \qquad \text{odd-N, } I = \ell - 1/2$$

*Zeits. f. Physik 106, 358 ('37)

In Fig. I.7 (purely schematic, not to scale) the curves represent I.17 and I.18. The dots are maximum observed $\mu_Z$'s for many nuclei. One might hopefully say that there is a tendency toward grouping near the two lines. Odd-N data is similar.*

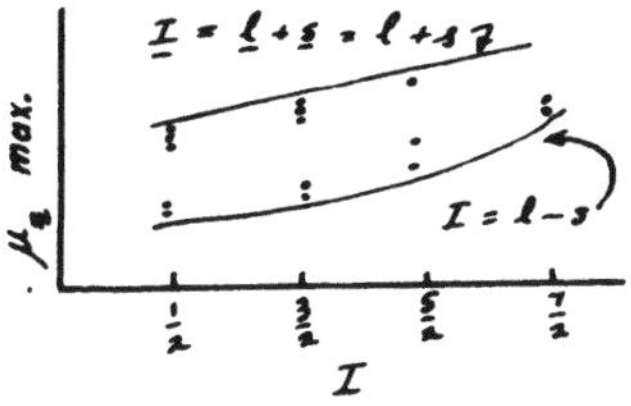


FIG. I.7, ODD PROTON

## E. ELECTRIC QUADRUPOLE MOMENT

For amplification, see Mattauch, p. 38; also Rosenfeld, "Nuclear Forces," p. 392.

The hyperfine structure of Europium, for example, shows lines that can be accounted for only by assuming that the nucleus has a permanent electric quadrupole moment, **Q**.

Rabi has shown that even deuterium possesses a quadrupole moment such that it appears as a spheroid elongated (prolate) along the I axis. $Q(\text{deut}) = 0.00273 \times 10^{-24}\text{cm}^2$.

The concept of quadrupole moment stems from classical electrostatic potential theory. Let us assume that the nuclear charge is rotating about $\underline{I}$. Then, no matter what its distribution, it must, on time average, appear cylindrically symmetric about $\underline{I}$. Because, outside the nucleus, $\nabla^2\phi = 0$, we can expand $\phi$ in terms of Legendre polynomials,

$$\phi(r,\theta) = \frac{1}{r}\sum_{n=0}^{\infty} \frac{a_n}{r^n} P_n(\cos\theta) \qquad \text{I.20}$$

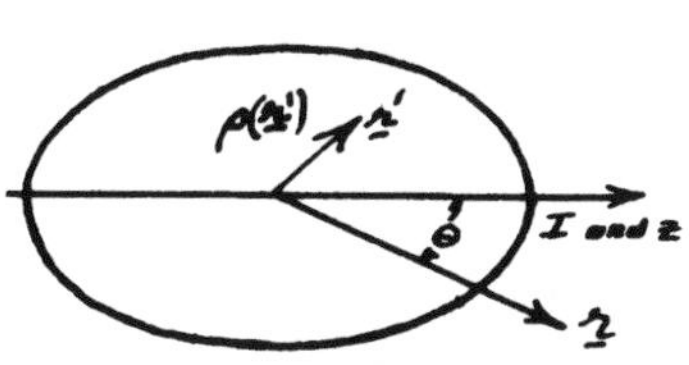


Nuclear elongation, much exaggerated

$\frac{2}{e}$ times the coefficient $a_2$ is called **Q**, the quadrupole moment. As shown in App. I.2 p. 21 it is

$$Q = \frac{1}{e}\int (3z'^2 - r'^2)\,\rho(\underline{r}')\,d\tau' \ \text{cm}^2 \qquad \text{I.21}$$

where the notation is shown in the Fig. and e is the charge on an electron.

The first term of I.20 is the ordinary Coulomb $\phi$. The next term, the permanent dipole, is not generated by nuclei, in the static case, because of parity considerations (Schiff, p. 158). The first static term giving a measure of the departure from spherical symmetry is the quadrupole term. A nucleus of the shape illustrated would have $Q > 0$. This is the more common occurence.

When $I = 0$, the nucleus has no preferred axis, the charge distribution appears spherically symmetric, and $Q = 0$. We shall now use quantum mechanics to prove that $Q = 0$ also when $I = 1/2$.

In quantum mechanics we must replace z' by $\underline{r}' \cdot \underline{e}_I = \underline{r}' \cdot \frac{\underline{I}}{|I|}$

$$z' = \frac{1}{|I|}(x'I_x + y'I_y + z'I_z)$$

*For a more complete discussion, see Rosenfeld, "Nuclear Forces," p. 394

$$z'^2 = \frac{1}{I(I+1)}\left( x'^2 I_x^2 + y'^2 I_y^2 + z'^2 I_z^2 + 2x'y'(I_xI_y + I_yI_x) + .. \right)$$

But because of the commutation relations for I, the three terms of the form

$$I_xI_y + I_yI_x = 0 \ ;$$

also

$$I_x^2 = I_y^2 = I_z^2 = 1/4$$

$$I(I + 1) = 3/4$$

therefore

$$z'^2 = \frac{4}{3}\ \frac{1}{4}\ (x'^2 + y'^2 + z'^2) + 0$$

$$= \frac{r'^2}{3}$$

so

$$Q = 0 \qquad \text{by I.21}$$

Experimentally, Q ranges from

$$Q = 7.0 \times 10^{-24}\ \text{cm}^2 \text{ for } {}_{71}\text{Lu}^{176}$$

to " $-0.6$ " " " ${}_{53}\text{I}^{127}$

A later value of $-0.46 \times 10^{-24}$ is given by Feld, Addendum to Nuclear Science Series Report No 2, May 1949.

but $Q > 0$ is more common.

Problem. Assume that $Lu^{176}$ and $I^{127}$ are ellipsoids of rotation, obtained by deforming, without changing its volume, a sphere of radius $R = 1.5 \times 10^{-13}\ A^{1/3}$. Calculate the ratio, a/b, of the axial to the equatorial semi-axes.

Solution. Using I.21, the integration over an ellipsoid is straightforward. We get

$$Q = \frac{\rho}{e}\ V\ \frac{2}{5}\ (a^2 - b^2) \qquad \text{I.22}$$

where

$$V = \frac{4}{3}\pi\ ab^2 = \frac{4}{3}\pi\ R^3$$

$\rho V$ is the charge of the nucleus, eZ, so I.22 gives

$$\boxed{Q = Z\frac{2}{5}\ (a^2 - b^2)} \qquad \text{I.23}$$

Elimination of

$$b^2 = R^3/a$$

would appear to give a cubic equation. Before we bother to solve it, however, let us hopefully assume that the nucleus is very close to spherical, so that

$$a = b(1+d) \qquad d \ll 1 \qquad \text{I.24}$$

We shall drop terms in $d^2$ and $d^3$; solve for d. If d turns out indeed to be small, our assumption was justified, and we have saved much algebra. Inserting I.24, I.23 goes to

$$\frac{Q}{Z} = \frac{2}{5}\ 2b^2 d$$

Thus for Lu, setting $b^2 \approx R^2 = 7.02 \times 10^{-25}$

$$d = \frac{5Q}{4ZR^2} = 0.18 \pm 0.18^2$$

So $(a/b)_{Lu} = 1.18 \pm 0.03$; $(a/b)_I = 0.97$ with negligible error

## F. RADIOACTIVITY AND ITS GEOLOGICAL ASPECTS*

Presumably, at the beginning of the universe ($\sim 3$ billion years ago) all isotopes were formed. The only radioactive isotopes and chains now present are those originating with isotopes whose half life, $\tau$, is appreciable compared with $10^9$ y.

Some naturally occurring radioactive isotopes are known that are not parts of chains:

$_{19}K^{40}$ ($\beta^-$, $\beta^+$, or K capture; $\tau = 2.4 \times 10^8$ y);

$_{37}Rb^{87}$ ($\beta^-$, $\tau = 6.6 \times 10^{10}$ y); $_{62}Sm^{148}$ ($\alpha$, $1.4 \times 10^{11}$ y);

$_{71}Lu^{176}$ ($\beta^-$, $2.4 \times 10^{10}$ y); $_{75}Re^{187}$ ($\beta^-$, $3 \times 10^{12}$ y)

Also, minute amounts of $C^{14}$ are formed by cosmic radiation.

However, the three most important radioactive families are:

| A*** | Name | Parent | Half life |
|---|---|---|---|
| (4n - 2) | Uranium-Radium | $_{92}U^{238}$ | $4.49 \times 10^9$ years |
| (4n) | Thorium | $_{90}Th^{232}$ | $1.39 \times 10^{10}$ " |
| (4n - 1) | Uranium-Actinium | $_{92}U^{235}$ | $7.07 \times 10^8$ " |

A nucleus decays down the chain by $\alpha$ and $\beta^-$ processes. When the half-lives of these competing processes are comparable, there is "branching", but the daughter products all eventually loop back into one another. These chains end at Pb.

Naturally occurring uranium contains $U^{235}$ and $U^{238}$ in the ratio of 1:140, and all their decay products, which are in radioactive equilibrium, often called "secular equilibrium." The ratio of the activities of $U^{235}$ and $U^{238}$ is 4%.

By measuring the ratio of $Pb^{206}$ to $U^{238}$ in rocks, we can estimate the age of the earth, which appears to be about 2.5 billion years. Astronomy leads in two or three different ways to the assumption that the age of the universe is about 3 billion years. The earth thus appears to have been formed early in the life of the universe.**

### Terrestrial Distribution of Uranium

One gram of natural U produces 0.95 erg $sec^{-1}$ when in equilibrium with its decay products. On the average, there are five parts per million of U in the earth's crust. One g Th produces 0.27 erg $sec^{-1}$. The average abundance of Th is about $10^{-5}$. Thus a gram of crust produces $7.5 \times 10^{-6}$ erg $sec^{-1}$.

The known conductivity and temperature gradient of the earth's crust allow us to calculate the earth's energy loss due to conduction, which is $6 \times 10^{12}$ cal $sec^{-1}$. Assuming a steady state, this requires a generation of heat, throughout the whole earth, of only $4 \times 10^{-8}$ erg $g^{-1}$ $sec^{-1}$.

*Cork, "Radioactivity and Nuclear Physics"

**See, for example, NRC Bulletin No. 80, "The Age of the Earth."

***Since all the transitions involve a $\Delta A = 0$ or 4, one can characterize the families with these formulas.

This factor of about 200 is not very satisfactorily accounted for, although there is some geochemical evidence that U should tend to concentrate in the crust of the earth rather than in the core.

## G. MEASUREMENT AND BIOLOGICAL EFFECTS OF RADIOACTIVITY

**ONE CURIE** is defined as $3.71 \times 10^{10}$ disintegrations per second. This is roughly the number of disintegrations per sec. in one gram of pure radium. The strength of a source is usually given in terms of the activity of the parent nucleus; thus, if a source contains a parent nucleus in equilibrium with $n$ of its daughters, then there will actually be $(n + 1) \times 3.71 \times 10^{10}$ disintegrations per sec. There may also be gamma radiation.

**ONE ROENTGEN** (r) is that amount of x- or $\gamma$-radiation which will, on passing through pure air under standard conditions, produce one esu of positive and one esu of negative ions $cm^{-3}$. One r liberates 83 erg $g^{-1}$ air and roughly the same per gram of water.

Typical soft tissue absorbs energy to about the same extent per gram as does water. A dose of one **ROENTGEN EQUIVALENT, PHYSICAL** (rep) is an irradiation by particles other than photons such that, again, 83 erg $g^{-1}$ is absorbed.

Dose*

Biological damage depends not only upon the total dose, but also upon the specific ionization or ion density along the path of the particle. Particles with higher density of ionization generally have greater biological effectiveness. It is customary to present the following table.

RELATIVE BIOLOGICAL EFFECTIVENESS

| Particle | RBE |
|---|---|
| x-rays, gammas, betas* | 1 |
| Protons | 5 |
| Alphas | 10 |
| Fast Neutrons | 10 |
| Thermal Neutons | 5 |

*In the case of alphas and weak betas, some extra protection is given by the dead, outermost ($\sim$0.1 mm) layers of the skin.

Tolerance Dose:

The latest proposal (Jan. 1949) is that the limit be 0.3 rep, divided by the RBE, per week, for total body irradiation. Thus, for example, the limit would be 0.3 r per week for gammas, but only

*Goodman, Vol. II, Chap. 16; also Siri (see p. 239 of this book)
Nucleonics 4, 42 (1949) and 3, 60 (1948)
"Safety Code for the Industrial Use of x-rays", Am. Standards Assoc. N. Y., 1946
Lea, D.E. "Action of Radiation on Living Cells" 1947
Tompkins, P.C. "Lab. Handling of Radioactive Material" AEC, MDDC 1414 and 1527
Cantril and Parker, "The Tolerance Dose" AEC MDDC 1100

0.03 rep per week for fast neutrons.

It might be healthy, however, always to allow for much larger safety factors than called for by the table, which may imply that more is known about the dangers of radiation than is actually the case. RBE varies with the species of organism or cell, the kind of biological effect studied, dose rate, and probably other factors.

Protons, fast and slow neutrons, really have quite similar RBE. According to experience to date, for effects on mammals, this varies from 3 to 30.

For local exposure, e.g., the hands, a limit of about 1.5 rep per week is proposed.

## A Few Sundry Facts

Cosmic radiation is about 1 mr per day, or roughly 2.5% of the proposed weekly dose.

One curie of gamma emitter produces on the order of 1r $hr^{-1}$ at 1 meter. 8 in. concrete or 2 in. Pb reduce this to one tenth.*

In terms of neutron flux, assuming total body irradiation, 40-hr week,

Fast neutrons: allowable flux, 100 $cm^{-2}$ $sec^{-1}$
Slow neutrons: " " 2000 " "

***********************

## APPENDIX I.1: MAGNETIC MOMENT FOR CLOSED-SHELL-PLUS-ONE NUCLEI

In this section, * does not mean footnote but, instead, denotes a quantum mechanical vector.

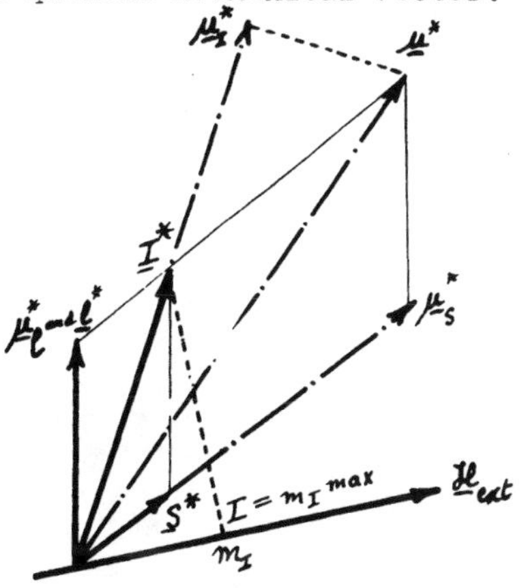


FIG. I.7

mechanical vector

magnetic vector

Interpretation of Fig. I.7. $\underline{I}^*$ is of such a length that, along $\mathcal{H}_{ext}$, it has a maximum observable component of I, a half integer.

If, for example, I were 3/2, then $\underline{I}^*$ may have (2I + 1) different components, given by

$$m_I = \frac{3}{2} \quad \frac{1}{2} \quad -\frac{1}{2} \quad -\frac{3}{2}$$

The length of $\underline{I}^*$ is given by

$$I^{*2} = I(I+1)$$

or

$$I^* = \sqrt{I(I+1)}$$

The same applies to the other angular momentum vectors, $\underline{l}^*$ and $\underline{s}^*$.

*Experimental data for shielding against high energy gammas is found in papers by Westendorp and Charleton, Jour. Appl. Phys. 16, 590 ('45), and Blocker et al., Phys. Rev. 79, 419 ('50).

The magnetic vectors are treated differently. In quantum mechanics

$$\mu_{\text{operator}} = k s_{\text{operator}}$$

Therefore, for the vector model, we must write

$$\underline{\mu}^* = k\underline{s}^* = k\sqrt{s(s+1)}\,\underline{e}_s$$

where

$$\mu_{\text{observed}} \equiv \mu = (\mu_z)_{\max} = k(s_z)_{\max} \equiv ks$$

Thus, for a proton, $2.79 = \frac{k}{2}$

$$k_{\text{proton}} = 5.58; \quad k_{\text{neutron}} = -3.82$$

Note that $(\mu^*)^2 \neq \mu(\mu+1)$

5.58 and -3.82 are analogous to the anomalous gyromagnetic ratio, -2.0, for the electron. Because the spin part of the total I contributes this anomalously large magnetic moment, (notice the figure) the total $\underline{\mu}^*$ is not parallel to $\underline{I}^*$.

If there is no $\mathcal{H}_{ext}$, $\underline{I}^*$ will stay fixed in space, and $\underline{\mu}^*$ will rotate about $\underline{I}^*$ with a high frequency, $\omega$. The time average of $\mu^*$ will be the component of $\mu^*$ along $I^*$.

We call this

$$\underline{\mu}_I^* = \underline{\mu}_s^* \cdot \frac{\underline{I}^*}{I^*} + \underline{\mu}_l^* \cdot \frac{\underline{I}^*}{I^*}$$

When $\mathcal{H}_{ext}$ is applied, $\underline{I}^*$ will start to precess about $\underline{\mathcal{H}}_{ext}$ with a frequency $\omega_{\text{Larmor}} << \omega$. (see footnote)

The projection of $\underline{\mu}_I^*$ along $\underline{\mathcal{H}}_{ext}$, i.e. the observable $\mu$, will be:

$$\mu_{obs} = \mu_I^* \frac{m_I}{I^*} = \mu_I^* \frac{m_I}{\sqrt{I(I+1)}}$$

We shall now calculate this quantity.

$$s^{*2} = \frac{3}{4}; \qquad l^{*2} = l(l+1)$$

$$\cos(\underline{s}^*, \underline{I}^*) = \frac{s^{*2} + I^{*2} - l^{*2}}{2s^*I^*}$$

$$\cos(\underline{l}^*, \underline{I}^*) = \frac{l^{*2} + I^{*2} - s^{*2}}{2\,l^*I^*}$$

Working in nuclear magnetons, $\underline{\mu}_s^* = k\underline{s}^*$, $\underline{\mu}_l^* = \begin{cases} \underline{l}^* & \text{for protons} \\ 0 & \text{for neutrons} \end{cases}$

$$\mu_s^* \cos(\underline{s}^*, I^*) = k\frac{s^{*2} + I^{*2} - l^{*2}}{2I^*}$$

---

FOOTNOTE: $\hbar\omega \sim \underline{l}\cdot\underline{s}$ interaction; i.e. $\omega \sim 10^{17}$

On the other hand, $\omega_L = \frac{e\mathcal{H}}{2Mc} \sim \frac{10^{-20}}{10^{-24}}\mathcal{H}$

Since the largest attainable $\mathcal{H} < 10^5$ gauss, the two frequencies are vastly different.

$$\mu_l^* \cos(l^*, I^*) = \frac{l^{*2} + I^{*2} - s^{*2}}{2I^*}$$

For the odd-P case

$$\mu_I^* = \frac{l^{*2} + I^{*2} - s^{*2} + k(s^{*2} + I^{*2} - l^{*2})}{2I^*}$$

$$\mu_{obs} = \mu_I^* \frac{m_I}{I^*} = \frac{l^{*2} + I^{*2} - s^{*2} + k(s^{*2} + I^{*2} - l^{*2})}{2I^{*2}} m_I .$$

Now write the denominator $2I(I + 1)$; $\mu_{obs}$ max. will be given when $m_I$ takes on the value I.

$$\mu_{obs} \text{ max.} = \frac{l^{*2} + I^{*2} - s^{*2} + k(s^{*2} + I^{*2} - l^{*2})}{2(I + 1)} \qquad \text{I.25}$$

nuclear magnetons

We shall now use I.25 to get the first of the four expressions on p. 14, I.17

$$I = l + 1/2; \quad l = I - 1/2$$

$$\mu_{obs} \text{ max} = \frac{(I - \frac{1}{2})(I + \frac{1}{2}) + I(I+1) - 3/4 + k[3/4 + I(I+1) - (I - \frac{1}{2})(I + \frac{1}{2})]}{2(I+1)}$$

$$= \frac{I^2 - \frac{1}{4} + I^2 + I - \frac{3}{4} + k(\frac{3}{4} + I^2 + I - I^2 + \frac{1}{4})}{2(I+1)}$$

$$= \frac{I^2(2) + I(1+k) - 1 + k}{2(I+1)}$$

$$= \frac{I^2 + 3.29\,I + 2.29}{I + 1} = I + 2.29 \quad \text{Nuclear Magneton}$$

The other equations are obtained in the same manner, remembering, of course, that the first three terms in the numerator of I.25, above, are multiplied by zero for the case of the neutron.

## APPENDIX I.2: ELECTRIC QUADRUPOLE MOMENT

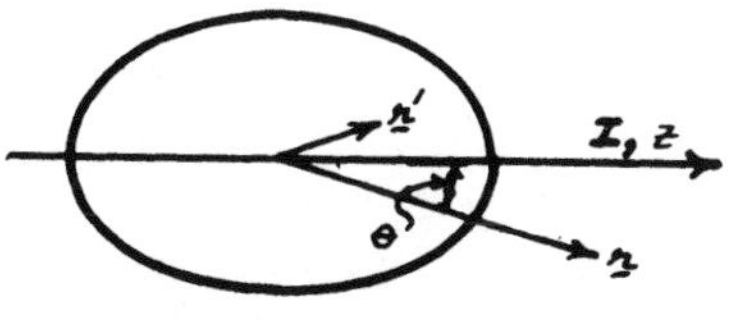


**FIG. I.8:** Whole nucleus; symmetric about I.

Look at Fig. I.8. The polynomials of I.20, p. 15, are unity for θ=0, so I.20 gives, along the z-axis,

$$\phi(z) = \frac{1}{z} \sum_{n=0}^{\infty} \frac{a_n}{z^n} \qquad \text{I.26}$$

If we can calculate $\phi$ along the z-axis, by any method at all, and expand it as a convergent series in integral powers of z, we can get the coefficients, $a_n$, by comparison. We are interested only in $a_2$.

By Coulomb's law

$$d\phi(r) = \frac{\rho(r')\, d\tau'}{|r - r'|} \qquad \text{I.27}$$

Denoting the unit vector, $\frac{\mathbf{r}}{r}$, by $\mathbf{e}_r$

$$|\mathbf{r}-\mathbf{r}'| = (r^2 + r'^2 - 2\mathbf{r}\cdot\mathbf{r}')^{1/2} = r\left\{1 - \left(\frac{2\mathbf{e}_r\cdot\mathbf{r}'}{r} - \frac{r'^2}{r^2}\right)\right\}^{1/2} \equiv r\{1-(x)\}^{1/2}$$

$$\frac{1}{\sqrt{1-x}} = 1 + \tfrac{1}{2}x + \tfrac{3}{8}x^2 + \dots \qquad x < 1$$

$$\text{So } \frac{1}{|\mathbf{r}-\mathbf{r}'|} = \frac{1}{r}\left\{1 + \frac{\mathbf{e}_r\cdot\mathbf{r}'}{r} + \frac{1}{r^2}\left(-\frac{r'^2}{2} + \frac{3}{8}\cdot 4(\mathbf{e}_r\cdot\mathbf{r}')^2\right) + \cdots \right\} \qquad \frac{r'}{r} < 1$$

For comparison with the $a_2$ term in I.26, we are interested only in the two terms in $1/r^3$.

$$d\phi^{(2)}(r) = \frac{\rho\, d\tau'}{2r^3}\left(3(\mathbf{e}_r\cdot\mathbf{r}')^2 - r'^2\right)$$

and when r lies along the z-axis

$$\phi^{(2)}(z) = \frac{1}{2z^3}\int \rho(\mathbf{r}')\left(3z'^2 - r'^2\right)d\tau'$$

which, by comparison with I.26, justifies I.21, p 15.

---

## APPENDIX I.3: MASS CORRECTION FOR NEUTRON EXCESS, $M_3$, p 6.

Equation VIII.46, p. 160, shows that nuclei can be described as cold Fermi-Dirac gases, with a Fermi-energy (for the neutrons)

$$\mu_0 = c_0\left(\frac{N}{A}\right)^{2/3}$$

Now, the total energy of the neutrons, measured from the bottom of the well, is *

$$E_0 = \tfrac{3}{5}N\mu_0$$

so

$$E_N = c_1\frac{N^{5/3}}{A^{2/3}} \qquad \text{and} \qquad E_P = c_1\frac{Z^{5/3}}{A^{2/3}}$$

Term $M_4$, p. 6, corrects explicitly for electrostatic energy, so we ignore it here. Then $E_{N+Z}(\text{min})$ will occur when $N=Z=A/2$. The mass correction will be proportional to

$$E_{N+Z} - E_{N+Z}(\text{min}) = c_1 A^{-2/3}\left[N^{\frac{5}{3}} + Z^{\frac{5}{3}} - 2\left(\frac{A}{2}\right)^{\frac{5}{3}}\right]$$

$$\text{Let } \Delta \equiv \frac{N-Z}{2} = \begin{cases} N - \frac{A}{2} \\ \frac{A}{2} - Z \end{cases}$$

$$E_{N+Z} - E_{N+Z}(\text{min}) = c_1 A^{-\frac{2}{3}}\left[\left(\frac{A}{2}+\Delta\right)^{\frac{5}{3}} + \left(\frac{A}{2}-\Delta\right)^{\frac{5}{3}} - 2\left(\frac{A}{2}\right)^{\frac{5}{3}}\right]$$

Taylor-expand the first two binomials as far as terms in $\Delta^2$:

$$E_{N+Z} - E_{N+Z}(\text{min}) = c_1 A^{-\frac{2}{3}}\left[2\,\frac{\frac{5}{3}\cdot\frac{2}{3}}{2}\left(\frac{A}{2}\right)^{-\frac{1}{3}}\Delta^2\right] \propto \frac{\Delta^2}{A} = \frac{\left(\frac{A}{2}-Z\right)^2}{A}$$

as assumed in $M_3$.

---

*Mayer and Mayer, "Statistical Mechanics," p 376.

The solutions, references, etc., are not due to Dr. Fermi.

1. Design a mass spectrograph to measure the mass difference between Hydrogen and Deuterium. Measure the separation of the close lines $(H_2)^+$ and $D^+$.

   References: Mattauch and Fluegge, "Nuclear Physics Tables" 1942; Harnwell and Livingood, "Experimental Atomic Physics" 1933; M.G. Inghram, chapter on Modern Mass Spectroscopy in "Advances in Electronics" 1948.

2. Use the semi-empirical mass formula to calculate the energy of α-particle emitted from $_{92}U^{235}$. Compare this with the observed value.

   Calculate the binding energy of a proton and a neutron in $U^{235}$.

   Answers: α-particle-- theoretical, 4.14 Mev; experimental, 4.52. BE(N) = 6.8; BE(P) = 4.85. See Metropolis, "Table of Atomic masses," Oak Ridge.

3. Design a molecular beam apparatus to determine the atomic magnetic moment of Na in the ground state, $^2S_{\frac{1}{2}}$.

   Reference: RGJ Fraser, "Molecular Beams" 1937.
   Consider: Temperature of furnace, slit dimensions, magnet dimensions, pressure, beam separation after splitting, width of beam.

4. Problem on relativity. A cosmic ray meson, mass = 216m, passes through two Geiger counters, 10 m apart. What error, dt, in the time, Δt, between the two pulses, is allowable, if we wish the uncertainty in energy to be less than 10%?. Consider the cases where the "energy" (i.e. kinetic energy) of the meson is 50, 100, 1000, Mev.

   Solution: Be sure to calculate

$$\frac{dT}{T} < 10\%$$

   where T is the kinetic, not the total, energy. For low-T particles, dE/E has little importance.

   Answers: For T = 50 Mev, dt = 1.29 $\times 10^{-9}$ sec.

| T (Mev) | dt | |
|---|---|---|
| 50 | 1.29 | $\times 10^{-9}$ sec. |
| 100 | 0.71 | " |
| 1000 | 0.0296 | " |

5. Design a 10 Mev Betatron. (Three assignments; one week's homework)

   Points to consider: Shape of pole pieces (taper, position of stable orbit), ampere turns required and power supply, frequency, lamination, vacuum, d.c. bias, injection, extraction.

   References: W. Bosley, "Betatrons," a review, <u>Jour</u>. <u>Sci</u>. <u>Inst</u>., <u>23</u>, 277 (1946)
   D.W. Kerst, "20 Mev Betatron," <u>Rev</u>. <u>Sci</u>. <u>Inst</u>. <u>13</u>, 387 (1942), also <u>Phys</u>. <u>Rev</u>. <u>60</u>, 47 and 63 (1941)
   W.F. Westendorf, "Use of d.c. in Induction Accelerators," <u>Jour</u>. <u>App</u>. <u>Phys</u>. <u>16</u>, 657 (1945). See also page 581.

For a general reference see M.S. Livingston, chapter on particle accelerators in "Advances in Electronics," Academic Press, 1948

6. Design a 200 Mev synchro-cyclotron (i.e. fm cyclotron). Also one week's homework.

Points to consider: Dimensions. frequency, frequency of modulation, radial decrease of $\mathcal{H}$, phase stability, voltage on dees, electrostatic focussing, injection, extraction, vacuum.

References: Chapter by Pickavance in "Progress in Nuclear Physics, 1" by O. Frisch, 1950. The Berkeley machine is described by Chew and Moyer, Am. Jour. Phys. 18, 125 (1950). The Chicago machine in the "170-in. synchro-cyclotron, Progress Report" Institute for Nuclear Studies, Univ. of Chicago, 1950.

7. Design a one Mev Cockroft-Walton accelerator.

References: Proc. Roy. Soc. Lond. A136 610, 619 ('32) To reduce the expense to 1/10 by using radiofrequency, see Rev. Sci. Inst. 20, 216 (1949).

8. Describe the precautions and apparatus necessary to carry out simple chemical operations upon a one curie sample.

One should not approach within about ten meters of the unshielded sample. Thus about 5"Pb would be a reasonable thickness of shield.

9. The activity of a sample is the total number of processes counted per unit time.

$$A = \Sigma A_i = \Sigma \lambda_i n_i$$

where $\lambda_i$ is the decay constant, and $n_i$ the number, of the i-th sort of disintegrating nucleus.

Plot, against time, the activity of a two-element radioactive chain.

Solution. The equations are

$$n_1 = n_1^0 e^{-\lambda_1 t}$$

$$\dot{n}_2 = -\dot{n}_1 - \lambda_2 n_2$$

*Write this in the form*

$$\frac{dn_2}{dt} + \lambda_2 n_2 = \lambda_1 n_1^0 e^{-\lambda_1 t}$$

$$\frac{dn_2}{dt} + P(t) n_2 = Q(t)$$

$$n_2 = e^{-\int P dt}\left[\int e^{\int P dt} Q\, dt + C\right]$$

$$n_2 = e^{-\lambda_2 t}\left[\int e^{\lambda_2 t}\lambda_1 n_1^\circ e^{-\lambda_1 t}\,dt + C\right]$$

$$= e^{-\lambda_2 t}\left[\frac{\lambda_1 n_1^\circ}{\lambda_2 - \lambda_1} e^{\lambda_2 t} e^{-\lambda_1 t} + C\right]$$

$$\boxed{A_2 = \lambda_2 n_2 = \frac{\lambda_2 A_1^\circ}{\lambda_2 - \lambda_1}\left[e^{-\lambda_1 t} - e^{-\lambda_2 t}\right]} \quad \text{and} \quad A = A_1 + A_2$$

It can be shown by straighforward substitution that the curves of parent and daughter activity cross at the exact time that the daughter activity is a maximum. This is illustrated in the curves below.

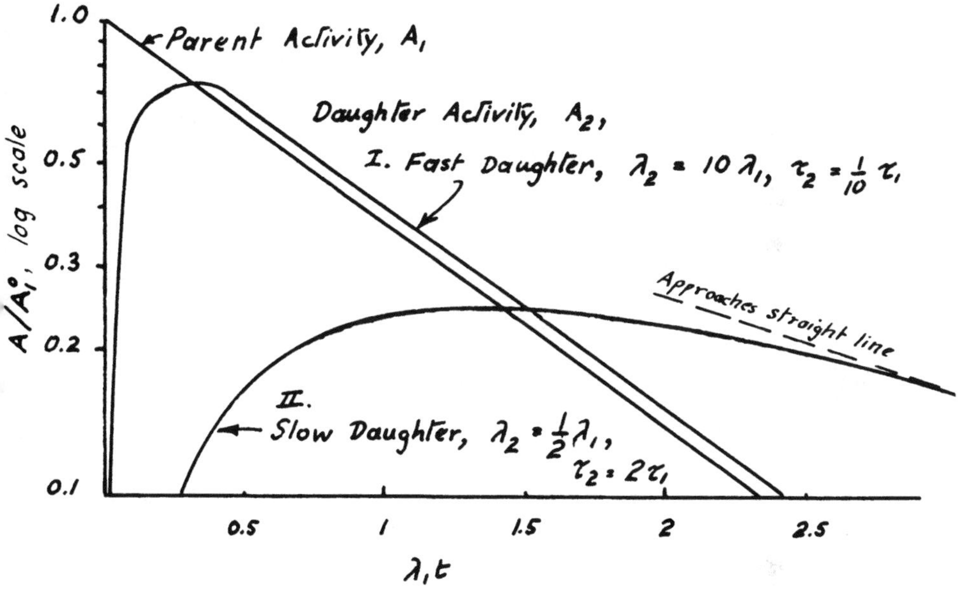


PARENT and DAUGHTER ACTIVITIES

Two statements can be made about the curves above:

1. The final decay rate depends upon the longer $\tau$; $\tau \equiv \frac{1}{\lambda}$.

2. The simultaneous $A_2$ max. and cross-over point occurs at a time roughly of the order of magnitude of the smaller $\tau$.

This is shown as follows: If the cross-over time is called $\theta$, then straightforward algebra gives

$$\theta = \frac{1}{\lambda_2 - \lambda_1} \ln \frac{\lambda_2}{\lambda_1}$$

Since the log. is a slowly varying function, $\theta$ varies roughly as the longer $\lambda$ ; that is, as the shorter $\tau$ .

10. At time $t = 0$, $U^{235}$ is stripped of all its decay products. Plot the buildup and decay of Actinium X ($_{88}Ra^{223}$).

Solution. The chain, with decay constants, $\lambda$, in $sec^{-1}$, is as follows ($\lambda = 0.693/T$, where T is the half life).

$$U^{235} \xrightarrow[3.1\times 10^{-17}]{\alpha} Th^{231} \xrightarrow[7.6\times 10^{-4}]{\beta^-} Pa^{231} \xrightarrow[6.8\times 10^{-13}]{\alpha} Ac^{227} \xrightarrow[1.6\times 10^{-9}]{\beta^-} Th^{227} \xrightarrow[4.2\times 10^{-7}]{\alpha} Ra^{223} \xrightarrow[7.4\times 10^{-7}]{\alpha}$$

The first daughter, Th, will grow into secular equilibrium with the U within a matter of days. This is illustrated by the "fast-daughter" (type I) curve of problem 9. After a few days we may consider the Th activity equal to the U activity, and go on to consider how the Pa grows in.

The Pa will build up in a period $\sim 10^{13}$ sec ($10^5$ years). We may thus neglect the comparatively short Th growth period, and write

$$\frac{d}{dt} n_{Pa} = \lambda_U n_U - \lambda_{Pa} n_{Pa}$$

If we now restrict ourselves to a time considerably less than $\tau_U$, we may consider $\lambda_U n_U$ a constant, equal to $\lambda_U n_U^\circ$. The solution to the differential equation is then

$$\lambda_{Pa} n_{Pa} = \lambda_U n_U^\circ \left(1 - e^{-\lambda_{Pa} t}\right) \qquad T_{Th} \ll t \ll T_U$$

Pa has a $T$ much longer than any of its daughters. Thus, as the Pa grows in gradually, so do all its daughters, in secular equilibrium. This is another way of saying that the AcX activity will always equal the Pa activity.

We have now completed the problem, except for a description of the disappearance of the whole chain (which is by then in perfect equilibrium) as the U decays.

During this time the AcX activity is given by

$$A_{AcX} = A_U^\circ e^{-\lambda_U t}$$

The complete curve is given below.

Notice that the asymptotic increase of the Pa activity after a time on the order of its half life is important in an analagous problem: irradiation. One accomplishes little by irradiating a sample for a time longer than, say, twice the half life under consideration.

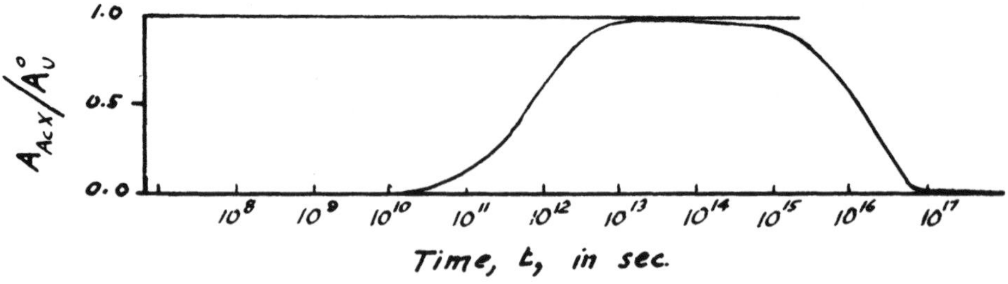

## CHAPTER II. INTERACTION OF RADIATION WITH MATTER

### A. ENERGY LOSS BY CHARGED PARTICLES

1. A charged particle moving through matter loses energy by electromagnetic interactions which raise electrons of the matter to excited energy states. If an excited level is in the continuum of states the electron is ionized; if not, the electron is in an excited bound state. In either case the increment of energy is taken from the kinetic energy of the incident particle. In the following section "ionization" will refer to both degrees of excitation.

Range = total distance traveled by the particle until its kinetic energy is 0. Before a formula for the range of a particle can be derived, the rate of energy loss per unit path must be calculated. The first such calculation is due to Bohr*, and is essentially classical, i.e., non-quantum mechanical.

2. Bohr Formula. Consider one electron of mass m at a distance b from the path of an incident particle having charge $ze$, mass M and velocity V.

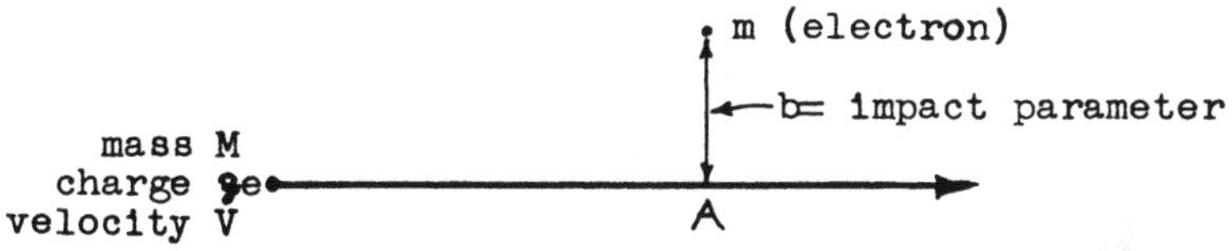


FIG. II.1

Assume the electron is free and initially at rest, and moves so slightly during the collision that the electric field acting on the electron due to the particle can be calculated at the initial location of the electron. The last assumption is not valid for an incident particle of velocity comparable to that acquired by the electron.

We shall calculate first the momentum acquired by an electron during a collision, and from this find the energy acquired. As the particle passes, the electrostatic force $\underline{F}$ changes direction. By symmetry the impulse $\int F_{\parallel}\,dt$ parallel to the path is zero, since for each position of the particle to the left of A, yielding a forward contribution to the impulse, there is a position at equal distance to the right of A giving an equal but opposite contribution.

The impulse $\perp$ to the path is $I_{\perp} = \int F_{\perp}\,dt$ . We first estimate the order of magnitude of $I_{\perp}$ :

$$I_{\perp} = \int F_{\perp}\,dt \approx \text{(electrostatic force)} \times \text{(time of collision)} \approx \frac{ze^2}{b^2} \times \frac{b}{V}$$

More exact computation: Consider a circular cylinder centered on the path and passing through the position of the electron, Fig. II.2. Let $\mathcal{E}$ be the electrostatic field intensity due to the particle $ze$. The electric flux is

$$\int_{\text{cyl.}} \mathcal{E} \cdot d\sigma = 4\pi z e \qquad \text{(independent of V)}$$

---

* Bohr, Phil. Mag. 24, 10 (1913), 30, 581 (1915)

by Gauss's theorem. If $\mathcal{E}_\perp$ = component of $\mathcal{E}$ $\perp$ to the path, then the flux = $\int_{-\infty}^{\infty} dx\, 2\pi b\, \mathcal{E}_\perp = 4\pi \mathfrak{z}e$. Therefore $\int_{-\infty}^{\infty} \mathcal{E}_\perp\, dx = \frac{2\mathfrak{z}e}{b}$.

The variation of $\mathcal{E}_\perp$ with time at the electron is the same function as would be found by keeping $\mathfrak{z}e$ fixed and observing at a point moving with velocity $V$ along the cylinder surface. Therefore

$$\int_{-\infty}^{\infty} \mathcal{E}_\perp(t)\,dt = \int_{-\infty}^{\infty} \mathcal{E}_\perp(x)\,\frac{dx}{V} = \frac{1}{V}\int_{-\infty}^{\infty} \mathcal{E}_\perp(x)\,dx = \frac{2\mathfrak{z}e}{Vb} \qquad \text{II.1}$$

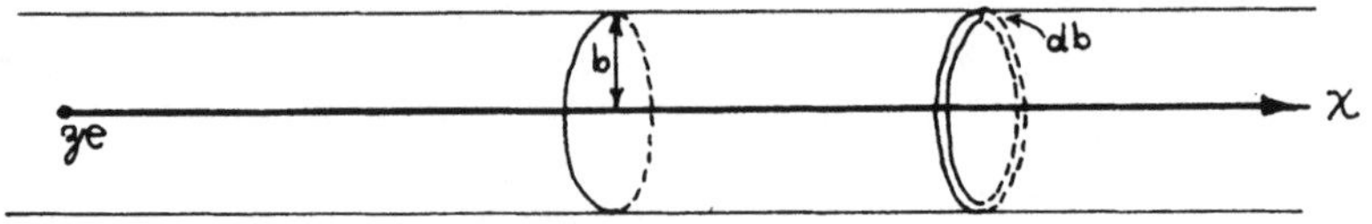


FIG. II.2

The impulse $I_\perp = \int_{-\infty}^{\infty} \mathcal{E}_\perp e\,dt = \frac{2\mathfrak{z}e^2}{Vb} = p$, where p is the momentum acquired by the electron. The energy acquired by one electron is then II.2

$$\frac{p^2}{2m} = \frac{2\mathfrak{z}^2 e^4}{mV^2 b^2} \qquad \text{II.3}$$

The number of collisions per unit path length such that b lies in the range b to b + db is equal to the number of electrons per cm. length in the shell bounded by cylinders of radii b and b + db (Fig. II.2). If $\mathfrak{N}$ = number of electrons per $cm^3$, this is $2\pi \mathfrak{N}$ bdb. The energy lost per cm. to such electrons is

$$dE(b) = \frac{4\pi \mathfrak{z}^2 e^4 \mathfrak{N}}{mV^2}\,\frac{db}{b} \qquad \text{II.4}$$

The total change in energy to all shells in the range $b_{min}$ to $b_{max}$ is

$$-\frac{dE}{dx} = \frac{4\pi \mathfrak{z}^2 e^4 \mathfrak{N}}{mV^2} \ln \frac{b_{max}}{b_{min}} \quad (\text{ergs cm}^{-1}) \qquad \text{II.5}$$

The limit $b_{max}$: The force on an electron as a function of time will be a pulse occupying a time $\tau = b/V$, non-relativistically. It can be shown that if $1/\tau$ is much less than the vibration frequency $\nu$ of an electron in an atom, then the electron absorbs no energy, i.e., the probability for transition to a higher state is small.

Relativistically, the electric field of the incident particle is contracted in the direction of motion and $\mathcal{E}_\perp$ is increased by a factor $1/\sqrt{1-\beta^2}$, $\beta \equiv V/c$. This "sharpens" the impulse given the electron; the duration $\tau$ of the pulse of force is now approximately $(b/V)\sqrt{1-\beta^2}$. The integral of the impulse is not changed, since it depends essentially on the product (field strength)X(duration).

$b_{max}$ is chosen so that $1/\nu > \tau \approx \frac{b\sqrt{1-\beta^2}}{V}$ is valid over the range integrated. Thus we may set

$$b_{max} = \frac{V}{\bar{\nu}\sqrt{1-\beta^2}} \qquad \text{(cm)} \qquad \text{II.6}$$

where $\bar{\nu}$ is an appropriate average frequency for electrons in the absorbing material.

---

Problem: Discuss the statement that for $1/\nu < \tau$, energy transfer to electrons is negligible (principle of adiabatic invariance).

Consider the component of the motion of the electron ⊥ to the path of the incident particle. Let the coordinate of the electron be y. $y = b + d \sin\nu t$. $\dot{y} = \nu d \cos\nu t$. $b \gg d$.

$$\Delta(\text{energy}) = \int (\text{y component of force}) \times (\text{velocity})\, dt$$

$$= \int_{-\infty}^{\infty} \frac{z e^2 \nu d \cos\nu t \cos\theta}{y^2 + V^2 t^2}\, dt$$ neglecting terms of order$(d/b)$ or higher.

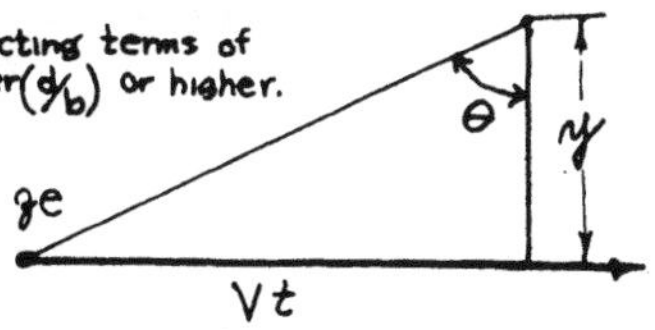


Since $\cos\theta \leqq 1$

$$\Delta(\text{energy}) < z e^2 \nu d \int_{-\infty}^{\infty} \frac{\cos\nu t}{(y^2 + V^2 t^2)}\, dt =$$

$$= \frac{z e^2 \nu^2 d}{V^2} \int_{-\infty}^{\infty} \frac{\cos\mu\, d\mu}{a^2 + \mu^2} \qquad \mu = \nu t, \quad a = \frac{y\nu}{V} \approx \frac{b\nu}{V} = \nu\tau$$

$$= \frac{z e^2 \nu^2 d}{V^2} \, \frac{\pi e^{-a}}{a}$$

$$= \frac{z e^2 \nu d \pi}{bV} e^{-\nu\tau} \longrightarrow 0 \quad \text{as } \nu\tau \rightarrow \gg 1$$

---

The limit $b_{min}$: (1) Classically, the maximum velocity that can be imparted to the electron (in head-on collision) is less than 2V.* The energy given cannot exceed $\frac{1}{2}m(2V)^2$. Therefore b cannot have values that imply a greater energy transfer per collision than $2mV^2$. As a function of b, the energy transferred per collision is $\frac{2z^2e^4}{mV^2b^2}$. Values of b smaller than the solution of $\frac{2z^2e^4}{mV^2b^2} = 2mV^2$ must be excluded in the integration. This determines $b_{min}$ as

$$(b_{min})_{cl.} \approx \frac{ze^2}{mV^2} \qquad \text{II.7}$$

(2) This classical treatment is valid only if the Coulomb field of the incident particle varies negligibly over the dimension $\lambda\!\!\!^{-}$ of the quantum mechanical wave packet representing the

---

* This is easiest to see in the rest system of the incident particle. Then the electron appears to collide with something like a rigid wall.

electron. $\lambda\!\!\!^{-}$ is approximately the de Broglie wavelength of the electron as seen from the incident particle. In a coordinate system in which the incident particle is at rest (this nearly coincides with the center of mass system, for a heavy incident particle), the electron has velocity of about V, assuming its orbital velocity is much less than V. The momentum of the electron in this coordinate system is $mV/\sqrt{1-\beta^2}$ and therefore $\lambda\!\!\!^{-} = \frac{\hbar\sqrt{1-\beta^2}}{mV}$ . Only values of $b > \lambda\!\!\!^{-}$ have meaning, and therefore another criterion for $b_{min}$ is

$$(b_{min})_{QM} \approx \frac{\hbar\sqrt{1-\beta^2}}{mV} \qquad \text{II.8}$$

The larger of $(b_{min})_{cl}$ and $(b_{min})_{QM}$ should be used in the integration*. For values of V where $b_{max} > b_{min}$, $(b_{min})_{QM} > (b_{min})_{cl}$ and therefore II.8 whould be used. Using II.8 in II.5:

$$-\frac{dE}{dx} = \frac{4\pi g^2 e^4 \mathfrak{N}}{mV^2} \ln \frac{mV^2}{\hbar\bar{\nu}(1-\beta^2)} \quad (\text{erg cm}^{-1}) \qquad \text{II.9}$$

where $\bar{\nu}$ is a suitable average of the oscillation frequencies of the electrons.

More precise calculation** leads to the following formula for heavy particles, i.e., not electrons:

$$\left.-\frac{dE}{dx}\right)_{ion.} = \frac{4\pi g^2 e^4 \mathfrak{N}}{V^2 m}\left(\ln\frac{2V^2 m}{\bar{I}(1-\beta^2)} - \beta^2\right) \quad \begin{matrix}(\text{erg cm}^{-1})\\ (\text{heavy particles})\end{matrix} \qquad \text{II.10}$$

where $\bar{I}$ is the average ionization potential of the electrons of the absorber, in ergs. The ln term $\approx 9$ for 1 mev protons in NTP air.

3. Electrons. There are two main reasons why II.10 cannot apply to electrons. (1) The derivation assumes that the incident particle is practically undeflected. But the incident particle acquires a transverse component of momentum per collision approximately equal to that given to an electron in the absorber, and if the incident particle is an electron, the transverse velocity corresponding to this momentum will not be negligible. (2) For collisions between identical particles exchange phenomena must be taken into account***. Bethe ** gives the following formula for energy loss by electrons:

$$\left.-\frac{dE}{dx}\right)_{ion.} = \frac{2\pi e^4 \mathfrak{N}}{V^2 m}\left[\ln\frac{V^2 m T}{2\bar{I}^2(1-\beta^2)} - \ln 2\left(2\sqrt{1-\beta^2} - 1 + \beta^2\right) + 1 - \beta^2\right] \quad (\text{electrons}) \qquad \text{II.11}$$

where $\bar{I}$ is the average ionization potential of the atoms of the absorber and $T$ = relativistic kinetic energy of the electron.

*In cutting the integral of II.4 off at $b_{min} \neq 0$, we have neglected a term $\int_0^{b_{min}} dE(b)$ . This is justified in "Lecture Series in Nuclear Physics", LA 24, Lecture XI, printed edition p. 27.
** Bethe, Handbuch der Physik, p. 519
***Mott, Proc. Roy. Soc. 125, 222, 126, 259 (1929)

An approximation for $\bar{I}$ is*

$$\bar{I} \approx (13.5\, Z) \times 1.601 \times 10^{-12} \qquad \text{(ergs)} \qquad \text{II.12}$$

The formula II.11 for electrons, applied to air, yields the values given in the table.

| Electrons in Air | |
|---|---|
| Energy, in e.v. | $-\frac{dE}{d\xi}$ in e.v. per $g/cm^2$ |
| $10^4$ | $19.5 \times 10^6$ |
| $10^5$ | 3.67 |
| $10^6$ | 1.69 |
| $10^7$ | 1.95 |
| $10^8$ | 2.47 |
| $10^9$ | 2.79 |
| $10^{10}$ | 3.48 |

4. Other particles. For incident particles of identical charge moving in like absorbers $-dE/dx$ is a function of V only. Therefore if $-dE/dx$ as a function of energy is known for, say, protons of mass $M_p$ it can be found for some other singly charged particle B by changing the energy scale so that the new energy values are $M_B/M_p$ times the old. The following table for protons absorbed in air enables filling out the table for (1) deuterons, mass 2, (2) $\mu$ mesons of mass $\sim 215m$. For a particle B of different charge as well as different mass, the above energy correction is made, but furthermore the ionization loss value is multiplied by $(z_B^2/z_p^2) = z_B^2$ since $z^2$ enters the formula. In this way column (3) for alpha particles is derived from that for protons. The unit for rate of energy loss in this table is ergs per gram $cm^{-2}$ and is denoted by

$$-\frac{dE}{d\xi} = -\frac{1}{\rho}\frac{dE}{dx} \qquad \text{II.13}$$

where $\xi$ is the thickness in $g\text{-}cm^{-2}$. For this table $-dE/dx$ in $erg\text{-}cm^{-1}$ is obtained from $-dE/d\xi$ by multiplying by the density of air, $\approx .0012\ g\text{-}cm^{-3}$ (STP).

| In Air | | | | | | | | In Lead | |
|---|---|---|---|---|---|---|---|---|---|
| PROTONS | | (1) Deuterons | | (2) $\mu$ mesons | | (3) Alphas | | Protons | |
| Energy, e.v. | $-\frac{dE}{d\xi}$ in $\frac{Mev}{g\text{-}cm^{-2}}$ | e.v. | $-\frac{dE}{d\xi}$ | e.v | $-\frac{dE}{d\xi}$ | e.v. | $-\frac{dE}{d\xi}$ | e.v. | $-\frac{dE}{d\xi}$ |
| $10^6$ | 300 | $2\times10^6$ | 300 | $.117\times10^6$ | 300 | $4\times10^6$ | 1200 | $10^6$ | 150 |
| $10^7$ | 47 | $2\times10^7$ | 47 | $.117\times10^7$ | 47 | $4\times10^7$ | 188 | $10^7$ | 27.5 |
| $10^8$ | 7.6 | $2\times10^8$ | 7.6 | $.117\times10^8$ | 7.6 | $4\times10^8$ | 30 | $10^8$ | 5. |
| $10^9$ | 2.3 | $2\times10^9$ | 2.3 | $.117\times10^9$ | 2.3 | $4\times10^9$ | 9.2 | $10^9$ | 1.6 |
| $10^{10}$ | 2.3 | $2\times10^{10}$ | 2.3 | $.117\times10^{10}$ | 2.3 | $4\times10^{10}$ | 9.2 | $10^{10}$ | 1.6 |

5. Other absorbers. If ionization loss were exactly proportional to the density of the absorber, then $-dE/d\xi$ ergs per $g\text{-}cm^{-2}$ for a given particle would not vary. But $-dE/d\xi$ depends on two further factors: (1) The number of electrons per atom; $Z$ of an atom does not increase as fast as the weight, thus $\mathcal{N}/\rho$ in the formula for $dE/d\xi$ is less for heavier elements. (2) $\bar{I}$, appearing in the log term, depends on the absorber (Eqn. II.12)**.

6. Range. Often in experiments the original data is the range and from this the energy is estimated. We have derived an

* Bloch, Zeits. f. Physik 81, 363 (1933)

**See also Livingston and Bethe, Rev.Mod.Phys. 9, p. 265.

equation of the form $-dE/dx = f(E)$. Integrating, we get

$$x = -\int_{E_0}^{0} \frac{dE}{f(E)} = \int_{0}^{E_0} \frac{dE}{f(E)} = \mathcal{R} \qquad \text{II.14}$$

where the limits of integration have been chosen so as to define X as the range of a particle with initial energy $E_0$. The integration may be performed numerically. For a rough approximation we assume $f(E) \propto 1/V^2 \propto 1/E$. Then $\mathcal{R} \propto E_0^2/2 \propto V^4$ nonrelativistically. More precise consideration shows that a better approximation is*

$$\mathcal{R} \propto E_0^{3/2} \propto V^3 \qquad \text{II.15}$$

Empirical range-energy formulas: A rough formula giving the range of alpha particles in air at 15°C. and atmospheric pressure is

$$\mathcal{R} \approx .32\,(\text{Mev})^{3/2}\ \text{cm.} \quad \text{(alphas in air)} \qquad \text{II.16}$$

This is correct to about 10%. It breaks down for relativistic velocities. The general nature of the range-energy relation is shown in Fig. II.3.

In 1938 Feather proposed for electrons the empirical relationship: $R = 0.543E - 0.160$ for $E > 0.7$ Mev, where R is the range in $gm/cm^2$, E is energy in Mev. Glendenin and Coryell (1946) have improved this and also worked out a low energy relation:

$$R = 0.542E - 0.133 \quad \text{for } E > 0.8 \text{ Mev} \qquad \text{II.16a}$$
$$R = 0.407E^{1.38} \quad \text{for } 0.15 \text{ Mev} < E < 0.8 \text{ Mev.} \qquad \text{II.16b}$$

(for Al, but is close for all other substances)

7. Polarization effects. In the derivation of II.10 no account has been taken of the influence on one electron due to the simultaneous motion of the other electrons near it. The electrons in a region move so as to diminish the electric field beyond that region. This partial shielding effect increases with increase in density of electrons. The change in -dE/dx due to this effect is usually small.**

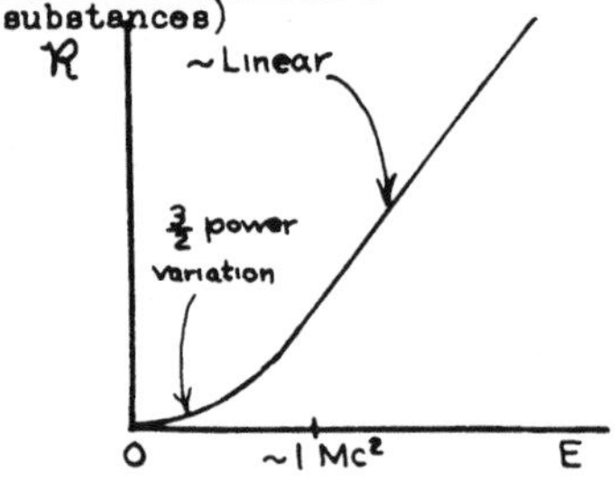


FIG. II.3

If the index of refraction, n, is not one, the velocity of light is less than c. In water, for example, $n \sim 1.5$ and velocity of light is $\sim 2/3$ c. If the incident particle has velocity $V > c/n$ its electric field is strongly perturbed and can be likened to a wake in water. Such a particle produces radiation known as Cerenkov radiation, after its experimental discoverer***

8. Nature of the equation for -dE/dx. Equation II.10 has the form

$$-\frac{dE}{dx} \propto \frac{n z^2}{V^2} f(V, I)$$

* Exponents for various particles and energies in Livingston and Bethe, p. 265.
**Treated in detail by Fermi, Phys. Rev. 57, 485 (1940)
***Cerenkov, Phys. Rev. 52, 378 (1937); theory in Schiff, l.c., p. 261.

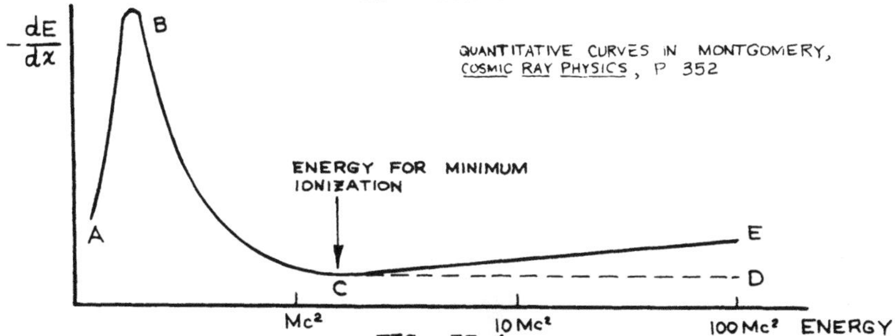


FIG. II.4

The curve BCD gives the $1/V^2$ dependence. At relativistic energies V changes little and CD is asymptotic to V = c. At relativistic energies, the log term in $(V^2/1-\beta^2)$ changes, and increases as $V \longrightarrow c$, giving the rise in the curve from C to E. At very low energies (region AB) equation II.10 breaks down because the particle has velocity comparable to that of the orbital electrons in the absorber, and the efficiency of energy exchange is much lower. The particle itself captures electrons and spends part of its time with reduced charge.

9. Ionization of a gas. If ionization is produced in a gas the ions may be collected by charged electrodes, and the amount of charge collected will be proportional to the number of ions produced. The change of potential of one of the electrodes will depend on the charge collected (and the external circuit) and therefore on the number of ions produced. This voltage pulse may be amplified linearly and measured quantitatively, as with an oscillograph. A gas chamber for this purpose is called an ionization chamber*.

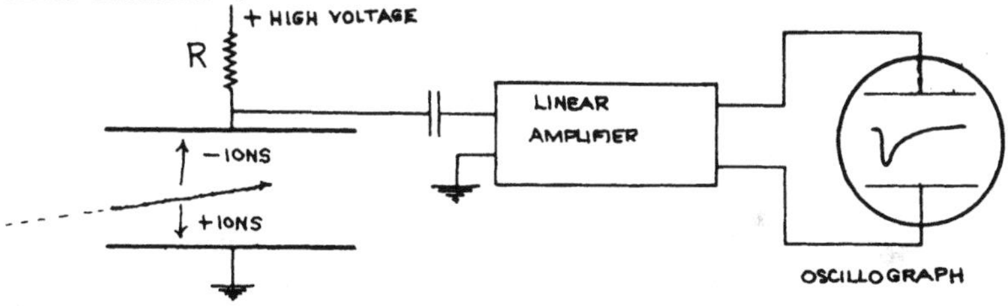


FIG. II.5

In the arrangement in Fig. II.5, electrons are collected at the top plate. A negative pulse, of duration determined by R and the capacity of the ionization chamber and associated circuit, is produced at the grid of the linear amplifier.

It turns out that there is a close proportionality between number of ions produced and total energy lost by the incident particle. For most gases one ion pair (electron plus ionized atom) is produced for each 32-34 e.v. lost by the particle, (see table on following page). Although empirically the result is a

* References on ionization chambers are: Korff, Electron and Nuclear Counters (Van Nostrand), Rossi and Staub, Ionization Chambers and Counters (McGraw-Hill).

simple proportionality between number of ions and energy spent, the explanation is very complicated. Theoretical prediction of the average energy per ion pair involves: (1) calculating the % of all primary collisions that lead to removal of an electron in order to know how much energy is "wasted" on non-ionizing excitation of the atom; (2) calculating what fraction of energy carried away by primary ionized electrons is used in producing secondary ionization. This problem has not been completely investigated.

Energy for one ion pair*

| Gas | Energy spent for one ion pair, e.v. |
|---|---|
| H | 33.0 |
| He | 27.8 |
| N | 35.0 |
| O | 32.3 |
| Ne | 27.4 |
| A | 25.4 |
| Kr | 22.8 |
| Xe | 20.8 |

> Problem: Design an ionization chamber and specify the characteristics of an associated linear amplifier so that the system is suitable for measuring the α energy difference between the $U^{238}$ (UI) and $U^{234}$ (UII) α decays.
>
> $$_{92}U^{238} \rightarrow \alpha + {}_{90}Th^{234}; \quad {}_{90}Th^{234} \rightarrow 2\beta^- + {}_{92}U^{234}; \quad {}_{92}U^{234} \rightarrow \alpha + {}_{90}Th^{230}$$
>
> The following are among the necessary considerations: (1) loss of energy of particles while still in the emitting substance, (2) gas to be used, (3) dimensions and electrostatic capacity of chamber, (4) gain and frequency response necessary in the amplifier, (5) rate of emission of particles by the emitter.

10. High energy β particles lose energy mainly by radiation. This effect is taken up later, in II, section C, 3.

## B. SCATTERING DUE TO A COULOMB FIELD

Scattering due to interation of charged particles with the Coulomb field of nuclei is distinguished from scattering in which the incident particle enters a nucleus. Only the former is treated in this section.

Scattering due to collisions with nuclei is observed for all charged particles in varying degree. An alpha track in a cloud chamber may have a single kink, indicating one large angle scattering event. Electrons are scattered much more frequently, and their tracks are as shown:

1. Classical calculation for single scattering. If the screening of the nuclear charge by nearby electrons is neglected, the force on an incident particle due to one nucleus is the Coulomb force $Zze^2/r^2$, where $Z$ is the charge of the nucleus and $z$ the charge of the particle. Assume that the nucleus is heavy compared to the particle so that the center of mass of the system is almost at the nucleus. Let b, the impact parameter, and θ, the angle of deflection be defined in Fig. II.6, p. 35.

For inverse square forces between particles, classical

*Rutherford, Chadwick and Ellis, Radiations from Radioactive Substances, p. 81.

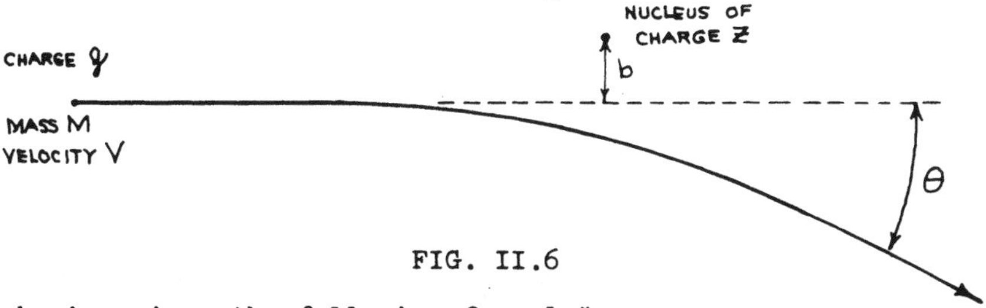


FIG. II.6

mechanics gives the following formula*:

$$\tan\frac{\theta}{2} = \frac{qZe^2}{MbV^2} \qquad \text{II.17}$$

This formula is valid at non-relativistic velocities, $V \ll c$. A relativistically correct version of II.19 for small angles θ is given in the paragraph containing II.20.

Exact quantum mechanical calculation gives the same formula provided the nuclear field is exactly a Coulomb field. Both classically and quantum mechanically the formula is valid only if the distance of nearest approach of the particle to the nucleus is larger than the nuclear radius.

The cross-section for scattering of the incident particle at an angle θ in the range dθ is defined to be the total area ⊥ to the initial path of the particle such that if the particle passes through this area it is deflected by an angle θ in dθ. Since for given particle and nucleus, b is a function of θ only, the area corresponding to a given θ lies at a certain radius b(θ), and has magnitude $d\sigma_\theta = 2\pi\, b(\theta)\, db$. Substituting for b in terms of the corresponding angle θ: $d\sigma_\theta = 2\pi b(\theta) b'(\theta) d\theta$. Divide by the element of solid angle $2\pi \sin\theta d\theta$ to find the cross section per unit steradian, and substitute for b(θ) its value from solving II.17 for b. Then the cross-section per unit solid angle at θ is

$$\boxed{\frac{d\sigma}{d\omega} = \frac{1}{4}\left(\frac{e^2 Z q}{MV^2}\right)^2 \frac{1}{\sin^4 \theta/2}} \qquad \text{II.18}$$

Note that most particles are scattered at small angles.

A relativistically correct equation for θ as a function of b when θ is small can be derived easily, using the same arguments used to derive II.3. Since now we deal with nuclear charge Z and incident particle of charge $q$, we must multipy II.2 by the nuclear charge in order to get the transverse impulse imparted to the incident particle in the collision. This gives

$$\Delta p = \frac{2Zqe^2}{Vb} \qquad \text{II.19}$$

If $p$ is the relativistic momentum of the incident particle, the angle of deflection is very nearly $\Delta p/p$ , if $\Delta p \ll p$.

$$\boxed{\theta \approx \frac{\Delta p}{p} = \frac{2Zqe^2}{Vbp} \qquad (\theta \text{ small})} \qquad \text{II.20}$$

If we put $p = MV$ (non-relativistic) this becomes identical to II.17 when θ is small and $\tan\theta/2 \approx \theta/2$. In these formulae

*See, for example, Lindsay, Physical Mechanics, p.76.

b is limited to distances from the nucleus within which the nuclear charge can be felt, i.e., has not been screened by nearby electrons.

2. Multiple Scattering. Particles, particularly electrons, are deflected many times in passing through a foil of metal. The net angle of deflection, denoted $\Theta$ , is the result of a statistical accumulation of single small scattering events. The detailed theory is complicated*. A simplified treatment will be given here. We assume that no paths are complete loops, as in b FIG.II.7. It is plausible and can be shown** that the values of $\Theta$ for many traversals are distributed about $\Theta = 0$ according to the gaussian law, i.e., probability for $\Theta$ in the range $d\Theta$ is

$$p(\Theta)d\Theta = \text{Const.}\ e^{-K\Theta^2} d\Theta \qquad \text{II.21}$$

For small individual scattering angles, $\theta_i$, $\overline{\Theta_p^2} = \sum_{i=1}^{p} \overline{\theta_i^2}$ II.22

where $\Theta_p$ is the net angle of deflection for p collisions and the bar means the average for many such traversals of the absorber.***

Since statistically the individual events do not differ,

$$\overline{\theta_i^2} = \overline{\theta^2} \qquad \text{II.23}$$

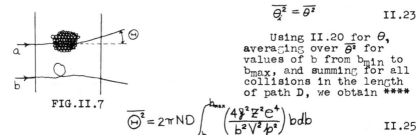

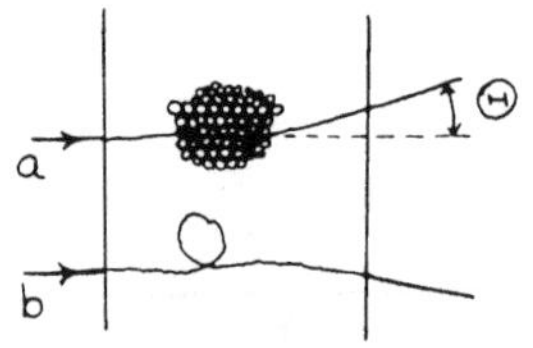

FIG.II.7

Using II.20 for $\theta$, averaging over $\overline{\theta^2}$ for values of b from $b_{min}$ to $b_{max}$, and summing for all collisions in the length of path D, we obtain ****

$$\overline{\Theta^2} = 2\pi N D \int_{b_{min}}^{b_{max}} \left(\frac{4z^2 Z^2 e^4}{b^2 V^2 p^2}\right) b\,db \qquad \text{II.25}$$

For thin foils, D differs little from the thickness of the foil. N = no. of atoms per cc.

Due to screening of the nuclear charge by electrons, the Z felt by a particle depends on b, therefore Z in the integral is a function of b. We take Z outside and adjust for the error by choice of $b_{max}$. Assuming the absorber is so thin that $p$ and $V$ do not change,

$$\overline{\Theta^2} = \frac{8\pi N D z^2 Z^2 e^4}{V^2 p^2} \ln \frac{b_{max}}{b_{min}} \qquad \text{II.26}$$

Choice of limits: $b_{max}$: The equation II.26 would be strictly correct if at distances beyond $b_{max}$ the screening of the nucleus were perfect; i.e., no scattering, and for distances within $b_{max}$ there were no screening at all, i.e., full value of Z were felt. No such boundary exists, since screening increases gradually with distance, but fortunately the log term

---

* E. J. Williams, Proc.Roy.Soc. A169 531 (1939); Rossi and Greisen, Rev. Mod. Phys. 13 249 (1941).

** Rossi and Greisen, l.c.

*** This is shown in Appendix II.1

**** Shown in greater detail in Appendix II.2

is not sensitive and we may put*

$$b_{max.} = \frac{\text{BOHR RADIUS}}{Z^{1/3}} = \frac{a_0}{Z^{1/3}} \qquad \text{II.27}$$

The factor $Z^{1/3}$ takes into account the variation of the function $Z_{eff.}(r)$ with $Z$. $Z_{eff.} \equiv Z$ — (electron charge within sphere of radius $r$).

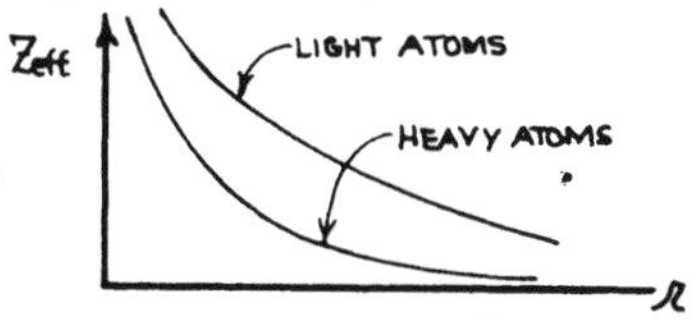


FIG. II.8

The limit $b_{min}$ effectively adjusts the maximum angle in a single scattering process. Since we are not counting any values of $\Theta > 1$, we may impose the rough restriction that $\theta < 1$. This gives

$$b_{min} = \frac{2 z Z e^2}{V p} \qquad \text{(from II.20)} \qquad 11.28$$

Other considerations may govern $b_{min}$ such as (1), the finite size of the nucleus: $b_{min} > A^{1/3}\, 1.5 \times 10^{-13}$, from equation 1.4 (page 6); (2), size of wave packets of particles in the collision**. The result for $\overline{\Theta^2}$ using our choices for $b_{min}$ and $b_{max}$ is

$$\boxed{\overline{\Theta^2} = \frac{8\pi N D z^2 Z^2 e^4}{V^2 p^2} \ln \frac{a_0 V p}{2 Z^{1/3} z e^2}} \qquad 11.29$$

The result is not sensitive to the choice of $b_{min}$ and $b_{max}$; the log term is of the order of magnitude of 10. Due to the $Z^2$ factor, scattering increases rapidly with weight of the elements in the absorber, even for equal atomic densities.

Classically $Vp = 2 \times$ (kinetic energy). In the extreme relativistic range $Vp \approx cp \approx$ kinetic energy. Therefore, roughly,

$$\sqrt{\overline{\Theta^2}} \sim \frac{Z}{\text{KINETIC ENERGY}} \qquad 11.30$$

This formula gives the most important characteristics of II.29. Evidently a proton and an electron of the same energy will be scattered about the same amount. But the range of the electron is much greater, and therefore, experimentally, large angle scattering of light particles is more prominent than of heavy particles.

Numerical formulae:

$$\boxed{\text{ELECTRONS IN LEAD} \quad \overline{\Theta^2} \approx \frac{6 \times 10^8\, D}{(\text{KILO e.v.})^2}} \qquad 11.31$$

For example, from II.31, a 1 Mev electron is deflected about one radian in passing thru $10^{-3}$ cm. of lead.

$$\boxed{\text{ELECTRONS IN AIR} \quad \overline{\Theta^2} \approx \frac{7000\, D}{(\text{KILO e.v.})^2}} \qquad 11.32$$

---

* This is the radius of an atom of charge $Z$ using the Thomas-Fermi model; cf. Lindsay, Physical Statistics, p.226.

** These are discussed in Williams, l.c., and more briefly, in Jánossy, Cosmic Rays, para. 173.

Formulas for range and absorption must be interpreted with scattering in the mind. The true path lenght may be much longer than the thickness of the absorber. The equations for range of particles applied to a homogeneous beam of non-scattered particles passing through an absorber would yield a curve of number of particles emerging vs. thickness like FIG. II.9a. But experimentally, FIG. II.9 b is more nearly correct for electrons, and the reason for the difference is straggling due to scattering.

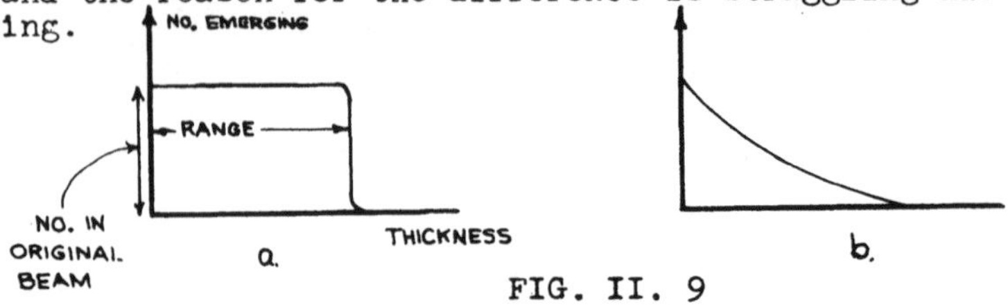


FIG. II. 9

There is a problem on multiple scattering in Appendix II.3

## C. Passage of Electromagnetic Radiation through Matter

There are three processes by which photons are either absorbed or lose energy:

(1) Photoelectric absorption
(2) Compton effect
(3) Pair formation.

1. Photoelectric absorption. The electrons bound to an atom may be grouped into shells according to their binding energies. The innermost shell, the K shell, contains the two 1s electrons and is bound with an energy of about Ry $(Z-1)^2$, where Ry is the Rydberg, about 13.5 e.v. The next shells are the L and M shells. Electrons in the L and M shells are bound with energies roughly Ry $(Z-5)^2/4$ and Ry $(Z-13)^2/9$, respectively. These last two numbers are very approximate, since the energy depends on screening and the shape of the orbit. The last filled orbits of the atom are the valence orbits producing optical spectra. Beyond this, the states are unfilled. Provided its energy is sufficient, a photon may remove an electron from any of the shells and leave it either in one of the previously unoccupied bound states, or in an ionized state. As the frequency $\nu$ of incident photons is increased from zero, they first become able to excite electrons in optical orbits of a few e.v. binding energy. This loss of energy is the first contribution to absorption of photons by photoelectric effect. As $\nu$ is further increased, electrons in deeper lying orbits may absorb energy. When $\nu$ reaches the binding energy of a particular shell of electrons, there is a sharp rise in absorption. A rough plot of the absorption as a function of $\nu$ is given in FIG. II.10.

The point at which the absorption changes abruptly due to, say, absorption by K electrons is called the K absorption edge, and is at the energy at which K electrons are lifted to energy levels $> 0$. Energy 0 is the energy of an ionized electron with zero velocity.

The L absorption edges are three-fold, since the L shell has sub-shells 2S, $2P_{1/2}$ and $2P_{3/2}$ having different energies. This energy splitting is due to (1) different screening of the nucleus for S and P orbits, and (2) electron spin-orbit interaction splitting of $P_{1/2}$ and $P_{3/2}$.

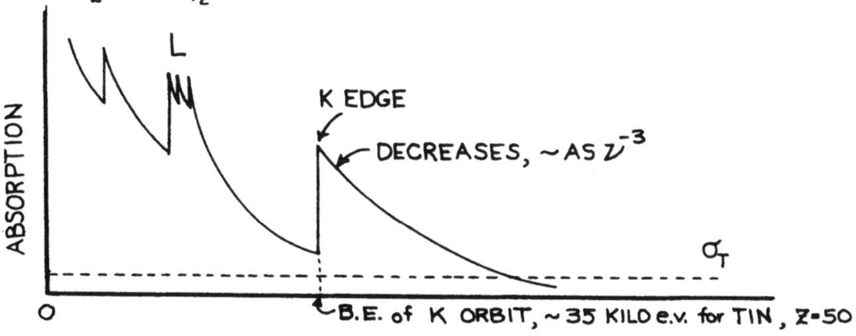


FIG. II.10

The cross-section for photoelectric absorption by one K shell electron (for transitions to the continuum of energy $> 0$) has been calculated by quantum mechanics, assuming a hydrogen-like wave function for the electron (1s). The result is*

$$\sigma_p(\text{cm}^2) \text{ per 1s electron} = \frac{128\pi}{3}\,\frac{e^2}{mc}\,\frac{\nu_K^3}{\nu^4}\,\frac{e^{-4\epsilon\cot^{-1}\epsilon}}{1-e^{-2\pi\epsilon}} \qquad \text{II.33}$$

where $\nu_K$ is the frequency of the K absorption edge, $\nu$ is the frequency of the photon and is greater than $\nu_K$, and $\epsilon \equiv \sqrt{\frac{\nu_K}{\nu-\nu_K}}$

The following simplified formula holds for $\nu$ near to $\nu_K$:

$$\sigma_p(\text{cm}^2) \text{ per 1s electron} = \frac{6.31\times10^{-18}}{Z^2}\left(\frac{\nu_K}{\nu}\right)^{8/3} \quad \nu \text{ near to } \nu_K \qquad \text{II.34}$$

The mass absorption coefficient $\mu$ is defined by the relation (fraction of photons not absorbed when passing through thickness $\xi$ in g-cm$^{-2}$ of absorber) $\equiv \exp(-\mu\xi)$. A similar equation defines the absorption coefficient $\chi$, i.e., the fraction passing through thickness x in cm is equal to $e^{-\chi x}$.

Problem: Derive the relation between $\mu$ and $\sigma$ per atom. Calculate $\mu$ for lead.

Consider a beam of $n_0$ photons-cm$^{-2}$-sec$^{-1}$ impinging on a section of absorber of thickness dx and cross-sectional area S. An area B given by B = $\sigma$ (cm$^2$) x N (cm$^{-3}$) x S (cm$^2$) x dx (cm) may be considered to absorb all photons hitting it. $B \le S$. The number of photons absorbed per second is $(B/S)n_0$. The emerging beam has density $n$ per cm$^2$ per sec.

$$n = n_0 - \frac{B}{S}n_0 \quad \therefore \frac{\Delta n}{n} = -\frac{B}{S} = -\sigma N\,dx \quad \therefore n = n_0\, e^{-\sigma N x}$$

The fraction not absorbed in a finite distance x (cm) is

$$\frac{n}{n_0} = e^{-\sigma N x} = e^{-\mu\xi} \qquad \text{by definition.}$$

* Heitler, Quantum Theory of Radiation, 2nd ed. p. 124

But $\xi = \chi\rho$ where $\rho$ is the density in g-cm$^{-3}$. Therefore

$$\mu = \frac{\sigma N}{\rho}\,(\text{cm}^2\text{-g}^{-1}) = \frac{\sigma N_0}{A} \text{ in terms of } N_0 = 6 \times 10^{23} \qquad \text{II.35}$$

$\sigma$ may be calculated by equation II.33, multiplied by the number of K electrons per atom, i.e., 2, and $\mu$ found by II.35. This calculation ignores the contribution of the L,M,N etc. shells. According to Rutherford, Chadwick & Ellis, (l.c., p. 464) the K shell accounts for ~ 80% of absorption for photons of energy ~ $\frac{1}{2}$ Mev. Using equation II.34, $\mu$ for lead at photon energy ~ .19 mc$^2$ is 5.76 cm$^2$-g$^{-1}$. Actually $\mu$ is ~ 8. In finding $\mu \approx 5.76$, we ignored absorption in other than K shell.

The sharpness of the absorption edges makes possible a method of measuring the frequency of $\gamma$ rays of low energy. The K edges vary for different elements, and one may determine e-tween which two K edges the unknown photon energy lies by observing the sudden change in absorption with change of K edge location of the absorber used.

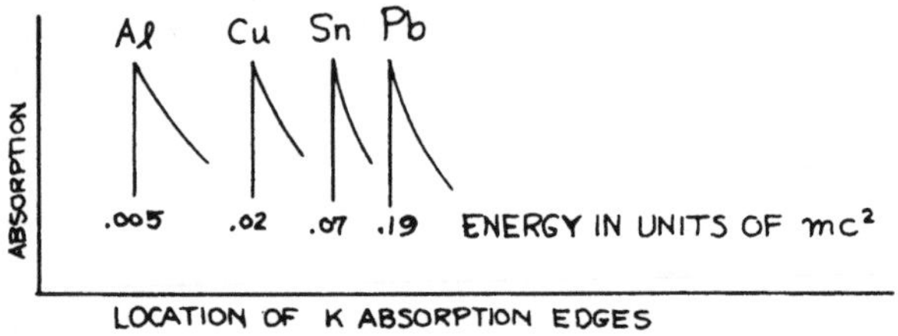


FIG. II.11

Problem: Given a source of 85 kilovolt $\gamma$ radiation. What absorbers would you use to bracket this energy and thus serve to determine it? Determine the thickness to be used. Give consideration to the availability of the chosen absorbers. Compounds may be used. The absorption of a compound is the sum of the separate absorptions due to the constituent elements.

2. Compton scattering. Photoelectric absorption decreases as $\nu$ increases. Soon after the K edge is passed the predominant process removing energy from a beam is scattering by electrons. This is not true absorption, since most of the energy is not absorbed but rather sent in a different direction. In the low energy region, $h\nu \ll mc^2$, scattering cross-section can be calculated classically and is given by the Thomson formula:

Assume a plane polarized electromagnetic wave impinges on an electron. The electron is set in motion and therefore radiates. Assume the electron is free, a good approximation if $\nu$ of the photon is much larger than the frequency of the electron. If $\underline{A}$ is the acceleration of the electron,

$$e\underline{\mathcal{E}} = m\underline{A}$$

$$\therefore\ \underline{A} = \frac{e\underline{\mathcal{E}}}{m}$$

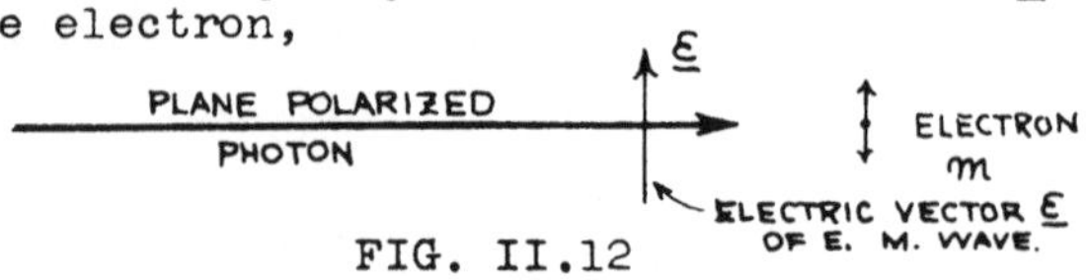


FIG. II.12

The instantaneous power radiated from a moving electron is*

$$\frac{dE}{dt} = \frac{2}{3}\frac{e^2}{c^3}A^2 \qquad \text{II.36}$$

In this case the average power is $\overline{\frac{dE}{dt}} = \frac{2}{3}\frac{e^4}{c^3 m^2}\overline{\mathcal{E}^2}$

This energy taken from the incident wave and re-radiated may be expressed in terms of a cross-section $\sigma_T$ per electron defined so that

$$\sigma_T \times (\text{intensity of incident beam}) = \begin{pmatrix}\text{total energy radiated per} \\ \text{unit time by the electron}\end{pmatrix}$$

$$\sigma_T(\text{cm}^2) \times I\ (\text{erg-cm}^{-2}\text{-sec}^{-1}) = \frac{dE}{dt}\ (\text{erg-sec}^{-1}) \qquad \text{II.37}$$

This definition is equivalent to saying that all the energy incident upon a surface $\sigma_T$ is absorbed and re-radiated. The intensity of the incident beam is

$$I = \frac{\overline{\mathcal{E}^2}}{4\pi}c \ (\text{erg-cm}^{-2}\text{-sec}^{-1})$$

Therefore

$$\sigma_T = \frac{\frac{dE}{dt}}{I} = \frac{8\pi}{3}\left(\frac{e^2}{mc^2}\right)^2 = \frac{8\pi}{3}(r_e)^2 = 0.66 \times 10^{-24}(\text{cm}^2)(\underline{\text{per electron}}) \qquad \text{II.38}$$

where $r_e$ is the classical radius of the electron, $2.8 \times 10^{-13}$ cm.

The Thomson cross-section is evidently independent of frequency. II.38 breaks down when $h\nu$ is near or above $mc^2 \approx \frac{1}{2}$ Mev. Then the phenomenon must be treated as Compton scattering. In the Compton effect, an incident photon $h\nu$ collides with an electron. Compared with the energy of $h\nu$, the binding energy of the electron is negligible. The scattered photon suffers a change of energy which can be calculated by imposing conservation of relativistic energy and relativistic momentum on the collision** The result is

$$\lambda_{\text{scattered}} - \lambda_{\text{incident}} = \frac{h}{mc}(1-\cos\theta) \qquad \text{II.39}$$

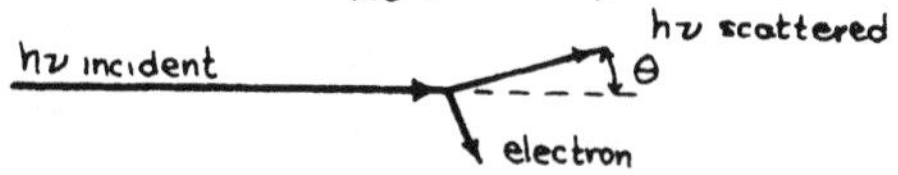


The frequency of the scattered photon is less than that of the primary. The factor $h/mc \equiv \lambda_c$ is the Compton wavelength, $0.024 \times 10^{-8}$ cm.

The cross-section for Compton scattering is given by the Klein-Nishina formula***:

$$\boxed{\sigma_C(\text{cm}^2)\text{ per electron} = 2\pi r_e^2\left\{\frac{1+\alpha}{\alpha^2}\left[\frac{2(1+\alpha)}{1+2\alpha} - \frac{1}{\alpha}\ln(1+2\alpha)\right] + \frac{1}{2\alpha}\ln(1+2\alpha) - \frac{1+3\alpha}{(1+2\alpha)^2}\right\}, \quad \alpha = \frac{h\nu}{mc^2}} \qquad \text{II.40}$$

*Abraham-Becker, Theorie der Electrizität, Vol. II, p. 75

**This is done, for example, in Richtmeyer and Kennard, Introduction to Modern Physics, 3rd ed. p. 533.

***Derived in Heitler, l.c., p.149

Problem: Find asymptotic expressions for II.40 for low energies and for high energies. Plot II.40 roughly.

The asymptotic expressions are:

Low energies: $\sigma_c = \sigma_T \left\{ 1 - 2\alpha + \frac{26}{5}\alpha^2 + \cdots \right\}, \quad \alpha \ll 1$ II.41

High energies: $\sigma_c = \frac{3}{8}\sigma_T \frac{1}{\alpha}\left( \ln 2\alpha + \frac{1}{2} \right) \quad \alpha \gg 1$ II.42

The Klein-Nishina formula is plotted below:*

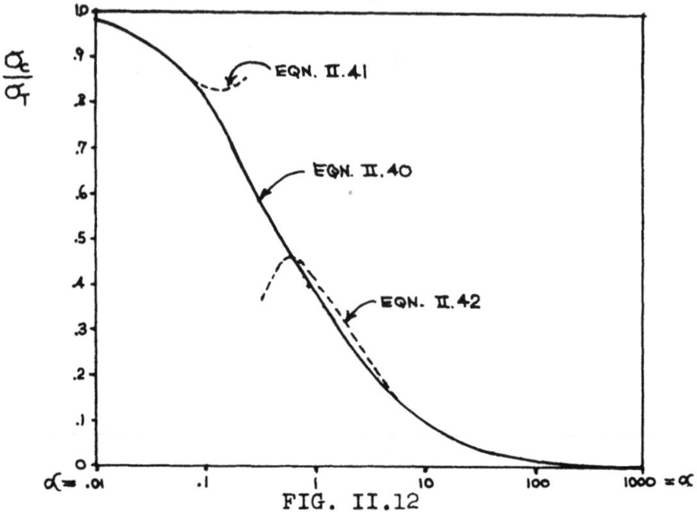


FIG. II.12

Angular distribution. For low energies where the Thomson formula II.38 holds, the angular distribution of radiated energy is that due to a radiating classical electron, and is proportional to $\sin^2\psi$, where $\psi$ is the angle between $\underline{A}$ and the direction of observation, Fig. II.13. If the primary radiation is not polarized, an average over the azimuth angle $\phi$ must be taken, and the result is that the maxima of scattered intensity are symmetrical forward and backward. When $h\nu$ approaches and exceeds $mc^2$ the radiation is predominantly forward.

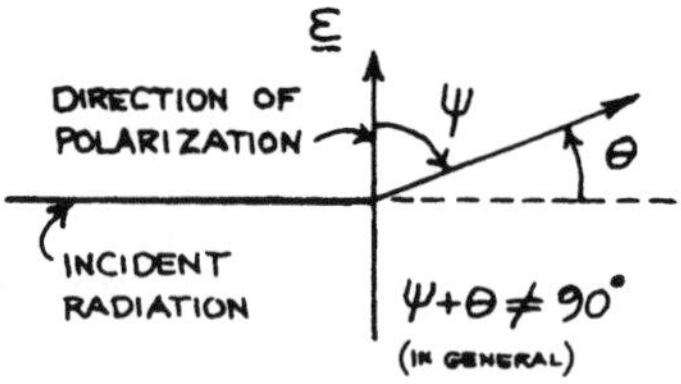


FIG. II.13

The general formula for angular distribution of scattered intensity is the following:

$$\frac{I(R,\psi,\theta)}{I_{primary}} = r_e^2 \frac{1}{R^2} \frac{\sin^2\psi}{[1+\alpha(1-\cos\theta)]^3} \left\{ 1 + \frac{\alpha^2(1-\cos\theta)^2}{2\sin^2\psi[1+\alpha(1-\cos\theta)]} \right\} \qquad \text{II.43}$$

where $I(R,\psi,\theta)$ is the intensity at $R$ in the direction $\theta,\psi$; $\alpha$ is $h\nu/mc^2$; and $I_{primary}$ is the intensity of plane polarized incident radiation. The Klein-Nishina formula II.40 would result from integrating this equation.

* From Jánossy, Cosmic Rays, (Oxford).

3. Radiation loss by fast electrons (bremsstrahlung). This paragraph should come under II. A, sub-section 10, but is placed here because our calculation makes use of the Thomson formula for $\sigma_T$ , derived in the last sub-section.

Consider an electron of velocity $V \approx c$ passing a nucleus Ze. We calculate the energy given off by the electron in the form of photons when it is accelerated by the field of the nucleus.

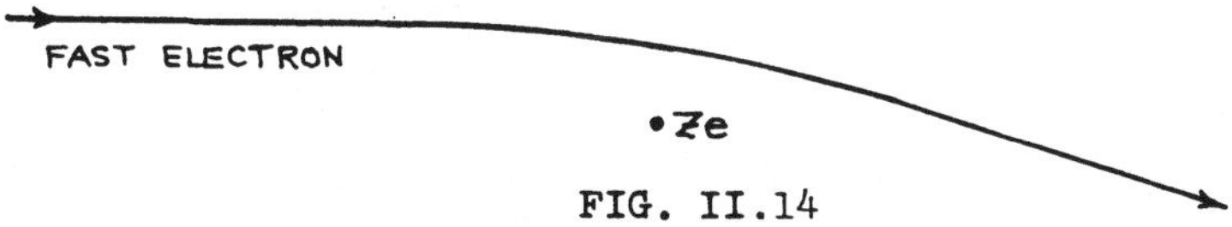


FIG. II.14

In the rest system of the electron, the nucleus Ze moves with velocity almost c. The electric field $\mathcal{E}$ of the nucleus is contracted, and the associated magnetic field $\mathcal{H}$ is $\perp$ to $\mathcal{E}$ and of about the same magnitude. Therefore the moving nucleus looks to the electron like a plane electromagnetic wave. This will be discussed more fully shortly. The wave of photons representing the nucleus suffers Compton scattering by the electron. These scattered photons, when viewed from the rest system of the nucleus (the laboratory system) appear as the photons emitted by the incident electron. According to this brief summary, the calculation has the following parts:

(a) Lorentz transform to the rest system of the electron.

(b) Calculate the characteristics of the pulse of electromagnetic radiation that represents the moving nucleus.

(c) Find the density of photons corresponding to the pulse of electromagnetic radiation.

(d) Calculate the probability of scattering of these photons by the electron.

(e) Transform back to the laboratory frame.

Throughout this calculation we leave out numerical factors. Their inclusion would obscure the ideas involved. The correct numerical factor is inserted at the end.

(a) Let $\beta = V/c \approx 1$; $\gamma \equiv 1/\sqrt{1-\beta^2}$ is large. The energy of the electron in the laboratory frame is $\gamma mc^2$.

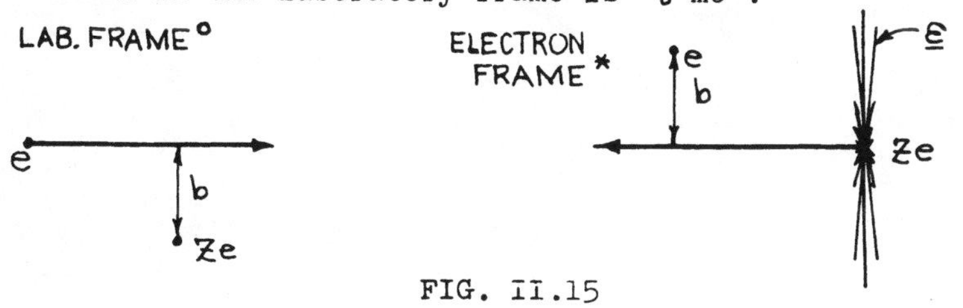


FIG. II.15

In the lab frame, denoted by $^{\circ}$, the electric field of the nucleus is $|\mathcal{E}^{\circ}| = Ze/b^2$. In the electron frame, denoted $*$, its perpendicular component $\mathcal{E}_{\perp}^{*}$ is increased by the factor $\gamma$;

$$\mathcal{E}_{\perp}^{*} = \mathcal{E}_{\perp}^{\circ}\gamma = Ze\gamma/b^2$$

The magnetic field $\mathcal{H}^*$ seen by the electron in its own rest system is $\perp$ to both $\mathcal{E}_{\perp}^{*}$ and the velocity vector. Denote this field by $\mathcal{H}_{\phi}^{*} = |\mathcal{H}_{\phi}^{*}|$. Then* $\mathcal{H}_{\phi}^{*} = \mathcal{E}_{\perp}^{\circ}\gamma\beta = \frac{Ze}{b^2}\gamma\beta \approx \frac{Ze}{b^2}\gamma$.

Therefore $\mathcal{H}_{\phi}^{*} = \mathcal{E}_{\perp}^{*}$, as in a plane electromagnetic wave.

(b) The region in which the electromagnetic field is not zero has a length, in the electron frame, of about $b/\gamma$, the factor $\gamma$ coming from the Lorentz contraction.

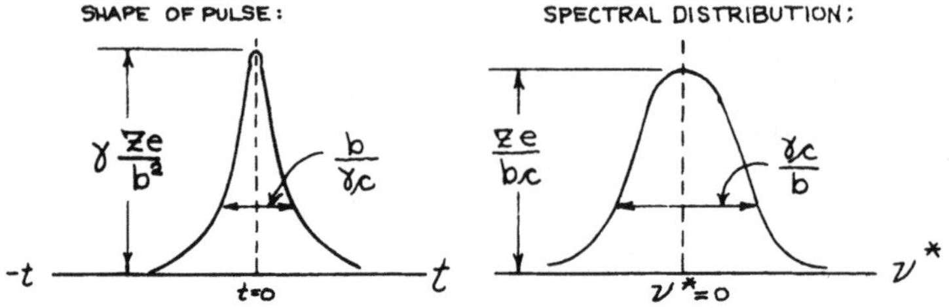


t=0 when the electron is in wave-front plane representing nucleus

FIG. II.16

The spectral analysis of this pulse may be done by assuming for simplicity that it is a Gaussian of time width $b/\gamma c$:

$$\mathcal{E}_{\perp}^{*} = \mathcal{H}_{\phi}^{*} \propto e^{-\frac{t^2\gamma^2c^2}{2b^2}} \qquad \text{II.44}$$

Then the amplitude distribution of component frequencies is another gaussian, of width $\gamma c/b$ **.

The energy per unit area in the disc-shaped electromagnetic disturbance is a function of the distance b. Energy per unit volume is given by $\frac{\mathcal{E}^2+\mathcal{H}^2}{8\pi} \approx \mathcal{E}^2 \approx \left(\frac{Ze\gamma}{b^2}\right)^2$. The volume per unit area of the disc equals the thickness $= b/\gamma$ (cm). Therefore the energy per unit area is

$$\left(\frac{Ze\gamma}{b^2}\right)^2 \times \frac{b}{\gamma} = \frac{Z^2e^2\gamma}{b^3} \quad (\text{erg-cm}^{-2})$$

(c) Approximate to the gaussian spectral distribution by a rectangle of width $\gamma c/b$ ***. This means that the energy carried by photons in the frequency range $\Delta\nu^*$ is simply proportional to

---

*Abraham-Becker, Vol. II, p.48.

**Stratton, Electromagnetic Theory, p. 290.

***It will turn out later, (just before II.54) that only frequencies in the electron system less than $\nu_e^* = mc^2/h$ will be integrated over, and that $\nu_e^* = \frac{mc^2}{h} << \gamma c/b$. Therefore the portion of the rectangle used will be at a relatively flat part of the gaussian, as shown in Fig. II.17.

$$\frac{\Delta\nu^*}{\nu^*_{MAX.}} = \frac{\Delta\nu^* b}{\gamma c} \;;$$

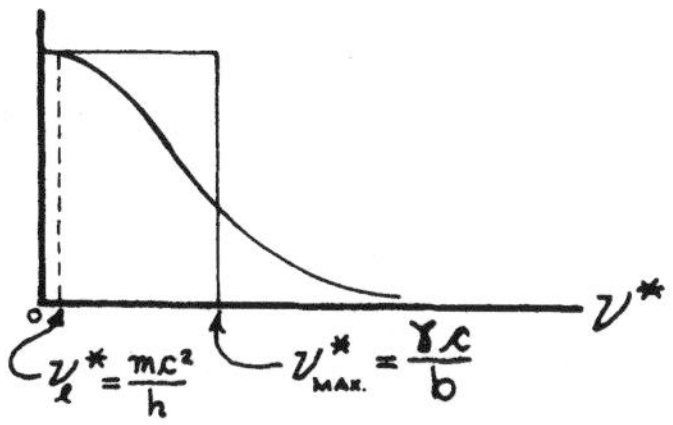


FIG. II.17

therefore the energy carried by photons in this range $\Delta\nu^*$, per $cm^2$ at distance b, is

$$\frac{\Delta\nu^* b}{\gamma c}\,\frac{Z^2 e^2 \gamma}{b^3}$$

The number of photons per $cm^2$ in the range $\Delta\nu^*$ is equal to this $\times$(energy per photon)$^{-1}$ which is $(h\nu^*)^{-1}$. Therefore the number of photons per $cm^2$ in the frequency range $\Delta\nu^*$ at the distance b is given by:

$$\frac{Z^2e^2\gamma}{b^3}\,\frac{\Delta\nu^* b}{\gamma c}\,\frac{1}{h\nu^*} = \begin{cases} \dfrac{Z^2e^2}{ch\,b^2}\,\dfrac{\Delta\nu^*}{\nu^*} & \text{up to } \nu^* = \dfrac{\gamma c}{b} \\ 0 & \text{for } \nu^* > \dfrac{\gamma c}{b} \end{cases}$$

(d) The mean number of scattering processes for photons of frequency $\nu^*$ occurring if the distance is b is:

$$\sigma_c \times \text{(density (per cm}^{-2}\text{) of photons of freq. } \nu^*)$$

Therefore

$$\begin{pmatrix}\text{No. of scattering events}\\ \text{for freq. } \nu^*\text{, as a function}\\ \text{of b, and for one nucleus}\end{pmatrix} = \frac{Z^2e^2}{ch\,b^2}\,\frac{\Delta\nu^*}{\nu^*}\,\sigma_c(\nu^*) \qquad \text{II.45}$$

Let the total cross-section per nucleus be $\Sigma(\nu^*)\Delta\nu^*$ = integral of II.45 over the area of the photon packet accompanying the nucleus. Therefore

$$\Sigma(\nu^*)\Delta\nu^* = \sigma_T(\nu^*)\frac{Z^2e^2}{ch\nu^*}\Delta\nu^*\int\frac{b\,db}{b^2} = \sigma_c(\nu^*)\frac{Z^2e^2\Delta\nu^*}{ch\,\nu^*}\ln\frac{b_{max.}}{b_{min.}}\;. \qquad \text{II.46}$$

$b_{max}$ is subject to limitations of which these are the most important: (1) Beyond $b'_{max} = c\gamma/\nu^*$ the frequency $\nu^*$ (for which we are computing $\Sigma(\nu^*)$) is not present in the packet. (2) The nucleus is screened by nearby electrons, and is not felt at distances greater than about $b''_{max} = a_0/Z^{1/3}$ (cf. eqn. II.27). At high velocities $\gamma$ is large, and $b'_{max} > b''_{max}$. We assume this is the case and use

$$b_{max} = \frac{a_0}{Z^{1/3}} \qquad \text{II.47}$$

$b_{min}$ is subject to several limitations. Which is dominant depends on the particular case. The most important is that $b_{min}$ must be larger than the quantum mechanical wave packet representing the electron. The "optimum" wavelength for the electron is the Compton wavelength $\lambda_c$ *. We may let the electron have velocity $c/\sqrt{2}$ with respect to the new system, this giving it de Broglie wavelength $\lambda_c$, and still have the nucleus move with velocity almost c, due to the relativistic addition law for velocities. Taking $b_{min} = \lambda_c$, we get:

$$\frac{b_{max.}}{b_{min.}} = \frac{a_0}{\lambda_c Z^{1/3}} = 137\,Z^{-\frac{1}{3}} \qquad \text{II.48}$$

*This is the smallest extension that can be attributed to an electron without ambiguity arising from quantum electrodynamics. This point is mentioned in Heitler, l.c., p. 193.

More precise calculation leads to 183 $Z^{-\frac{1}{3}}$ for II.48. Therefore

$$\Sigma(\nu^*)\Delta\nu^* = \frac{Z^2e^2}{ch}\frac{\Delta\nu^*}{\nu^*}\sigma_c \ln\frac{183}{Z^{1/3}} = \frac{Z^2e^2}{ch} r_e^2 \frac{\Delta\nu^*}{\nu^*}\ln\frac{183}{Z^{1/3}} \qquad \text{II.49}$$

(e) Transform back to the lab$^o$ system. The frequency of the photon changes from $\nu^*$ to $\nu^o$. From special relativity*

$$\nu^o = \frac{1+\beta\cos\theta^*}{\sqrt{1-\beta^2}}\nu^* \qquad \text{II.50}$$

Therefore

$$\frac{\Delta\nu^o}{\nu^o} \approx \frac{\Delta\nu^*}{\nu^*} \qquad \text{II.51}$$

Introducing this result into II.49, we get:

$$\Sigma(\nu^o)\Delta\nu^o = 4\frac{\Delta\nu^o}{\nu^o}\frac{Z^2e^2}{c\hbar} r_e^2 \ln\frac{183}{Z^{1/3}} \qquad \text{II.52}$$

The factor 4 $h/\hbar$ results from exact derivation. The cross-section per nucleus for scattering at frequency $\nu^o$ in the lab$^o$ system is

$$\boxed{\Sigma(\nu^o)\Delta\nu^o = \frac{4}{137}Z^2\frac{\Delta\nu^o}{\nu^o} r_e^2 \ln\frac{183}{Z^{1/3}}} \qquad \text{II.53}$$

where $e^2/c\hbar$, the fine-structure constant, has been replaced by 1/137.

Before integrating to find the total energy loss by the electron, we examine the limits of validity of this result. In II.49 the maximum value for $\sigma_c$ was used. This breaks down at $h\nu^* = mc^2$. This defines the limiting value of $\nu^*$, namely $\nu_l^*$. Corresponding to $\nu_l^*$ are two values of $\nu^o$, one for forward and one for backward scattering in the electron frame. (Thomson scattering is symmetrical, forward and back.) The larger value of $\nu^o$ results for a numerator in II.50 of about two, i.e.,

$$\nu^o_{max.} = \nu_l^*\frac{1+\beta\cos\theta^*}{\sqrt{1-\beta^2}} \approx \gamma\nu_l^*$$

Therefore the use of the Thomson scattering formula II.38 causes the theory to break down when $h\nu^o/\gamma = mc^2$, or

$$\nu^o_{max.} = \frac{mc^2\gamma}{h} \qquad \text{II.54}$$

But this says $\nu^o_{max} = \frac{\text{energy of electron}}{h}$, which clearly must be so, because the electron cannot radiate a photon of energy greater than its initial energy.

The expression II.53 diverges at low frequencies. But the energy carried by low frequency photons is finite. The total energy loss to photons of frequency $\nu^o$ per path length dx is equal to:

(energy per photon) X (no. of nuclei/cm$^3$) X (cross-section $\Sigma(\nu^o)\Delta\nu^o$) X dx

Therefore $-(dE)_{RAD.}$, the total average energy loss per path length dx for all frequencies is given by:

$$(dE)_{RAD.} = -dx\int_{\nu^o=0}^{\nu^o=\nu^o_{max}} h\nu^o N\,\Sigma(\nu^o)\,d\nu^o \qquad \text{II.55}$$

* cf. Abraham-Becker, Vol. II, p. 312.

$$\left(\overline{dE}\right)_{RAD} = -4Z^2 \frac{NE}{137} r_e^2 \ln\frac{183}{Z^{1/3}}\, dx \qquad \text{II.56}$$

where $E = h\nu^\circ_{max} =$ energy of electron.

Since an electron may give an appreciable amount of its energy to a single photon, the actual energy loss may vary widely from the average value of II.56.

$(\overline{dE/dx})_{rad}$ is evidently proportional to energy. It is therefore important in cosmic ray physics where particles have very high energies.

We may write II.56 in terms of $l_R$, the radiation length.

Define
$$l_R \equiv \frac{1}{\frac{4Z^2N}{137} r_e^2 \ln\frac{183}{Z^{1/3}}} \qquad \text{II.57}$$

Then
$$\left(\overline{dE}\right)_{RAD} = -E\frac{dx}{l_R} \quad \text{or} \quad E = E_0 e^{-\frac{x}{l_R}} \qquad \text{II.58}$$

For $x = l_R$, 1/2.7-- of the energy is dissipated, on the average.

$$l_R \approx \frac{C}{Z^2N} \qquad \text{where } C \text{ is nearly constant.} \qquad \text{II.59}$$

Roughly, $l_R \propto \frac{1}{ZAN} \propto \frac{1}{Z\rho}$

Radiation loss per g-cm$^{-2}$ traversed is greater for heavy elements. The ratio of radiative energy loss to ionization loss, for electrons, is given roughly by*

$$\frac{\left.\frac{dE}{dx}\right)_{RAD.}}{\left.\frac{dE}{dx}\right)_{ION.}} = \frac{Z\times(\text{Mev})}{800} \qquad \text{II.60}$$

| Material | $l_R$ | Energy for $\left.\frac{dE}{dx}\right)_{RAD} = \left.\frac{dE}{dx}\right)_{ION.}$ |
|---|---|---|
| Air, NTP | 330 m. | 120 Mev |
| Aluminum | 9.7 cm. | 52 |
| Lead | .517 cm. | 7.0 |

(Thickness of the atmosphere is equivalent to 8 km. of NTP air, or about 20 $l_R$)

4. Pair formation. We return to the discussion of absorption of electromagnetic radiation in matter. The most important energy loss by very high energy electromagnetic radiation is by pair formation. In this process a photon disappears and a positron and an electron appear. This is to be understood only by Dirac's relativistic wave mechanics. The following discussion is simplified and qualitative**.

According to the relativistic theory of the electron, an electron has energy $\pm\sqrt{(mc^2)^2 + p^2c^2}$. This equation permits

*This appears, inverted, in Lecture XIII of Los Alamos Report #24, p. 32 of printed version. Also inverted in page 72 of Rasetti.
** The theory is given, for example, in Heitler, l.c.

negative energy values. The energy spectrum of a free electron looks like Fig. II.18.

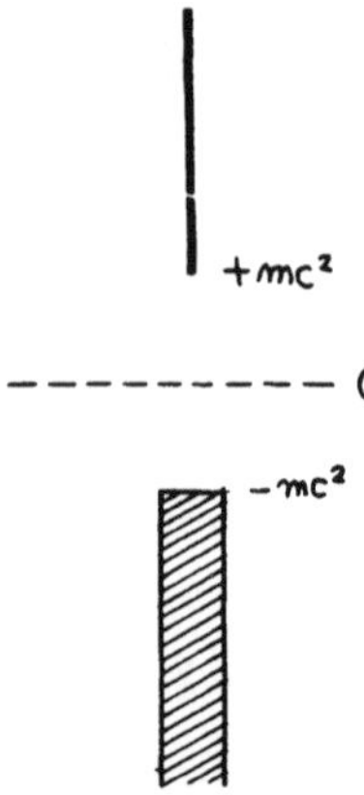


FIG. II.18

In Dirac's theory, practically all negative states are filled at all points in space. A vacuum is then a sea of electrons in negative energy states. The presence of this charge is not observed because it is uniformly distributed.

A photon of sufficiently high energy may lift an electron from a negative energy state to a positive energy state. The energy threshold for the photon is $2mc^2$, since for a free electron there are no states between $-mc^2$ and $+mc^2$. Physically, this means that the photon must supply enough energy to create two particles of mass m. Momentum must be conserved and this requires either that the negative energy electron be near a nucleus or an electron, i.e., not free, or that two photons coming from different directions coalesce and lift an electron from a negative energy state*. If the electron is near a nucleus it may occupy discrete states just below $+mc^2$. These are within a few e.v. of 510,000 e.v. Strictly, then, the threshold for pair formation near a nucleus is $2mc^2$ - (binding energy of electron). This is of no importance because binding energy $\ll mc^2$ and because transitions from negative energy states to the discrete part of the spectrum are improbable and not yet observed.

The result for cross-section for pair production near a nucleus is calculated in Heitler** and is

$$\sigma_{PAIR}(\mathrm{cm}^2) = \frac{Z^2}{137} r_e^2 \left(\frac{28}{9} \ln \frac{183}{Z^{1/3}} - \frac{2}{27}\right) \qquad h\nu \gg mc^2 \qquad \text{II.61}$$

In the extreme relativistic range, $\sigma_{PAIR}$ is independent of energy For other energies the situation is more complicated, and is described in Heitler.

(A simplified procedure similar to that used to compute the radiation loss by fast electrons may be used to compute the cross-section for pair formation at very high energies, if the process of collision of photons in a vacuum ($h\nu + h\nu \rightarrow e^+ + e^-$) is known This is outlined. Consider a photon of $h\nu \approx 1000\ mc^2$ or so. Transform to a new coordinate system in which the nucleus moves very fast, almost, but not quite, c. The photon seen in the new system has reduced frequency. The nucleus looks like a wave of photons (just as in subsection 3). Give the nucleus such a velocity that its photons have the same (reduced) frequency as the incident photon seen in the new coordinate system. The process is the simpler one of collision of photons in a vacuum, $h\nu + h\nu \rightarrow e^+ + e^-$.)

*It is shown in Appendix II.4 that momentum cannot be conserved in pair formation by one isolated photon.
**Heitler, l.c., p. 200

We may define a <u>mean free path for pair production</u>

$$l_P \equiv \frac{1}{N\sigma_{PAIR}} \qquad \text{II.62}$$

Then the decrease in intensity $n$ of a beam of photons will follow the equations:

$$\frac{dn}{n} = -\frac{dx}{l_P} \qquad n = n_0 e^{-\frac{x}{l_P}} \qquad \text{II.63}$$

$$l_P = \frac{1}{\frac{28}{9}\frac{Z^2N}{137}r_e^2 \ln\frac{183}{Z^{1/3}}} \qquad \text{II.64}$$

This is analogous to the equations for radiation loss by an electron, II.58 and II.57

$$\frac{dE}{E} = -\frac{dx}{l_R} \qquad E = E_0 e^{-\frac{x}{l_R}} \qquad \text{(II.58)}$$

$$l_R = \frac{1}{4\frac{Z^2N}{137}r_e^2 \ln\frac{183}{Z^{1/3}}} \qquad \text{(II.57)}$$

Evidently

$$l_R = \frac{7}{9} l_P \qquad \text{II.65}$$

5. The phenomena of radiation and pair formation are responsible for the shower phenomenon in cosmic rays*. The shower is initiated by an electron having energy of the order of billions of e.v. By radiation it produces photons of comparable energy. These photons produce electron-positron pairs, each particle of which radiates. The energy is soon divided among many particles. When the energy of a particle reaches the point where ionization loss predominates, the shower stops.

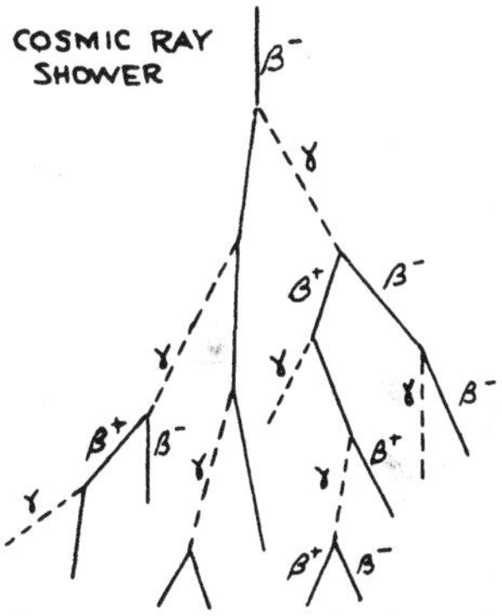


FIG. II.19

Mesons have mass about 200 m. The factor $(e^2/mc^2)^2$ appears in the formula for radiation loss. For a meson, this factor is $(1/200)^2$ as large as for electrons, hence energy loss by the meson by this process is much smaller. A meson looses energy mainly by ionization. For a meson of some billions of e.v. ionization loss is small, and the meson can penetrate several meters of lead.

6. <u>Summary</u>. The phenomena of absorption of photons can be summarized in the following graphs.

---

* The theory of showers is given in an article by Rossi and Greisen, <u>Rev. Mod. Phys.</u> <u>13</u>, 249 (1941)

GENERAL FEATURES OF ABSORPTION AS A FUNCTION OF ENERGY:

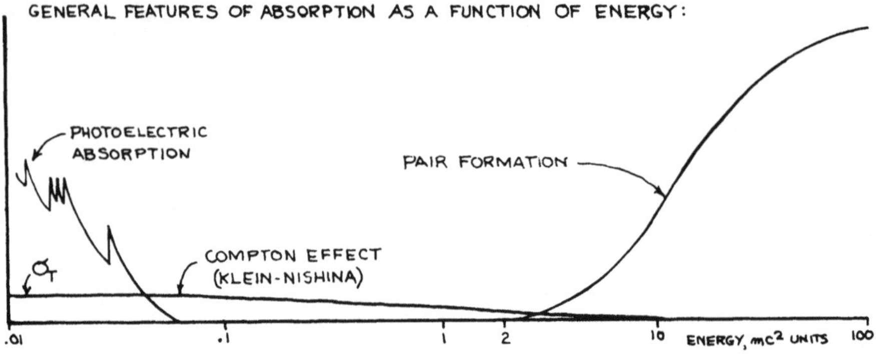


FIG. II. 20

ABSORPTION COEFFICIENT FOR PHOTONS IN LEAD, AS A FUNCTION OF ENERGY:

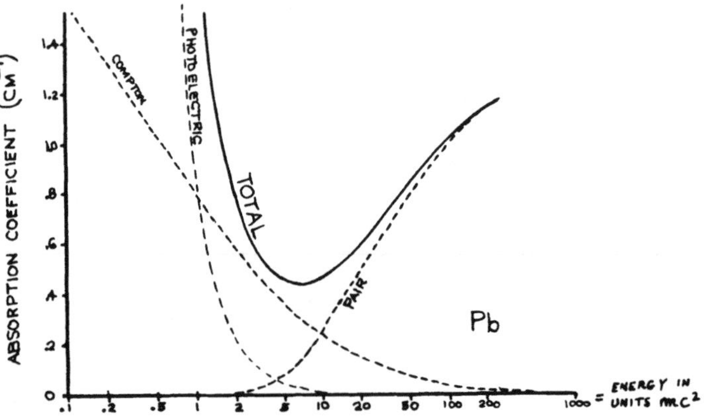


FIG. II. 21

| MATERIAL | Location of minimum in total absorption | Thickness ξ, in g/cm² for 1/2.7 = 1/e reduction in intensity |
|---|---|---|
| Aluminum | ~35 MEV | ~47 g/cm² at ~35 Mev |
| Copper | ~10 | ~31 " " ~10 Mev |
| Lead | ~ 3½ | ~25 " " ~3½ Mev |

Problem: Suppose a cosmic ray electron of energy 1000 mc² passes through a lead plate 1 mm. thick. What is the probability that the electron leaves the plate accompanied by a pair whose total energy (electron plus positron) is larger than 500 mc²? Assume the particles change direction negligibly.

Use II.53, writing it in terms of $\ell_R$: $\Sigma(\nu^\circ) = \frac{1}{\ell_R N \;\nu^\circ}$

To get the cross-section for all the nuclei in thickness dx, multiply by N dx, getting $\Delta\sigma_{total} = \frac{dx}{\ell_R}\frac{\Delta\nu^\circ}{\nu^\circ}$.

* From Heitler, p. 216.

The cross section for photons in the range 500 $mc^2$ to 1000 $mc^2$ is

$$\sigma_{total} = \frac{dx}{\ell_R}\int_{\nu=\frac{500\,mc^2}{h}}^{\nu=\frac{1000\,mc^2}{h}} \frac{\Delta\nu'}{\nu'} = \frac{.693}{\ell_R}\,dx$$

This is the probability that a photon of energy > 500 $mc^2$ will be produced in the thickness dx.

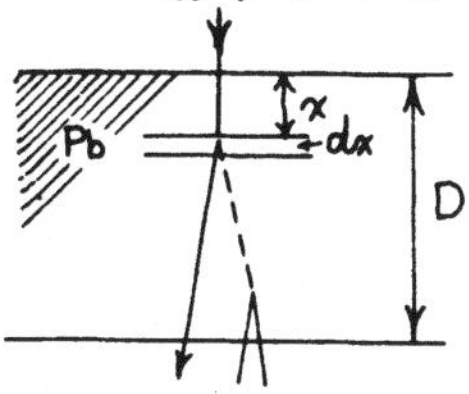


The probability that a photon produced at $x$ will then form a pair within the plate is equal to $\frac{D-x}{\ell_P}$, if $D-x \ll \ell_P$. This is from II.63. The pair produced will have all the energy of the photon. The total probability that both events will occur is

$$P = \frac{.693}{\ell_R\ell_P}\int_0^D (D-x)\,dx = \frac{.693\,D^2}{2\,\ell_R^2\,9/7}$$

using II.65. For lead in this energy region, $\ell_R \approx .6$. Therefore the probability is 0.0075.

Appendix II.1 We will show that $\overline{\Theta_p^2} = \sum_{i=1}^{p} \overline{\theta_i^2}$, where $\theta_i$ is the scattering angle for a single process and the mean is taken over many traversals of the absorber. Let $\Theta_i$ be the net accumulated angle between the initial direction of incidence and the particle's velocity vector just before the $i+1$ collision of the traversal in question. A shorter derivation appears on the next page.

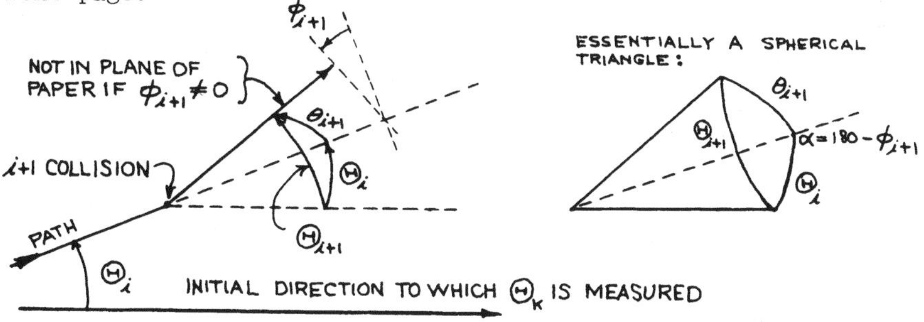


FIG. II.22

From the law of cosines in spherical trigonometry we have

$$\cos\Theta_{i+1} = \cos\Theta_i\cos\theta_{i+1} - \sin\Theta_i\sin\theta_{i+1}\cos\phi_{i+1} \qquad \text{II. 66}$$

First average over the random variable $\phi$, the azimuthal angle of scattering. The last term in II.66 vanishes. This gives

$(\cos\Theta_{i+1}) = (\cos\Theta_i\cos\theta_{i+1})$, where ( ) denotes average with respect to $\phi$. This is a recursion formula which gives

$$(\cos\Theta_p) = \left(\prod_{j=1}^{j=p}\cos\theta_j\right) \qquad \text{II. 67}$$

Assume that the $\theta_i$ and $\Theta_p$ are small. Then expand the cosines and retain the terms up to $\theta^2$ only.

$$1 - \frac{\overline{\Theta_p^2}}{2} \approx \prod_{j=1}^{p}\left(1 - \frac{\overline{\theta_j^2}}{2}\right) \approx 1 - \frac{1}{2}\sum_{j=1}^{p}\overline{\theta_j^2} \qquad \text{Thus } \overline{\Theta_p^2} = \sum_{j=1}^{p}\overline{\theta_j^2}$$

Shorter derivation of II.22:

We make use of the fact that small angles are vectors.

Then $\underline{\Theta}_P = \sum_{j=1}^{P} \underline{\theta}_j$

$$\Theta_P^2 = \sum_{j=1}^{P} \theta_j^2 + \sum_{i \neq j} \underline{\theta}_i \cdot \underline{\theta}_j$$

In averaging over many traversals, $\underline{\theta}_i$ is positive as often as it is negative, thus

$$\overline{\Theta_P^2} = \sum_{j=1}^{P} \overline{\theta_j^2}$$

Appendix II.2. Deriving II.25, $\overline{\Theta^2} = 2\pi ND \int_{b_{min}}^{b_{max}} \left(\frac{4z^2Z^2e^4}{b^2V^2p^2}\right) b\,db$ in greater detail. P, the average number of collisions in one traversal is given by

$$P = \int_{b_{min}}^{b_{max}} 2\pi b DN\, db$$

If $f$ is a function of $x$ and $\psi(x)$ is the normalized probability density for $x$, then the mean of $f$ is

$$\overline{f} = \int f(x)\,\psi(x)\,dx \qquad \text{II. 69}$$

applied to calculate $\overline{\theta^2}$, $\overline{\theta^2} = \int [\theta(b)]^2 \psi(b)\,db$.

The probability density for $b$, $\psi(b)$ is proportional to cross-sectional area at the distance b, $2\pi b\ db$, to the nuclear density N, and to the path length D. Therefore the normalized $\psi(b)$ is

$$\psi(b) = \frac{2\pi b DN}{\int_{b_{min}}^{b_{max}} 2\pi b DN\, db} = \frac{2\pi b DN}{P} \qquad \text{II. 70}$$

Therefore

$$\overline{\Theta^2} = P\,\overline{\theta^2} = \not{P}\,\frac{\int [\theta(b)]^2\, 2\pi b DN\, db}{\not{P}} \qquad \text{II. 71}$$

Finally

$$\overline{\Theta^2} = 2\pi ND \int_{b_{min}}^{b_{max}} \left(\frac{4z^2Z^2e^4}{b^2V^2p^2}\right) b\,db \qquad \text{(II.25)}$$

Since $\theta(b) = \dfrac{2zZe^2}{bVp}$ from II.20

Appendix II.3. Equation II.26 can be derived by considering the projections of the angles $\theta_i$ on some chosen plane and deriving a value for $\overline{\Psi^2}$, the mean square of the resultant projected angle. Then the unprojected, true scattering angle can be found from the result that $\frac{1}{2}\overline{\Theta^2} = \overline{\Psi^2} \quad (\theta_i \ll 1)$ II.72

The factor ½ comes from averaging $\cos^2\phi$, the angle being random. A similar relation holds between the actual linear deviation of the point of exit of the particle measured from the geometrical shadow of the point of entrance, and the projection of that displacement on some chosen plane; i.e.

$$\frac{1}{2}\overline{r^2} = \overline{x^2} \qquad \text{II.73}$$

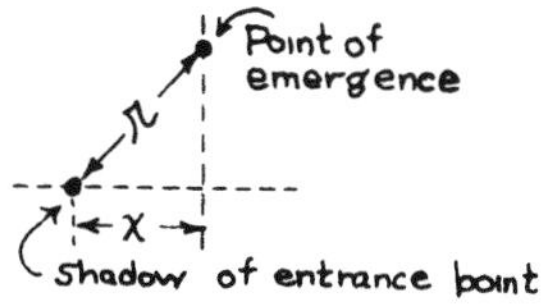


(plane of paper normal to incident direction. Looking toward source of particles)

In a cloud chamber photograph it is the projected angle of scattering that is shown.

This approach (use of projected angles) is used by Jánossy.

Problem: Electrons of energy $10^8$ e.v. (from cosmic rays) pass through a lead plate 1 cm. thick. Calculate the mean square of the linear deviation of the emergent points from the geometrical shadow point on the last surface of the lead plate. Assume the path is almost straight.

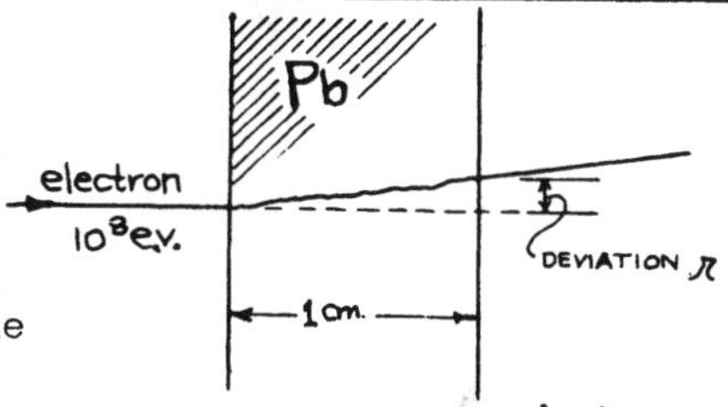


Let the collisions take place at distances $y_i$ from the last surface and result in deflections $\theta_i \ll 1$. The azimuth angle is $\phi_i$. The projected angle of deflection at $y = 0$ is $\Psi = \sum \theta_i \cos\phi_i$. The projection of linear displacement at $y = 0$ is $x = \sum_i y_i \theta_i \cos\phi_i$.

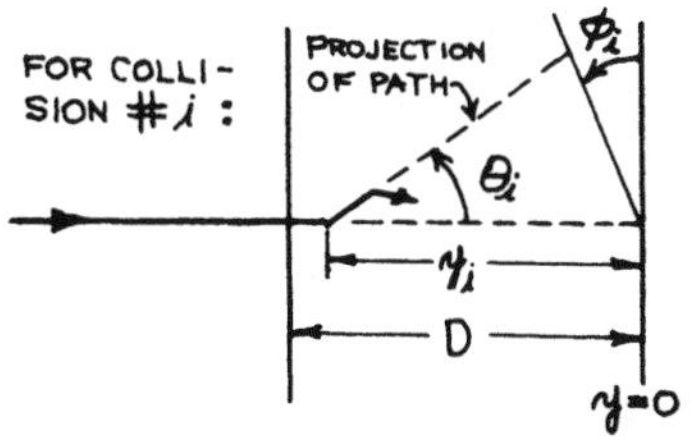


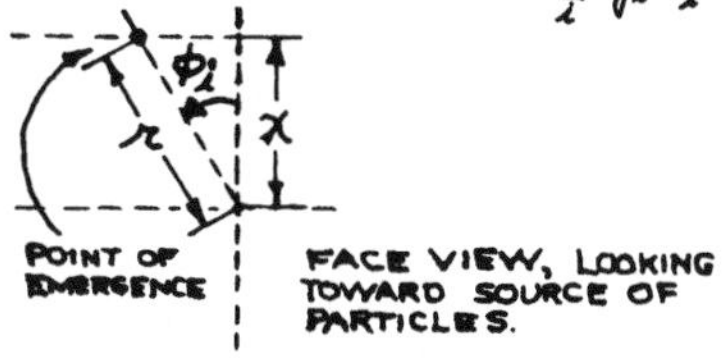


$$x^2 = \sum_{ij} y_i y_j \theta_i \theta_j \cos\phi_i \cos\phi_j\,, \quad \text{but} \quad \overline{\cos\phi_i \cos\phi_j} = \tfrac{1}{2}\delta_{ij}$$

$$\overline{x^2} = \tfrac{1}{2}\sum_i \overline{y_i^2}\,\overline{\theta_i^2} \quad \text{since } y_i \text{ and } \theta_i \text{ are independent.}$$

$$\overline{y_i^2} = \overline{y^2} = \frac{1}{D}\int_0^D y^2\,dy = \frac{D^2}{3}$$

$$\therefore\ \overline{x^2} = \frac{D^2}{6}\,\overline{\theta^2}\,P \qquad P = \text{average no. of collisions}$$

$$\overline{x^2} = \frac{D^2}{6}\,\overline{\Theta^2} \quad \text{by II.70}$$

The unprojected linear $\overline{(\text{displacement})^2}$, $\overline{r^2}$, will be twice this.

$$\overline{r^2} = \frac{1}{3}D^2\overline{\Theta^2} \qquad (= 0.02 \text{ cm. using II.31})$$

Appendix II.4. Proof that momentum cannot be conserved in creation of an electron-positron pair by a single isolated photon. By conservation of energy:

$$h\nu = \sqrt{(mc^2)^2 + p_+^2 c^2} + \sqrt{(mc^2)^2 + p_-^2 c^2} \qquad \text{II.74}$$

$p_+$ = momentum of $\beta^+$
$p_-$ = " " $\beta^-$

By conservation of momentum:

$$(h\nu)^2 \leqq (p_+ + p_-)^2 c^2 \qquad \text{II.75}$$

(The equality holds if $\beta^+$ and $\beta^-$ go off in the same direction)

From II.75 $(h\nu)^2 \leqq p_+^2 c^2 + p_-^2 c^2 + 2 p_+ p_- c^2$

From II.74 $(h\nu)^2 = p_+^2 c^2 + p_-^2 c^2 + 2(mc^2)^2 + \sqrt{[(mc^2)^2 + p_+^2 c^2][(mc^2)^2 + p_-^2 c^2]} \times 2$

The radical term is larger than $2 p_+ p_- c^2$, so both cannot be satisfied.

References: Books and articles referred to more than once in this chapter are listed.

Abraham-Becker, Theorie der Electrizität, Vol. II (Teubner, Edwards)

Heitler, Quantum Theory of Radiation, 2nd ed. (Oxford)

Jánossy, Cosmic Rays (Oxford)

Livingston and Bethe, Rev. Mod. Phys., 9, 245 (1937) (experimental nuclear physics)

Rutherford, Chadwick and Ellis, Radiations from Radioactive Substances, (Cambridge)

Rossi and Greisen, Rev. Mod. Phys. 13, 249 (1941)

## Chapter III

### ALPHA PARTICLE EMISSION

#### A. Penetration of Rectangular Barriers

A quantum mechanical explanation was applied to the problem of alpha emission independently by Gamow and by Gurney and Condon in 1928. This explanation makes use of the ability of a quantum mechanical particle to penetrate a potential barrier. First the problem of the one dimensional rectangular barrier will be considered.

Classically an incident particle of energy E (see Fig. III.1), coming from the left, could not penetrate into region II, since $U_0 > E$. It would be completely reflected. However, quantum mechanically, the particle, which is an incident wave, is partially reflected and partially transmitted.

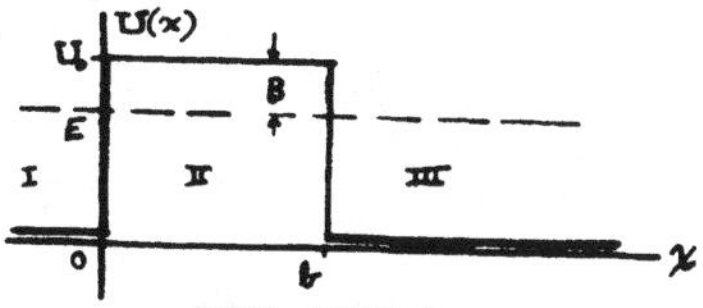


FIG.III.1

Let $\psi_{in} = \alpha e^{i(kx-\omega t)}$ be the incident particle travelling to the right $k \equiv \frac{p}{\hbar}$

then $\psi_r = \beta e^{i(-kx-\omega t)}$ is the reflected wave

and $\psi_{III} = \gamma e^{i(kx-\omega t)}$ is the transmitted wave.

The probability of penetration or the

$$\text{transparency} = \frac{\text{transmitted intensity}}{\text{incident intensity}} = \frac{|\gamma|^2}{|\alpha|^2}$$

Considering the time independent part of $\psi$,

$$\psi_I = \alpha e^{ikx} + \beta e^{-ikx}$$

$$\psi_{III} = \gamma e^{ikx}$$

These are solutions of the time independent wave equation:

$$\frac{d^2\psi}{dx^2} + \frac{2M}{\hbar^2}(E)\psi = 0$$

In region II the wave equation becomes

$$\frac{d^2\psi}{dx^2} - \frac{2M}{\hbar^2}(U_0 - E)\psi = 0$$

Let $B \equiv U_0 - E$

Then $\psi_{II} = K e^{\sqrt{\frac{2M}{\hbar^2}B}\,x} + L e^{-\sqrt{\frac{2M}{\hbar^2}B}\,x}$

Now the constants $\beta, \gamma$, K, and L will be determined by making use of the fact that the wave function must be well behaved physically.

Thus $\psi$ and $\frac{d\psi}{dx}$ are continuous across the boundaries x = 0 and x = b.

Thus $\alpha + \beta = K + L$ since $\psi_I(0) = \psi_{II}(0)$

$$ik(\alpha - \beta) = \sqrt{\tfrac{2M}{\hbar^2}B}\,(K - L) \qquad \psi_I'(0) = \psi_{II}'(0)$$

$$K e^{\sqrt{\frac{2M}{\hbar^2}B}\,b} + L e^{-\sqrt{\frac{2M}{\hbar^2}B}\,b} = r e^{ikb} \qquad \psi_{II}(b) = \psi_{III}(b)$$

$$\sqrt{\tfrac{2M}{\hbar^2}B}\left(K e^{\sqrt{\frac{2M}{\hbar^2}B}\,b} - L e^{-\sqrt{\frac{2M}{\hbar^2}B}\,b}\right) = ikr e^{ikb} \qquad \psi_{II}'(b) = \psi_{III}'(b)$$

Eliminating $\beta$ between the first two of these equations gives

$$\alpha = \tfrac{1}{2}\left(1 + \tfrac{q}{ip}\right)K + \tfrac{1}{2}\left(1 - \tfrac{q}{ip}\right)L \qquad q \equiv \sqrt{2MB}$$

Solving the last two of these equations for K and L gives

$$K = \tfrac{r}{2}\, e^{ikb}\left(1 + \tfrac{ip}{q}\right) e^{-\frac{qb}{\hbar}}$$

$$L = \tfrac{r}{2}\, e^{ikb}\left(1 - \tfrac{ip}{q}\right) e^{\frac{qb}{\hbar}}$$

Thus

$$\alpha = \tfrac{r}{4}\, e^{ikb}\left\{\left(1 + \tfrac{q}{ip}\right)\left(1 + \tfrac{ip}{q}\right) e^{-\frac{qb}{\hbar}} + \left(1 - \tfrac{q}{ip}\right)\left(1 - \tfrac{ip}{q}\right) e^{\frac{qb}{\hbar}}\right\}$$

Since $p = \sqrt{2ME}$, $\frac{p}{q} = \sqrt{\frac{E}{B}} \sim 1$

$$\alpha \sim r e^{ikb} e^{\frac{qb}{\hbar}} \qquad \text{for } \frac{qb}{\hbar} \gg 1$$

$$\frac{|r|^2}{|\alpha|^2} \sim e^{-\frac{2qb}{\hbar}}$$

$$\text{transparency} \approx e^{-2\sqrt{\frac{2M}{\hbar^2}(U - E)}\; b} \qquad \text{III.1}$$

Notice that as h → 0, this approaches the classical limit of zero transparency. Also the transparency is greater for particles of smaller mass. This explains why electrons have little trouble in going from one atom to another in diatomic molecules and metals. See FIG III.2.

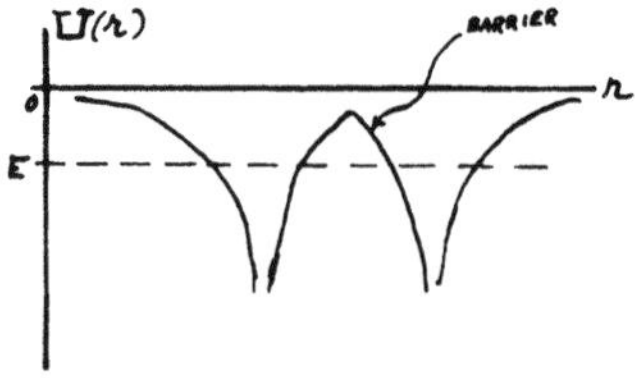

FIG III.2 Diatomic molecule with one electron

B. Barrier of Arbitrary Shape

The order of magnitude of the transparency of an arbitrarily shaped barrier can be obtained by finding the average "height" and treating as a rectangular barrier.

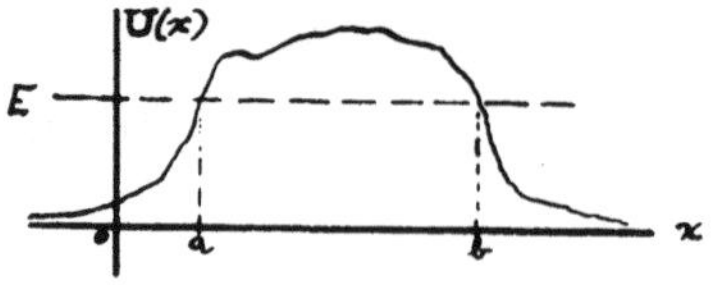

FIG III.3

III.1 then gives

$$\text{transparency} \sim e^{-2\int_a^b \sqrt{\frac{2M}{\hbar^2}\left[U(x)-E\right]}\,dx} \qquad \text{III.2}$$

Problem: A very slowly moving car of 1 ton (kinetic energy considered almost zero) encounters a sinusoidal bump in the road which is 1 ft. high and 100 ft. long. Classically, the car can't get past. However III.2 shows that there actually is a finite probability (w) that the car can overcome the bump.

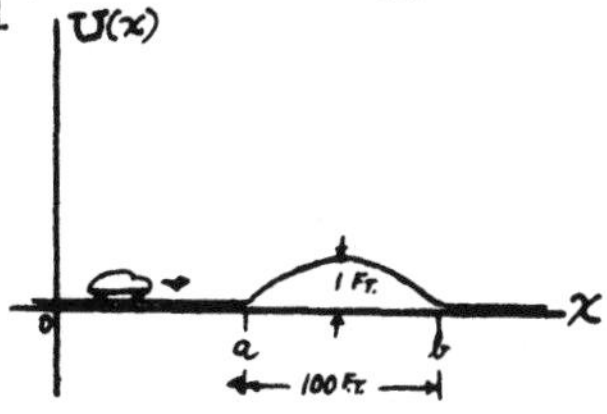


$M = 10^6$ gm.

$b - a = 3050$ cm.

$$w \sim e^{-2\sqrt{\frac{2M}{\hbar^2}}\int_a^b \sqrt{U(x)}\,dx}$$

$$U(x) = Mg \times 12 \times 2.54 \sin \frac{\pi(x-a)}{3050}$$

$$\int_a^b \sqrt{U(x)}\,dx \approx 5.3 \times 10^8$$

$$w \sim e^{-14.3 \times 10^{38}}$$

$$w \sim 10^{-6.2 \times 10^{38}}$$

In the following figure, III.2 is used to calculate the transparency for protons penetrating the coulomb barrier of nuclei of A = 1, 8, 90, and 238. These are plotted as a function of proton energy divided by barrier energy.

| CURVE | A | Z | BARRIER E |
|---|---|---|---|
| a | 1 | 1 | 0.5 Mev |
| b | 8 | 4 | 1.9 |
| c | 90 | 40 | 8.6 |
| d | 238 | 92 | 14.2 |

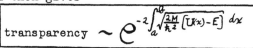

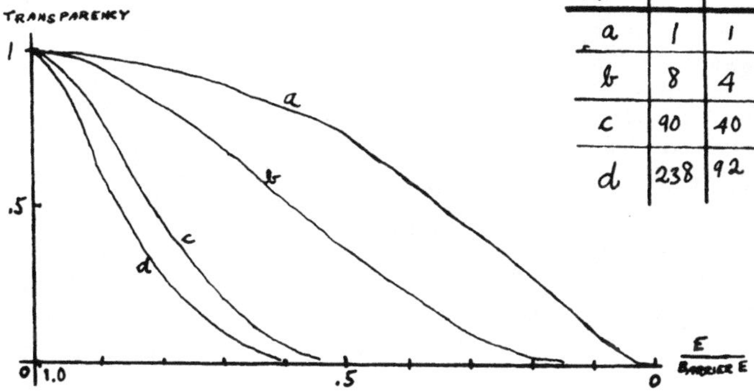


Transparency to Protons of Nuclear Coulomb Barriers

## C. Semi-classical Application to Alpha Decay

III.2 will now be applied to calculate the probability of an alpha particle penetrating the potential barrier due to a nucleus. First, assumptions must be made concerning this barrier. As an alpha particle is brought from $r=\infty$, the main interaction force is the electrostatic repulsion which gives the potential $U(r) = \frac{zZe^2}{r}$ . Let R be the distance where the alpha particle first "contacts" the nucleus. Then it encounters the strong short range attractive forces until it is within the nucleus. Once the alpha particle is in the nucleus it is considered as a free particle. The depth of this well is not very great for the alpha active nuclei, since the lowest energy state must give a positive energy eigenvalue. The height of the barrier is much greater, however. For uranium it is 29 Mev if equation I.4, page 6, is used for R.

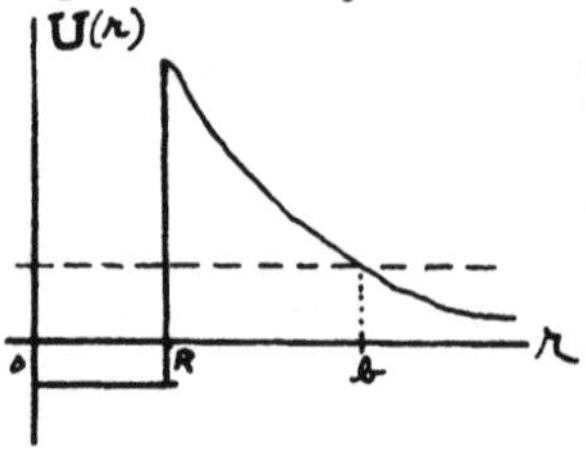


FIG III.4

probability / sec. of escape = rate of hitting barrier x transparency or

$$\frac{1}{\tau} \approx \frac{V_{in}}{R} \times e^{-2\int_R^b \sqrt{\frac{2\mu}{\hbar^2}(U-E)}\,dr}$$

where $\mu$ is the reduced mass of alpha particle and nucleus

$$\frac{V_{in}}{R} \sim 10^{21} \qquad \text{since } V_{in} \sim 10^9 \text{ and } R \sim 10^{-12}$$

$$\frac{1}{\tau} \sim 10^{21} e^{-G}$$

where $G \equiv \sqrt{\frac{8\mu zZe^2}{\hbar^2}} \int_R^b \sqrt{\frac{1}{r} - \frac{E}{zZe^2}}\,dr \doteq \sqrt{\frac{8\mu zZe^2 b}{\hbar^2}} \left[\cos^{-1}\left(\sqrt{\frac{R}{b}}\right) - \sqrt{\frac{R}{b} - \frac{R^2}{b^2}}\right]$ III.3

$$\boxed{\frac{1}{\tau} \sim 10^{21} \exp\left\{-\left(\frac{8\mu zZe^2 b}{\hbar^2}\right)^{\frac{1}{2}}\left[\cos^{-1}\left(\sqrt{\frac{R}{b}}\right) - \sqrt{\frac{R}{b} - \frac{R^2}{b^2}}\right]\right\}}$$

For a large barrier G has the following approximate form:

$G = \frac{2\pi zZe^2}{\hbar V}$ where V is the final velocity of the $\alpha$ particle

Problem: Using formula I.8, page 7, for the masses, determine which of the following nuclei are unstable to alpha decay and determine their lifetimes.

| A | Z |
|---|---|
| 100 | 44 |
| 150 | 62 |
| 200 | 80 |
| 250 | 97 |
| 300 | 113 |

For the first two nuclei listed, the mass of the daughter exceeds the mass of the parent, hence they are stable. Of the next three, Z=80 has the longest lifetime and will be calculated in detail here.

$M(200, 80) = M(196, 78) + He^4 + Q$

$Q = M(200,80) - [M(196,78) + He^4]$

$M(200,80) = 200.04474$
$M(196,78) = 196.03826$
$He^4 = 4.00390$

$Q = 0.00258 \text{ amu} = 3.84 \times 10^{-6}\text{ergs}$

$\frac{zZe^2}{b} = 3.84 \times 10^{-6}\text{ergs}$

$b = 9.52 \times 10^{-12}\text{cm}$

$R = 1.5 \times 10^{-13}(196)^{\frac{1}{3}} = 8.7 \times 10^{-13}\text{cm}$ by I.4

III.3 gives for G:

$$G = \frac{1}{\hbar}\sqrt{8\times\left(\frac{4\times196}{4+196}\right)\times1.66\times10^{-24}\times2\times78\times(4.80\times10^{-10})^2\times(9.52\times10^{-13})}\times[1.198]$$

$$= 127 \times 1.198$$

$$= 152$$

$$\frac{1}{\tau} \sim 10^{21}e^{-152}$$

$$\tau \sim 10^{43}\text{sec}$$

$\tau \sim 3 \times 10^{35}$ yrs. since $3.15 \times 10^7$ sec = 1 yr.

Thus this nucleus would appear stable.

## D. Virtual Level Theory of Alpha Decay

Now a complete quantum mechanical presentation will be given, including the solution for the alpha particle inside the nucleus. This leads to the virtual level theory of alpha decay. As practice in quantum mechanics, the work will be presented here in some detail.

It will be of help first to investigate the analogous rectangular barrier problem which has an exact solution.

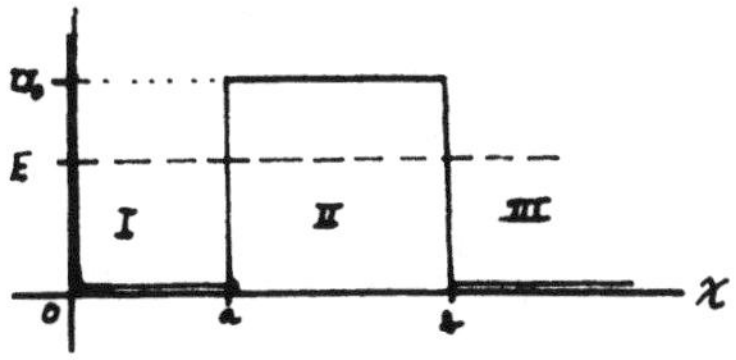


FIG III.5

Problem: At x = 0 there is an infinite barrier. The only finite solution in this region is zero

since $\frac{d^2\psi}{dx^2} = \infty$ for $x \leq 0$.

Let $p^2 = \frac{2M}{\hbar^2}E$ $\quad q^2 = \frac{2M}{\hbar^2}(U_0 - E)$

In regions I and III $\quad \frac{d^2\psi}{dx^2} + p^2\psi = 0$

$$\psi \propto e^{\pm ipx}$$

Since $\psi$ must be zero in the infinite potential region and $\psi$ must be continuous, we have

$$\psi_I = e^{ipx} - e^{-ipx}$$ where the arbitrary constant is chosen as one.

Notice that this represents equal waves travelling in both directions to give a standing wave steady-state solution. This distinction between a steady-state solution and an incident particle as before is to be emphasized.

In general $\quad \psi_{II} = c_+ e^{q(x-a)} + c_- e^{-q(x-a)}$

$$\psi_{III} = c_1 e^{ip(x-b)} + c_2 e^{-ip(x-b)}$$

The continuity conditions at the boundaries give

$$e^{ipa} - e^{-ipa} = C_+ + C_- \qquad \text{since } \psi_I(a) = \psi_{II}(a) \qquad \text{III.4}$$

$$\frac{ip}{q}(e^{ipa} + e^{-ipa}) = C_+ - C_- \qquad \psi_I'(a) = \psi_{II}'(a)$$

Let G = 2q(b-a)

$$C_1 + C_2 = C_+ e^{\frac{G}{2}} + C_- e^{-\frac{G}{2}} \qquad \psi_{III}(b) = \psi_{II}(b)$$

$$C_1 - C_2 = \frac{q}{ip}(C_+ e^{\frac{G}{2}} - C_- e^{-\frac{G}{2}}) \qquad \psi_{III}'(b) = \psi_{II}'(b)$$

$$C_1 = \frac{1}{2}\left\{(1+\frac{q}{ip})C_+ e^{\frac{G}{2}} + (1-\frac{q}{ip}) C_- e^{-\frac{G}{2}}\right\}$$

$$C_2 = \frac{1}{2}\left\{(1-\frac{q}{ip}) C_+ e^{\frac{G}{2}} + (1+\frac{q}{ip}) C_- e^{-\frac{G}{2}}\right\}$$

$$\psi_{III} = \frac{1}{2} C_+ e^{\frac{G}{2}}\left[(1+\frac{q}{ip}) e^{ip(x-b)} + (1-\frac{q}{ip}) e^{-ip(x-b)}\right] + \frac{1}{2} C_- e^{-\frac{G}{2}}\left[(1-\frac{q}{ip}) e^{ip(x-b)} + (1+\frac{q}{ip}) e^{-ip(x-b)}\right]$$

$$\psi_{III} = C_+ e^{\frac{G}{2}}\left[\cos p(x-b) + \frac{q}{p}\sin p(x-b)\right] + C_- e^{-\frac{G}{2}}\left[\cos p(x-b) - \frac{q}{p}\sin p(x-b)\right]$$

$$\frac{|\overline{\psi_{III}}|^2}{|\overline{\psi_I}|^2} = \left(\frac{1}{4}+\frac{q^2}{4p^2}\right)\left(C_+^2 e^{G} + C_-^2 e^{-G}\right) + \left(\frac{1}{4}-\frac{q^2}{4p^2}\right)C_+C_-$$

Thus the average intensity of $\psi_{III}$ is usually much greater than that of $\psi_I$. However, we are interested in the case where $\psi_{III}$ is as small as possible.

III.4 gives $C_+ = i\sin pa + \frac{ip}{q}\cos pa$.

When $\tan pa = -\frac{p}{q}$, $C_+ = 0$ or $\tan\sqrt{\frac{2ME_0}{\hbar^2}}\,a = -\sqrt{\frac{E_0}{U_0 - E_0}}$

The values of E which satisfy the above equation give the smallest possible intensity outside. These values of E are known as the virtual levels. It is seen the outside intensity decreases very sharply in the energy region of the virtual level.

In FIG 6, on the following page, the first curve shows the general situation when $C_+ \neq 0$. The second curve is for the virtual level energy. The final curve shows the behavior of $\psi$ for an energy slightly greater than the virtual level energy.

This same virtual level approach will now be used with the W.K.B. method in solving the problem of the potential barrier experienced by the alpha particle.

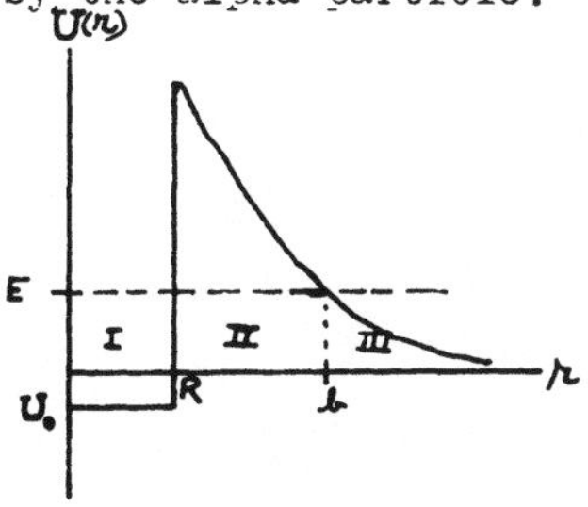


FIG III.7

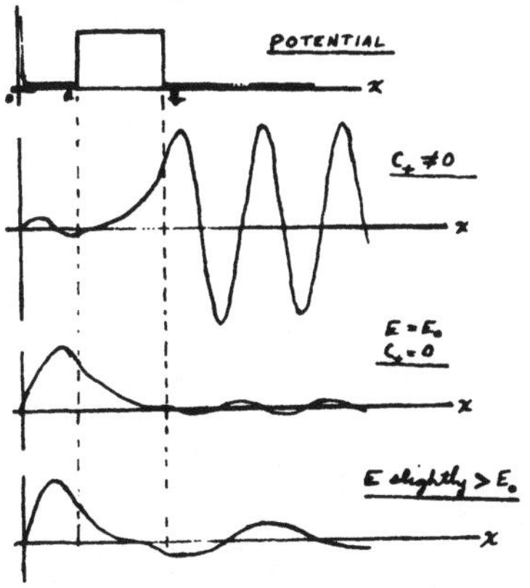


FIG III.6

FIG 7 gives the potential barrier experienced by the alpha particle. Since this is a three dimensional problem, spherical coordinates will be used which give III.5 for the radial part of the wave function.

The complete wave equation is

$$\nabla^2\psi(r,\theta,\phi) + \frac{2\mu}{\hbar^2}\left[E - \frac{2Ze^2}{r}\right]\psi(r,\theta,\phi) = 0 \qquad \text{for regions II and III}$$

Let $\psi(r,\theta,\phi) \equiv \psi(r)\,Y_{lm}(\theta,\phi)$

then $$\frac{d^2}{dr^2}\left[r\psi(r)\right] + \left[\frac{2\mu}{\hbar^2}\left(E - \frac{2Ze^2}{r}\right) - \frac{l(l+1)}{r^2}\right]r\psi(r) = 0 \qquad \text{III.5}$$

First the case where the angular momentum of the alpha particle with respect to the nucleus is zero (s-state, $l = 0$) shall be considered.

$$\frac{d^2u}{dr^2} + \frac{2\mu}{\hbar^2}\left(E - \frac{2Ze^2}{r}\right)u = 0 \qquad \text{where } u \equiv r\psi(r) \qquad \text{III.6}$$

Let $\gamma^2 \equiv \frac{2\mu}{\hbar^2}(E - U_0)$ for region I

$g_{II}^2(r) \equiv \frac{2\mu}{\hbar^2}\left(\frac{2Ze^2}{r} - E\right)$ for region II

$g_{III}^2 \equiv \frac{2\mu}{\hbar^2}\left(E - \frac{2Ze^2}{r}\right)$ for region III

$k^2 \equiv \frac{2\mu}{\hbar^2}E$

Since these are all positive quantities, $\gamma$, $g_{II}$, $g_{III}$, and $k$ are all real.

$u_I = \sin\gamma r$ since $\psi_I(0) = \left.\frac{u_I(0)}{r}\right|_{r=0}$ must be finite

The W.K.B. method gives for the general solution in II

$$u_{II} = C_+ \frac{1}{\sqrt{g_{II}}} e^{\int_R^r g_{II} dr} + C_- \frac{1}{\sqrt{g_{II}}} e^{-\int_R^r g_{II} dr}$$

The W.K.B. method gives for the connecting asymptotic functions across r = b (Schiff, page 184)

$$\frac{1}{\sqrt{g_{II}}} e^{+\int_r^b g_{II} dr} \rightarrow \frac{1}{\sqrt{g_{III}}} \cos\left[\int_b^r g_{III} dr + \frac{\pi}{4}\right]$$

$$\frac{1}{\sqrt{g_{II}}} e^{-\int_r^b g_{II} dr} \rightarrow \frac{2}{\sqrt{g_{III}}} \cos\left[\int_b^r g_{III} dr - \frac{\pi}{4}\right]$$

Thus

$$\frac{1}{\sqrt{g_{II}}} e^{-\left[\int_R^b g_{II} dr + \int_b^r g_{II} dr\right]} \rightarrow \frac{1}{\sqrt{g_{III}}} e^{-\frac{G}{2}} \cos\left[\int_b^r g\, dr + \frac{\pi}{4}\right]$$ where G is defined by III.3

$$\frac{1}{\sqrt{g_{II}}} e^{+\int_R^r g_{II} dr} \rightarrow \frac{2}{\sqrt{g_{III}}} e^{+\frac{G}{2}} \cos\left[\int_b^r g_{III} dr - \frac{\pi}{4}\right]$$

This gives for $u_{III}$

$$u_{III} = \frac{1}{\sqrt{g_{III}}} C_- e^{-\frac{G}{2}} \cos\left[\int_b^r g_{III} dr + \frac{\pi}{4}\right] + \frac{2}{\sqrt{g_{III}}} C_+ e^{+\frac{G}{2}} \sin\left[\int_b^r g_{III} dr + \frac{\pi}{4}\right]$$

At large r

$$\frac{\overline{u_{III}}^2}{\overline{u_I}^2} = \frac{1}{k}\left[C_-^2 e^{-G} + 4C_+^2 e^{+G}\right] \qquad \text{III.7}$$

In the appendix to this chapter, page 67, $C_-$ and $C_+$ will be calculated in the region near a virtual level with the result

$$C_-^2 \approx \frac{\gamma^2}{4g(R)} \quad \text{where} \quad g(R) \equiv \sqrt{\frac{2\mu}{\hbar^2}(U(R)-E)}$$

$$C_+^2 \approx \frac{\mu^2 R^2 g(R)}{4\gamma^2 \hbar^4} \mathcal{E}^2 = g(R) \frac{\pi^2 \mathcal{E}^2}{4(\delta E)^2} \qquad \text{III.8}$$

where $\mathcal{E}$ is the energy separation from the nearby virtual level and $\delta E$ is the separation of two adjacent virtual levels.

III.7 becomes

$$\frac{\overline{u_{III}^2}}{\overline{u_I^2}} \approx A^2 e^{-G} + B^2 \mathcal{E}^2 e^{G}$$
$$\sim e^{-G} + \left(\frac{\mathcal{E}}{\delta E}\right)^2 e^{G} \qquad \text{III.9}$$

where $A^2 \equiv \frac{\gamma^2}{4k g(R)}$ $\qquad B^2 \equiv \frac{\mu^2 R^2 g(R)}{\gamma^2 k \hbar^4}$

Thus except for a small energy range about the virtual level, the amplitude outside is much larger than that inside the nucleus. Even at the virtual level

$$\int_{III} \left(\frac{u_{III}}{r}\right)^2 d\tau \approx 4\pi \int_b^\infty u_{III}^2\, dr = \infty \qquad \text{while} \quad \int_I \left(\frac{u_I}{r}\right)^2 d\tau$$

is finite. Thus there is zero probability that the particle be within the nucleus if the particle is to be in an exact energy state. However, we are concerned with an initial state function $\psi(r,t)$ which at t = 0 gives zero probability or zero amplitude for r > b.

Such a state function can be expanded in terms of the known eigen functions $u_n(r)$, where $u_n(r)$ is $u_I(r)$ in region I, $u_{II}(r)$ in II, and $u_{III}(r)$ in III, all corresponding to the same energy $E_n$.

This will turn out to give a narrow energy spectrum about the virtual energy. The distribution will turn out to be such that $\psi_{inside}(r,t)$ will decrease exponentially with time. The rate of decrease of total probability inside will be the reciprocal of the mean life of the particle inside.

In summary the procedure will be:

(1) Determination of eigen functions in energy region about the virtual level.
(2) Expansion of $\psi(r,0)$ in terms of these eigen functions.
(3) Determination of rate of decrease of $\psi_{inside}(r,t)$.

(1) Actually there is a continuous spectrum of eigen values and functions. The problem will first be treated as one of discrete levels by limiting region III to r< L. Later L will be made to approach infinity, which approaches the continuous spectrum situation.

The region is limited to r < L by placing an infinite barrier at r = L. L is taken large enough so that:

$$\int_R^L u_{III}^2 \, dr >> \int_0^R u_I^2 \, dr$$

For outside, $u_{III}$ is effectively a sine wave (of n half-wave lengths) so that:

$$\int_0^L u_n^* u_n \, dr \approx \frac{L}{2}$$

From now on $u_n$ will be considered as normalized which gives it the amplitude of $\sqrt{2/L}$ outside.

Now $$\int_0^L u_n^* u_n \, dr \approx 1$$

III.9 gives for the amplitude inside:

$$\text{amplitude inside} = \frac{\sqrt{\frac{2}{L}}}{\sqrt{A^2 e^{-G} + B^2 \mathcal{E}_n^2 e^{G}}}$$

The shape of the function $u_n$ up to r = d (see FIG 8) is essentially the same for all energy levels near the virtual level. In FIG 8 the smooth line represents $u_n$ which has an $E_n$ slightly below the lowest virtual level, $E_0$, and the dotted line is for an energy slightly above $E_0$.

Let f(r) (of amplitude one) represent all these $u_n$'s for the region r< d.

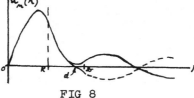

FIG 8

Then $$u_n(r) = \frac{\sqrt{\frac{2}{L}}}{\sqrt{A^2 e^{-G} + B^2 \mathcal{E}_n^2 e^{G}}} f(r) \quad \text{for } r< d \qquad \text{III.10}$$

If $\lambda_n$ is the wavelength of $u_n$ outside: $n\frac{\lambda_n}{2} \approx L$

$$\frac{1}{\lambda_n} = \frac{n}{2L} = \frac{p_n}{h}$$

$$p_n = \frac{\pi\hbar}{L}n$$

$$E_n = \frac{1}{2\mu}\frac{\pi^2\hbar^2}{L^2}n^2 \qquad \text{III.11}$$

$$E_{n+\Delta n} - E_n = \frac{1}{\mu}\frac{\pi^2\hbar^2}{L^2}n\,\Delta n \qquad \text{III.12}$$

(2) Expansion of $\psi(r,0)$ in terms of $u_n$.
Up to now f(r) has been defined only for r < d. For r > d it will be defined = 0.

Then $\psi(r,0) = f(r)$ so that the alpha particle has zero probability to be outside the nucleus at t = 0.

$$\psi(r,0) = \sum_n C_n u_n$$

$$C_n = \int_0^\infty u_n^* f(r)\,dr = \int_0^d u_n f(r)\,dr \qquad \text{since } f(r) = 0 \text{ for } r > d$$

$$C_n = \frac{\sqrt{\frac{2}{L}}}{\sqrt{A^2e^{-G}+B^2e^{G}\mathcal{E}_n^2}}\int_0^d f^2(r)\,dr \qquad \text{by III.10}$$

$$\int_0^d f^2(r)\,dr \approx \frac{R}{2}$$

$$\psi(r,t) = \sum_n C_n u_n e^{-\frac{i}{\hbar}E_n t}$$

$$\psi_{inside}(r,t) = \sum_n \left\{\frac{\frac{2}{L}}{A^2e^{-G}+B^2e^{G}\mathcal{E}_n^2}\,\frac{R}{2}\right\} f(r)\, e^{-\frac{i}{\hbar}E_n t} \qquad \text{using } u_n(r) \text{ for } r < d \text{ from III.10}$$

$$= \frac{R}{L} f(r)\, e^{-\frac{i}{\hbar}E_0 t}\sum_n \frac{e^{-\frac{i}{\hbar}\mathcal{E}_n t}}{A^2e^{-G}+B^2e^{G}\mathcal{E}_n^2}$$

This summation over n approaches $\int \frac{e^{-\frac{i}{\hbar}\mathcal{E}_n t}}{A^2e^{-G}+B^2e^{G}\mathcal{E}_n^2}\,dn$ as $L \to \infty$.

With the help of III.12 the variable of integration can be transformed:

$$\Delta n = \frac{\mu L^2}{n\pi^2\hbar^2}\Delta E$$

Since the integrand has large values only over a small range of n and n is very large, n may be taken outside the integral.

$$\psi_{inside} = \frac{R}{L} f(r)\, e^{-\frac{i}{\hbar}E_0 t}\frac{\mu L^2}{\pi^2\hbar^2 n}\int_{-\infty}^{\infty}\frac{e^{-\frac{i}{\hbar}t\mathcal{E}}}{A^2e^{-G}+B^2e^{G}\mathcal{E}^2}\,d\mathcal{E}$$

$$= \frac{R}{\pi\hbar}\sqrt{\frac{\mu}{2E_0}}\, f(r)\, e^{-\frac{i}{\hbar}E_0 t}\int_{-\infty}^{\infty}\frac{e^{-i\alpha x}}{A^2(1+x^2)}\,d\left[\frac{A}{B}e^{-G}x\right] \qquad \text{using III.11 to eliminate n}$$

where $\alpha = \frac{t}{\hbar}\frac{A}{B}e^{-G}$

$$\beta = A^2 e^{-G}$$

$$x = \frac{B}{A}e^{G}\varepsilon$$

Since $\int_{-\infty}^{\infty}\frac{e^{-i\alpha x}}{1+x^2}dx = \pi e^{-\alpha}$,

$$\psi_{inside}(r,t) = \left[\frac{R}{\pi\hbar}\sqrt{\frac{\mu}{2E_0}}\frac{1}{AB}\, e^{-\frac{i}{\hbar}E_0 t}\pi\right] e^{-\left(\frac{A}{\hbar B}e^{-G}\right)t} f(r)$$

(3) $\psi^2_{inside}(r,t)$ decreases to $\frac{1}{e}$ of $\psi^2(r,0)$ at

$$\tau = \frac{\hbar B}{2A}e^{G} \qquad \text{which is} \sim \frac{\hbar}{\delta E}e^{G}$$

$$\boxed{\tau = \frac{\mu R\, g(R)}{\hbar \gamma^2}e^{G}}$$ after substituting the values of A and B from page 62 III.13

$$\tau = \frac{R\sqrt{2\mu}\sqrt{U(R)-E_0}}{2E_{in}}e^{G}$$ Where $E_{in} = E_0 - U_0$ is the kinetic energy inside the nucleus.

$$\frac{1}{\tau} = \frac{2}{R}\sqrt{\frac{E_{in}}{U(R)-E_0}}\sqrt{\frac{E_{in}}{2\mu}}e^{-G}$$

$$= \frac{V_{in}}{R}\sqrt{\frac{E_{in}}{U(R)-E_0}}e^{-G}$$ since $V_{in} = \sqrt{\frac{2E_{in}}{\mu}}$

$$\frac{1}{\tau} \sim \frac{V_{in}}{R}e^{-G}$$ since $\sqrt{\frac{E_{in}}{U(R)-E_0}} \sim 1$

This is the same result as III.3 which used a semi-classical approach.

Comparison with Experiment

This theoretical result checks with experiment as far as orders of magnitude are concerned. The theoretical value of the lifetime may differ from the experimental by such a factor as $10^3$. Thus the formula is not very useful for computing lifetimes. However, since $\tau$ is very sensitive to the radius R which appears in the exponent, a rough knowledge of $\tau$ will give a much more exact knowledge of R. In fact the constant in the formula $r = 1.5 \times 10^{-13} A^{\frac{1}{3}}$ is determined by averaging the radii of alpha emitters which are calculated using this theory. This is done in Bethe B. section 68. No correction for the radius of the alpha particle is made in this determination. Bethe gives two sets of values, the "old" and the "new" which differ by about 40 percent. The "old" values are more in favor today. Bethe's "new" values are much larger because he considers that for most of the time the alpha particle does not exist separately in the nucleus. Thus the probability of formation of the alpha particle within the nucleus must be considered. This is the main weak point of the Gamow theory. This probability of pre-existance of the alpha particle in the nucleus prompts Bethe to change the factor $10^{21}$ ($= \frac{v}{R}$) to $10^{15}$, which gives a smaller barrier or larger radius for the same lifetime.

Another consideration which has been neglected thus far is the possibility of the alpha particle being emitted with angular momentum. The same mathematical treatment may be used if $\frac{\hbar^2}{2\mu}\frac{l(l+1)}{r^2}$ is added to the potential $\frac{zZe^2}{r}$ as can be seen by III.5. This will increase the barrier and half-life. See FIG 9. The effect is not very great since an $l$ as large as 5 changes $\tau$ only by a factor of 10.

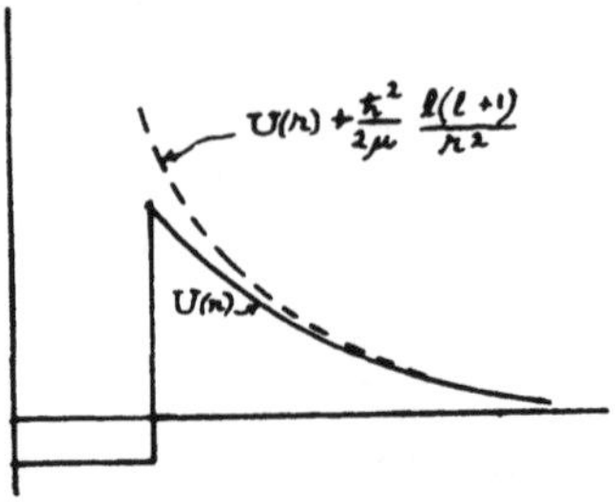


FIG III.9

The Geiger-Nuttall Law

In 1911 Geiger and Nuttall discovered the experimental relationship between members of the same radio-active family:

$\frac{1}{\tau} = \text{Const} \times R^{57.5}$ where R is the range in air III.14

For the 3 radio-active families this const. is

$10^{-42.3}$ for uranium family

$10^{-44.2}$ for thorium

$10^{-46.3}$ for actinium

Since in the energy region for alphas (4 to 10 Mev) $R \approx aV^3$ where $a = 9.67 \times 10^{-28}$ (see Ch.II, page 32), the Geiger-Nuttall law gives a relationship between $\frac{1}{\tau}$ and the velocity. This relation checks with that given by the Gamow theory (III.3) over the limited energy range which is observed.

## E. Alpha Ray Spectra

Many alpha emitters show a fine structure in the energy spectrum of the emitted alphas; with highest intensity for the maximum energy alpha emitted, and with rapid decrease in intensity for decrease in energy. FIG 10 shows the alpha spectrum for ThC. For this example it is seen that the highest energy alpha is an exception. The Gamow theory gives the ratio of intensity of a given energy group to that of the maximum energy group as $e^{-170\frac{\delta E}{(E_0)^{3/2}}}$

$\delta E$ is the energy difference and $E_0$ is the maximum energy. The general trend of group energies supports this relation. The individual departures from this relation may be explained by giving different probabilities for the pre-existance of the alpha particle within the nucleus.

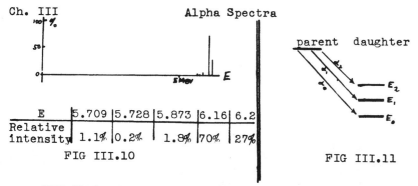

| E | 5.709 | 5.728 | 5.873 | 6.16 | 6.2 |
|---|---|---|---|---|---|
| Relative intensity | 1.1% | 0.2% | 1.8% | 70% | 27% |

FIG III.10

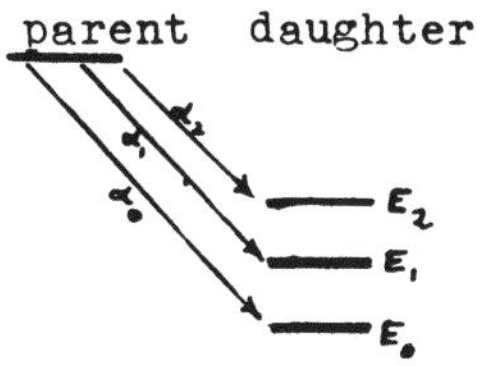


FIG III.11

FIG 11 is an energy level diagram illustrating the phenomenon of discrete energy groups. $\alpha_0$ is the maximum energy alpha The excited nuclear states of the daughter decay rapidly by gamma emission to the ground state, $E_0$, where $h\nu = E_1 - E_0$, etc. The energies of some of these gamma rays have been observed and do check.

## Long Range Alpha Emission

Two nuclei, RaC' and ThC', give an alpha spectrum which contains low intensity groups of alphas up to 2 Mev above the main group. The intensities of these long range groups are from $10^{-4}$ to $10^{-6}$ of the total intensity. This is because both RaC' and ThC' are products of a beta disintegration which leaves them in excited states. These excited states have a choice of decaying by a gamma ray to the ground parent state, or else by alpha to the daughter nucleus. FIG 12 shows the situation for ThC'. Since the gamma lifetimes are on the order of $10^{-5}$ times smaller than the alpha lifetimes for the excited parent states, the long range alpha intensities are very low.

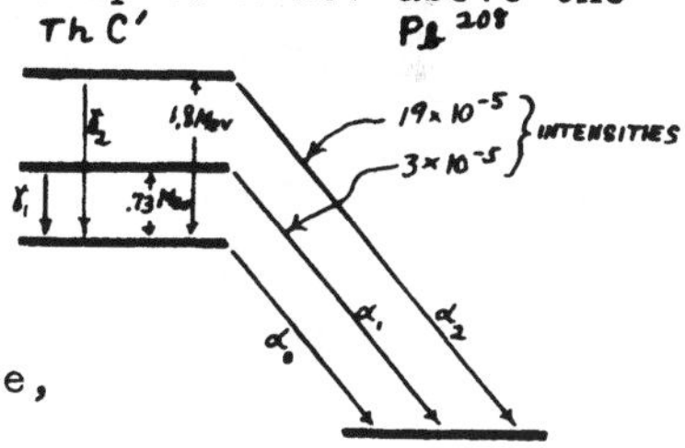


Half lives: $\alpha_0 = 3 \times 10^{-7}$ sec.
$\gamma_1 = 10^{-14}$
$\gamma_2 = 6 \times 10^{-17}$

Decay of ThC'

FIG III.12

## F. Appendix to Ch. III

The calculation of $C_+$ and $C_-$ as used in III.8 will be made here. See page 62. The continuity conditions across the boundary at r = R give

$$C_+ + C_- = \sqrt{g(R)}\,\sin\gamma R \qquad \text{since } u_{II}(R) = u_I(R)$$

$$C_+ - C_- = \sqrt{g(R)}\,\frac{\gamma}{g(R)}\cos\gamma R \qquad u'_{II}(R) = u'_I(R)$$

$$C_+ = \frac{\sqrt{g(R)}}{2}\left[\sin\gamma R + \frac{\gamma}{g(R)}\cos\gamma R\right]$$

$$C_- = \frac{\sqrt{g(R)}}{2}\left[\sin\gamma R - \frac{\gamma}{g(R)}\cos\gamma R\right]$$

At the virtual level $E = E_0$, $C_+ = 0$ or

$\tan \gamma R = -\frac{\gamma}{g(R)} = -\sqrt{\frac{E_0 - U_0}{U(R) - E_0}}$ which is less than unity for an unstable nucleus

Thus $\gamma R \approx n\pi$ where n is an integer III.15

$$C_- \approx -\frac{\sqrt{g(R)}}{2}\frac{\gamma}{g(R)}(-1)^n$$

$$\left.\frac{\partial C_+}{\partial E}\right|_{E_0} \approx \frac{\sqrt{g(R)}}{2}R\frac{\partial \gamma}{\partial E}\cos\gamma R$$

$$\approx \frac{\mu R\sqrt{g(R)}}{2\gamma\hbar^2}(-1)^n \quad \text{since} \quad \frac{\partial}{\partial E}\gamma^2 = \frac{2\mu}{\hbar^2}$$

$$\frac{\partial \gamma}{\partial E} = \frac{\mu}{\gamma\hbar^2}$$

This may also be expressed in terms of $\delta E$ by using III.15.

$$\delta(R\gamma) = \pi \text{ gives } \delta\gamma = \frac{\pi}{R}$$

but $\delta\gamma = \frac{\mu}{\gamma\hbar^2}\delta E$

thus $\delta E = \frac{\gamma\hbar^2\pi}{\mu R}$

So $\left.\frac{\partial C_+}{\partial E}\right|_{E_0} \approx \frac{\pi}{2}\sqrt{g(R)}\,(-1)^n\frac{1}{\delta E}$; also

$$C_+ \approx \left.\frac{\partial C_+}{\partial E}\right|_{E_0}\varepsilon$$

$$\left.\begin{aligned} C_+^2 &\approx \frac{\mu^2 R^2 g(R)}{4\gamma^2\hbar^4}\varepsilon^2 = \frac{\pi^2 g(R)}{4}\frac{\varepsilon^2}{(\delta E)^2} \\ C_-^2 &\approx \frac{\gamma^2}{4g(R)} \end{aligned}\right\} \qquad \text{III.16}$$

## CHAPTER IV BETA DECAY

### A. Introduction

$\beta$ decay is the process in which an electron or a positron is emitted by a nucleus. The term is extended to include absorption of electrons.

The most remarkable feature of $\beta$ decay phenomena is the apparent failure of energy conservation. In other nuclear processes, such as $\alpha$ decay, energy is clearly conserved. For example, if nucleus A decays to nucleus B, producing an $\alpha$, the energy equation is $E_A = E_B + E_\alpha$. If the nuclei are in excited states before and after, the energy equation may be different: $E_A' = E_B' + E_\alpha'$, but always energy isconserved. In $\beta$ emission, such an energy equation involving only the observed particles cannot be written. The reason is that the energies of $\beta$'s from a single type of process have a continuous distribution of values. Empirically, the relative number of $\beta$'s of a given energy, N(E), as a function of energy is a curve like FIG. IV.1. No $\beta$ is emitted having energy greater than some value $E_\beta^{max}$.

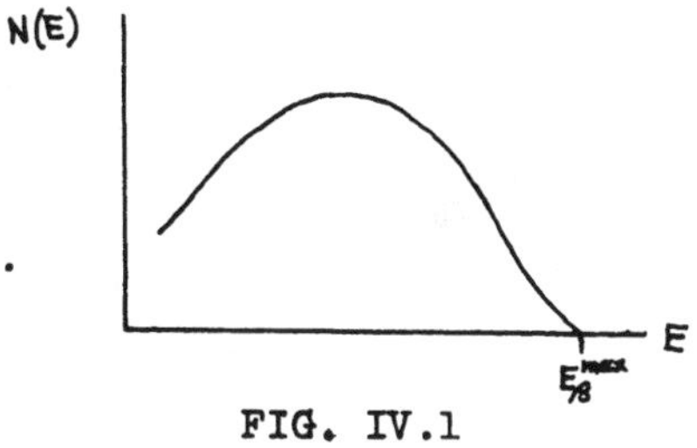

FIG. IV.1

A conceivable explanation that retains the energy conservation law is that the states of the final nucleus are very closely spaced, and the various $\beta$ energies correspond to various final states. But we must account for the extra energy of the final excited state. The only conceivable mode of decay to the ground state is by gamma emission. Since for some cases of $\beta$ emission there is no gamma radiation at all, and in any case no gamma radiation with a nearly continuous energy spèctrum, the hypothesis of many final states of the nucleus must be discarded.

There is no alternative to admitting that the final and initial states of the nucleus are definite, but that the $\beta$ may have any energy below $E_\beta^{max}$.

Experimentally, to within about 20 kev., $E_\beta^{max} = E_A - E_B$, showing that although energy may disappear, none is ever gained. The law of energy conservation might be forsaken, and replaced by: $E_{initial} \geq E_{final}$, a law that energy either is conserved or disappears, but never increases. Such a law would forbid perpetual motion machines of the first kind and accord with $\beta$ decay phenomena.

The most favored explanation of this apparent non-conservation of energy is the neutrino hypothesis first suggested by Pauli. It postulates that an additional particle, the neutrino, (or perhaps more than one) is produced in $\beta$ decay and carries away the missing energy. To accord with experiment, the neutrino, denoted by $\nu$, must be made very difficult to detect. This is done by postulating that it is electrically neutral, (conservation of charge also imposes this), and of very small mass. Under the neutrino hypothesis, the energy balance equation is: $E_A - E_B = E_\beta + E_\nu$.

## B. Examples of β Processes

Nuclei are almost certainly composed of neutrons and protons. (Reasons why electrons cannot exist in the nucleus are given in section D, p. 73 ) Therefore we may diagram the β process

$H^3 \xrightarrow{(12\ yrs)} He^3+\beta^-+\nu$, $E_\beta^{max}=0.019$ Mev. as follows:

Evidently the essence of the process is $N \rightarrow P+\beta^-+\nu$. $H^3$ is the lightest observed example of β decay.

An example of a second type of β decay is

${}_6C^{11} \xrightarrow{(20.5\ min)} {}_5B^{11}+\beta^++\nu+e^-$, $E_\beta^{max}=0.99$ Mev.

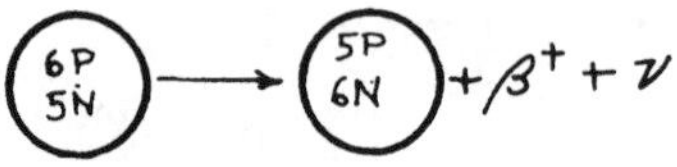


The net result is the transformation $P \rightarrow N + \beta^+ + \nu$. (The orbital electrons are not shown in these diagrams.)

A third type of β decay is exemplified by ${}_4Be^7 \xrightarrow{(52.9\ days)} {}_3Li^7+\nu-e_k$. The notation $-e_k$ means that an electron from the K shell of Be is absorbed into the ${}_4Be^7$ nucleus, and the result is a ${}_3Li^7$ nucleus with full complement of atomic electrons, but in an excited atomic state corresponding to a vacancy in the K shell.

The statements $N \rightarrow P+\beta^-+\nu$, $P \rightarrow N+\beta^++\nu$ describe the net transformation occurring in nuclei. The decay of a single neutron into a proton, or the reverse, has not been observed in a complete and conclusive way in experiment, although some evidence has been obtained, and experiments are in progress. One such experiment involves observing the decay protons and electrons from a beam of slow (few km/sec) neutrons. It is difficult to get a large number of neutrons per $cm^3$, even with high beam intensity and low velocity, and therefore the particles looked for are masked by the background. The latest estimate based on this type of experiment gives for the half-life of the neutron about 30 min, and definitely longer than 15 min.*

In principle, the $N \rightarrow P$ process may be performed with the help of a "catalyst". The process is this: We bombard $Fe^{56}$ with neutrons of a few Mev energy, and $Mn^{56}$ is produced by an (N,P) reaction:

$${}_0N^1+{}_{26}Fe^{56} \longrightarrow {}_1H^1+{}_{25}Mn^{56}$$

$Mn^{56}$ then decays:

$Mn^{56} \xrightarrow{(26\ hrs)} Fe^{56}+\beta^-+\nu$ giving back the $Fe^{56}$.

The net result is $N \longrightarrow P+\beta^-+\nu$. The $Fe^{56}$ serves as the catalyst.

## C. Energy Diagrams

β processes may be elucidated by diagrams showing the energy changes of the atom involved. Consider an atom having nuclear charge Z+1. Plot its total energy $E_{Z+1}$ on an arbitrary energy scale. In principle, $E_{Z+1}$ can be determined by weighing the neutral atom. Take the 0 of the scale so that the singly ionized atom (Z+1) plus the free electron at rest have total energy 0. (FIG. IV.3) This means that $E_{Z+1}$, the total energy of the neutral

* Snell and Miller, Phys.Rev. 74 1217 (1948)

atom, lies slightly below 0, by an amount equal to the first ionization potential of the atom. Now add to the diagram the energy level scheme for one electron, FIG. IV.2, and place this so that the last filled discrete state coincides with $E_{Z+1}$. Then the lower end of the continuum of states lies at E=0.

Now consider an atom of nuclear charge **Z** but of the same mass number. Its total energy $E_Z$, measured from the same 0 of energy, may be plotted on the same diagram, and the various possible β processes will be determined by the position of $E_Z$.

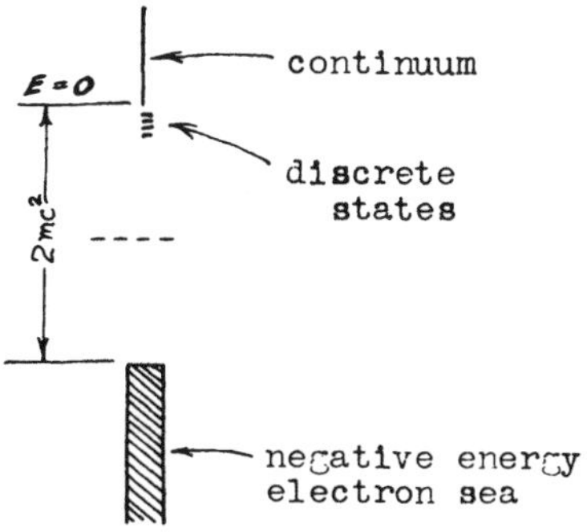


(This scheme is discussed in Ch. II (FIG.II.18))

FIG. IV.2

First suppose that $E_Z$ lies at position A, above $E_{Z+1}$. (See next page). If the nucleus (Z) converts by $\beta^-$ emission to (Z+1), energy is released. Some of this energy excess stays with the atom, because the new atom (Z+1) is produced in a singly ionized state. The rest is taken by the emitted $\beta^-$ and the neutrino. We have chosen the 0 of the energy scale so that the energy available for the $\beta^-$ and $\nu$ is just $E_Z$. The transition is represented diagrammatically by the arrows on the figure. The one-electron energy scheme now shows the possible states for the emitted $\beta^-$ particle. The $\beta^-$ can in principle go either to an empty discrete state or to the continuum. The latter is very much more likely, and the diagram shows such a transition. This first case is ordinary $\beta^-$ emission, as in $H^3 \rightarrow He^3 + \beta^- + \nu$. Positron emission is energetically impossible because $E_Z > E_{Z+1}$.

Secondly, suppose $E_Z$ lies below $E_{Z+1}$ but is greater than $-2mc^2$. Such a position is at B. Now the transition (Z)⟶(Z+1) is impossible because all electron states below the discrete valence levels are filled* and transition to electron states above this is energetically impossible. Therefore there is no state vacant to accept the $\beta^-$ from the (Z) nucleus. The transition (Z+1)⟶(Z) is energetically possible, but only if the nucleus (Z+1) can absorb an electron. The only filled electron states available are those in the discrete levels, so the absorbed electron must come from there. Usually the K shell supplies the electron, in which case the process is called K-capture. L-, M-,... capture is possible. The new atom (Z) is produced in an excited atomic state B* corresponding to a K (or L, etc.) shell vacancy. The energy difference B*-B leaves as gamma radiation. If the level B* corresponding to K shell vacancy is above $E_Z$ but B* for L excitation is below, then L, M, capture can occur, but not K capture, FIG.IV.4(a). If B* for minimum excitation (valence electron vacancy) is above $E_{Z+1}$, then electron capture cannot occur. If, in addition, B is below the level at which an electron can be created into a discrete state of atom (Z), then no transition is possible, FIG.IV.4(b). In this case (Z+1) and (Z) are both stable. Since the discrete energy levels occupy a small energy range compared to the actual energy differences between

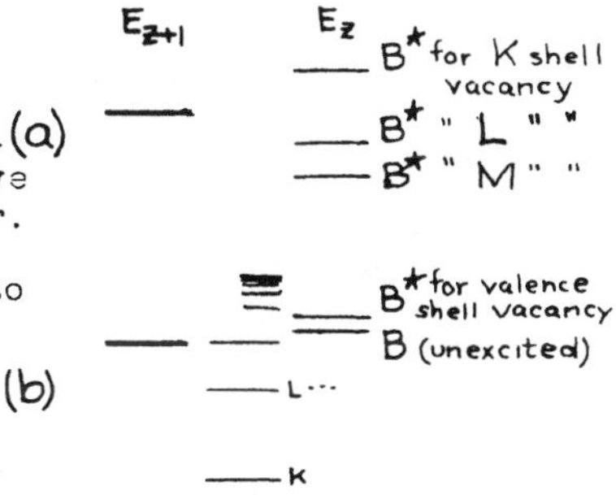


FIG. IV.4

* The Pauli exclusion principle operates.

isobars, such a situation is unlikely. In any other case a transition one way or the other is possible, and only one of the isobars (Z+1) and (Z) is stable.

The third case is for $E_Z$ less than 0 by $2mc^2$ or more. Position C, FIG. IV.5, is such a case. Evidently K-capture is possible. But now the (Z+1) nucleus may take an electron from the sea of electrons in negative energy states, thus producing a positron.

D. Theory of β Decay.

Present theories are not fully satisfactory. The presentation here is a summary of the most significant points, and is

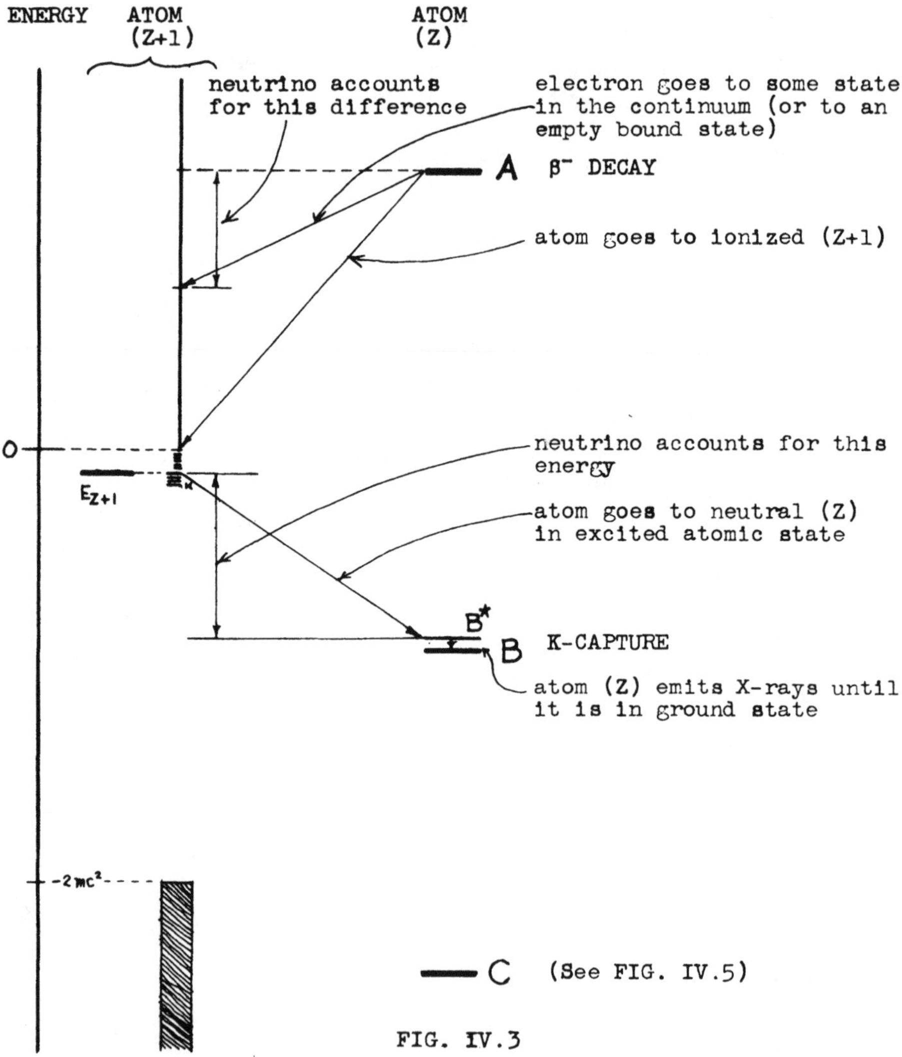


FIG. IV.3

oversimplified in the interest of clarity. The theory will give a) the rate of decay, b) the energy spectrum of the emitted particles.

Electrons do not exist in nuclei. In the early days of nuclear theory the neutron was unknown. The nucleus was thought to be made of protons and electrons, a nucleus $_Z()^A$ having A protons and A-Z electrons. After the discovery of the neutron, the nucleus was, and is, supposed to consist of protons and neutrons, so

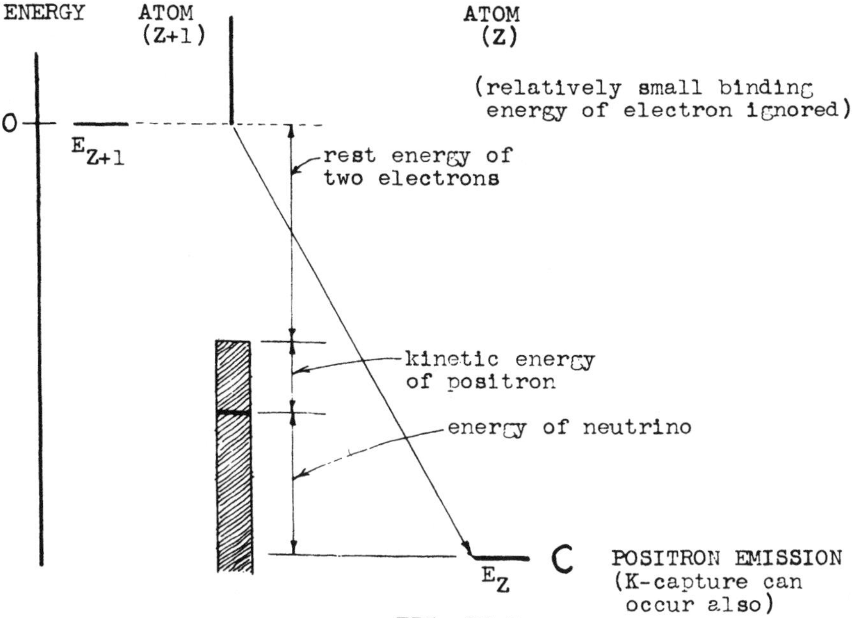


FIG. IV.5

that the nucleus $_Z()^A$ contains Z protons and (A-Z) neutrons. This change of view was brought about by compelling reasons why the nucleus could contain no electrons. Several are given here.

(a) The size of a nucleus is about $10^{-12}$ cm. An electron in the nucleus must be no larger than the nucleus. The electron momentum corresponding to such a wavelength is $p=\hbar/\lambda\!\!\!^{-} \approx \hbar/10^{-12} = 10^{-15}$. The corresponding electron energy is about 20 Mev. This is much larger than the energy of any β particle emitted by a nucleus. Such a high kinetic energy might be tolerated if the electron were in a deep potential well, but the Dirac theory predicts that electrons from negative energy states would spill into the potential well if it is more than $2mc^2$ deep, FIG.IV.6. (Note. If the potential for electrons differs by more than $2mc^2$, there is possibility for electrons in negative energy states to fall into positive energy states. Unless the variation by $2mc^2$ occurs in a very short distance, the probability of transition is very small. The intervening space acts as a barrier, FIG. IV.7. The creation of electrons and positrons in a potential

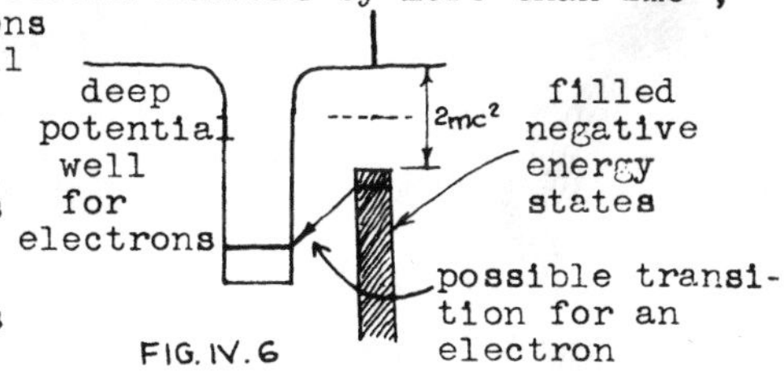


FIG. IV.6

field of large gradient has not been observed.

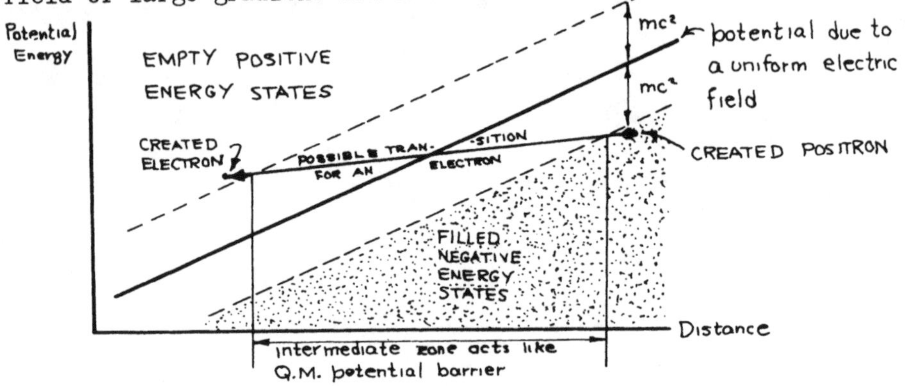


FIG. IV. 7

(b) Statistical arguments favor the proton-neutron theory of nuclear constitution. A nucleus having an odd number of elementary particles has Fermi statistics; a nucleus having an even number has Bose-Einstein statistics. The different hypotheses for the composition of nuclei lead to different numbers of particles. For the nucleus $_Z()^A$,

| Electrons-in-nucleus | Neutron hypothesis |
|---|---|
| A protons | Z protons |
| A-Z electrons | A-Z neutrons |
| 2A-Z elem. particles | A elem. particles |

For example, the $_7N^{14}$ nucleus has 14 particles under the neutron hypothesis, but 21 under the electron-in-nucleus hypothesis. Experiments in molecular spectroscopy* show that $_7N^{14}$ has Bose-Einstein statistics, therefore an even number of particles, confirming the neutron-in-nucleus hypothesis.

(c) The spin of the nucleus, whether integral or half-odd, depends on the number of elementary particles, and can be determined experimentally. The evidence again favors the neutron hypothesis.

Since the electron does not exist in the nucleus, it must be formed at the moment of its emission just as a photon is formed at the moment of its emission from an atom. The neutrino is created at the moment of emission, also. These particles are created into states represented by the wave-functions $\Psi_\beta$ and $\Psi_\nu$. Assume these are functions for plane waves with momenta $\underline{p}_\beta$ and $\underline{p}_\nu$, respectively,

$$\psi_\beta = N_\beta e^{i\,\underline{p}_\beta\cdot\underline{r}\,\hbar^{-1}}, \qquad \psi_\nu = N_\nu e^{i\,\underline{p}_\nu\cdot\underline{r}\,\hbar^{-1}} \qquad \text{IV.1}$$

where N is a normalization factor. $\Psi_\beta$ is actually more complicated than given here because it is affected by the nuclear charge Z. The plane wave $\Psi_\beta$ is a good approximation if the energy of the electron is much larger than Zx(Rydberg). For a low energy electron, say 200 Kev, near a nucleus of high Z, $\Psi_\beta$ is strongly perturbed.

The probability of emission will be assumed to depend on the

---

* See Chapter I, sec. D.

expectation value for the electron and the neutrino to be at the nucleus, i.e., on the factor $|\Psi_\beta(0)|^2|\Psi_\nu(0)|^2$. It also depends on other factors, whose nature is uncertain.

One factor is the square modulus of a matrix element $\mathcal{M}$ taken between the initial and final states of the nucleus. This matrix element is analogous to the matrix element in the theory of emission of photons. In photon emission the matrix element is definitely known and has the form (for dipole radiation)

$$\int \psi^*_{final}\,(\text{electric moment})\,\psi_{initial}\,d\tau$$

$\mathcal{M}$, in β theory, is not known. There are several possible forms. For the case N→P, the simplest is

$$\mathcal{M} = \int \psi_P^* \psi_N \, d\tau \qquad \text{IV.2}$$

assuming just one nucleon participates. $\Psi_N$ represents the initial state of the nucleon, $\Psi_P$ the final, i.e., proton, state of the nucleon. According to another form of the theory, $\mathcal{M}$ is a vector having x component

$$\mathcal{M}_x = \int \psi_P^* \sigma_x \psi_N \, d\tau \qquad \text{IV.3}$$

where $\sigma_x$ is the x component of a (relativistic) spin operator. Then

$$|\mathcal{M}|^2 = |\mathcal{M}_x|^2 + |\mathcal{M}_y|^2 + |\mathcal{M}_z|^2 \qquad \text{IV.4}$$

The choice of $\mathcal{M}$ determines the selection rules, discussed in section H.

The expression for the probability of emission includes also a constant factor $g^2$ which represents the strength of the coupling giving rise to emission, and is a universal constant. Experimentally,

$$g = 10^{-48} \text{ to } 10^{-49} \text{ g cm}^5 \text{ sec}^{-2} \qquad \text{IV.5}$$

Altogether, the probability of emission per unit time is*

$$\frac{2\pi}{\hbar}\left(|\psi_\beta(0)||\psi_\nu(0)||\mathcal{M}|\,g\right)^2 \frac{dn}{dE} \qquad \text{IV.6}$$

where dn/dE=energy density of final states, and 0 refers to the location of the nucleus. The $\Psi$ functions are normalized in a volume $\Omega$ so that $\int_\Omega \Psi^*\Psi \, d\tau = 1$. Therefore $N = 1/\sqrt{\Omega}$ IV.7

and

$$\psi_\beta = \frac{1}{\sqrt{\Omega}} e^{i\hbar^{-1}\underline{p}_\beta \cdot \underline{r}} \qquad \psi_\nu = \frac{1}{\sqrt{\Omega}} e^{i\hbar^{-1}\underline{p}_\nu \cdot \underline{r}} \qquad \text{IV.8}$$

It has meaning to say the nucleus is at $\underline{r} = 0$ only if $\Psi$ changes little over the dimension of the nucleus. The rapidity of variation of $\Psi$ is measured by $\lambda\!\!\!^- = \hbar/p \approx 10^{-11}$ cm, for a usual value of p. But the nucleus is about $10^{-12}$ cm in diameter. Therefore it is permissible to say that the nucleus is at $\underline{r} = 0$.

---

* This is analogous to the usual Q.M. formula for transition probability per unit time. This formula, "Golden Rule No. 2", is

prob. per second $= \frac{2\pi}{\hbar}|H_{21}|^2\frac{dn}{dE}$, where $H_{21} = \int \psi_2^* H' \psi_1 \, d\tau$.

$\Psi_1$ and $\Psi_2$ are the wave functions of the initial and final states, resp. This is derived, for example, in Schiff, Q.M., p. 193. It is discussed in more detail in Ch. VIII, sec. B.

For $\underline{r} = 0$

$$\psi_\beta(0) = \frac{1}{\sqrt{\Omega}} \qquad \psi_\nu(0) = \frac{1}{\sqrt{\Omega}} \tag{IV.9}$$

The number of plane wave states having magnitude of momentum between p and p + dp, with the particle anywhere in $\Omega$, is*

$$\frac{p^2 dp\,\Omega}{2\pi^2\hbar^3} \tag{IV.10}$$

Therefore
$$dn = \frac{p_\beta^2\,dp_\beta}{2\pi^2\hbar^3} \times \frac{p_\nu^2\,dp_\nu}{2\pi^2\hbar^3} \times \Omega^2 = \Omega^2 \frac{p_\beta^2 p_\nu^2}{4\pi^4\hbar^6} dp_\beta\,dp_\nu$$

$dp_\beta\,dp_\nu = J\,dp_\beta\,dE$ where J is the Jacobian. Using the relation $E = cp_\nu + E_\beta$, J is found to be 1/c.* (Mass of $\nu$ assumed zero for this derivation.)

Thus $$\frac{dn}{dE} = \Omega^2 \frac{p_\beta^2 p_\nu^2}{4\pi^4\hbar^6 c} dp_\beta$$

Using this to express IV.6, the probability of emission per unit time, $P(p_\nu, p_\beta)dp_\beta$, is

$$P(p_\nu,p_\beta)dp_\beta = \frac{2\pi}{\hbar}\left(\frac{1}{\Omega}|\mathcal{M}|\,g\right)^2 \frac{\Omega^2 p_\beta^2 p_\nu^2\,dp_\beta}{4\pi^4\hbar^6 c} \tag{IV.14}$$

Using the relation $p_\nu c = E_\nu = E_\beta^{max} - E_\beta$ to eliminate $p_\nu$, and writing p for $p_\beta$ from now on,

$$P(p)dp = \frac{g^2|\mathcal{M}|^2}{2\pi^3\hbar^7 c^3}\left(E_\beta^{max} - E_\beta\right)^2 p^2 dp \tag{IV.15}$$

Using the equation $E_\beta^{max} = \sqrt{m^2c^4 + c^2p_{max}^2}$ to define $p_{max}$, (IV.16) we get:

$$\boxed{P(p)dp = \frac{g^2|\mathcal{M}|^2}{2\pi^3\hbar^7 c^3}\left(\sqrt{m^2c^4 + c^2p_{max}^2} - \sqrt{m^2c^4 + p^2c^2}\right)^2 p^2\,dp} \tag{IV.17}$$

## E. Rate of Decay

The lifetime $\tau$ is found by integrating over all possible p.

---

* For simplicity, assume $\Omega$ is a cubical box of side L. $\Omega = L^3$. That the particle is confined to the box means that the potential rises to $\infty$ at the sides. Schrödinger's equation within the box is
$$-\nabla^2 u = \frac{2mE}{\hbar^2}u = (p_x^2 + p_y^2 + p_z^2)\hbar^{-2}$$

Solutions satisfying the boundary condition u = 0 at x.y.z = 0, and u = 0 at x,y,z = L are $u = \sin\frac{p_x}{\hbar}x \sin\frac{p_y}{\hbar}y \sin\frac{p_z}{\hbar}z$

where $p_x/\hbar$ is restricted to the values $n_x\pi/L$, etc. The number of states, n′, representing momentum less than p equals the number of combinations of $p_x, p_y, p_z$ such that $p_x^2+p_y^2+p_z^2 < p^2$, or, using the condition on the p's, $n_x^2+n_y^2+n_z^2 < p^2L^2/\pi^2\hbar^2$. $n_x, n_y, n_z > 0$ since $-$ values give no new independent solutions. The number of sets of $n_x, n_y, n_z$ satisfying the condition above equals 1/8 the number of points of a cubical lattice enclosed by a sphere of radius $\sqrt{\frac{p^2L^2}{\pi^2\hbar^2}}$. The 1/8 comes from restricting $n_x, n_y, n_z > 0$. Then the number of states,
$$n' = \frac{1}{8}\,\frac{4\pi}{3}\left(\frac{pL}{\pi\hbar}\right)^3. \tag{IV.18}$$

The number of states having momentum between p and p+dp is
$$\frac{dn'}{dp}dp = \Omega\,p^2dp/2\pi^2\hbar^3$$

---

* $$J = \begin{vmatrix} \frac{\partial p_\beta}{\partial p_\beta} & \frac{\partial p_\beta}{\partial E} \\ \frac{\partial p_\nu}{\partial p_\beta} & \frac{\partial p_\nu}{\partial E} \end{vmatrix} = \begin{vmatrix} 1 & \frac{\partial p_\beta}{\partial E} \\ 0 & \frac{1}{c} \end{vmatrix} = \frac{1}{c}$$

$$\frac{1}{\tau} = \frac{g^2|\mathcal{M}|^2}{2\pi^3\hbar^7c^3}\int_0^{p_{max}}\left(\sqrt{m^2c^4+p_{max}^2c^2}-\sqrt{m^2c^4+p^2c^2}\right)^2 p^2\,dp \qquad \text{IV.19}$$

Define $mc\eta = p$, $mc\eta_0 = p_{max}$. IV.20

Write the integral in terms of $\eta$ and $\eta_0$, and call it $F(\eta_0)$. The expression for the lifetime becomes

$$\frac{1}{\tau} = \frac{g^2|\mathcal{M}|^2 m^5c^4}{2\pi^3\hbar^7}F(\eta_0) \qquad \text{IV.21}$$

$$F(\eta_0) = \int_0^{\eta_0}\left(\sqrt{1+\eta_0^2}-\sqrt{1+\eta^2}\right)^2\eta^2\,d\eta \qquad \text{IV.22}$$

Straightforward integration leads to

$$F(\eta_0) = -\tfrac{1}{4}\eta_0 - \tfrac{1}{12}\eta_0^3 + \tfrac{1}{30}\eta_0^5 + \tfrac{1}{4}\sqrt{1+\eta_0^2}\,\ln\left(\eta_0+\sqrt{1+\eta_0^2}\right)$$

$$= -\tfrac{1}{4}\eta_0 - \tfrac{1}{12}\eta_0^3 + \tfrac{1}{30}\eta_0^5 + \tfrac{1}{4}\sqrt{1+\eta_0^2}\,\sinh^{-1}\eta_0 \qquad \text{IV.23}$$

Limiting forms for $F(\eta_0)$:

$\eta_0$ large compared to 5: $F(\eta_0) \longrightarrow \frac{1}{30}\eta_0^5$

$\eta_0$ small compared to .5: $F(\eta_0) \longrightarrow \frac{2}{105}\eta_0^7$ IV.24

(expanding the log term leads to powers of $\eta_0$ which cancel the powers of $\eta_0$ lower than $\eta_0^7$ in the expression for $F(\eta_0)$ )

Between these extremes, $F(\eta_0)$ has the values in the table:

| $\eta_0$ | $F(\eta_0)$ |
|---|---|
| 0.5 | 0.00012 |
| 1 | 0.0115 |
| 2 | 0.557 |
| 3 | 6.23 |
| 4 | 29.5 |

$F(\eta_0)$ calculated here is for a plane wave $\Psi$ for the electron. If the distortion of the wave function by the Coulomb field of the nucleus is taken into account a factor $f(Z,\eta_0)$ must be inserted into the un-integrated probability function $P(p)$. The integrated function $F(\eta_0)$ then depends on Z and should be written $F(Z,\eta_0)$. For small Z, $F(Z,\eta_0) \approx F(\eta_0)$ as defined before.

The final expression for $\tau$ is

$$\frac{1}{\tau} = g^2\frac{m^5c^4}{2\pi^3\hbar^7}|\mathcal{M}|^2\,F(Z,\eta_0) \qquad \text{IV.25}$$

where $|\mathcal{M}|^2$ is uncertain, but of order unity for "allowed" transitions. (Allowed and forbidden transitions will be discussed in sections H and J). From equation IV.25 it is seen that the theory predicts that $F\tau$ = constant if $|\mathcal{M}|^2$ does not change.

Problem: Look up the $\beta$ spectra of the following $\beta$ emitters: $H^3$, $He^6$, $C^{11}$, $N^{13}$, $O^{15}$, $F^{17}$, $C^{10}$, $C^{14}$; and fill in the table:

| Nucleus | Max. Energy | $\eta_0$ | $F(\eta_0)$ | $\tau$ | $F(\eta_0)\times\tau$ |
|---|---|---|---|---|---|
| | | | | | |

## F. Shape of Energy and Momentum Spectrum

Equation IV.17 for P, the probability of emission of an electron in the momentum range dp depends on p through the factor $\left[\left(\sqrt{m^2c^4+p_{max}^2c^2}-\sqrt{m^2c^4+p^2c^2}\right)^2 p^2\right]$ which has the form:

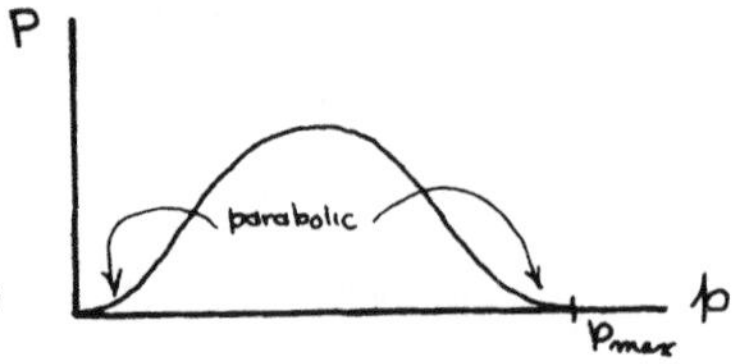


FIG. IV.8

The curve approaches P = 0 parabolically at 0, since there the expression $(\sqrt{\ } - \sqrt{\ })^2$ is almost constant. It approaches P = 0 parabolically at p = $p_{max}$ because there $p^2$ is almost constant and a Taylor expn. of the 2nd term about $p_{max}$. gives

$$c\sqrt{m^2c^2+p^2} = c\sqrt{m^2c^2+p_{max}^2} + c\frac{(p-p_{max})p_{max}}{\sqrt{m^2c^2+p_{max}^2}} + \cdots$$

$$\text{So} \lim_{p\to p_{max}} (\sqrt{\ } - \sqrt{\ })^2 = \frac{c^2(p-p_{max})^2}{m^2c^2+p_{max}^2} p_{max}^2$$

This plot must be corrected for the perturbation of the electron or positron wave function by the nuclear charge. $P_{\beta-}$ is larger than for Z = 0; $P_{\beta+}$ is smaller. The correction is greater for low energies. For negative electrons the correction near p = 0 is roughly proportional to 1/p; thus the corrected curve for negative electrons is linear near p = 0 (FIG. IV.9). For positrons the correction is in the other direction.

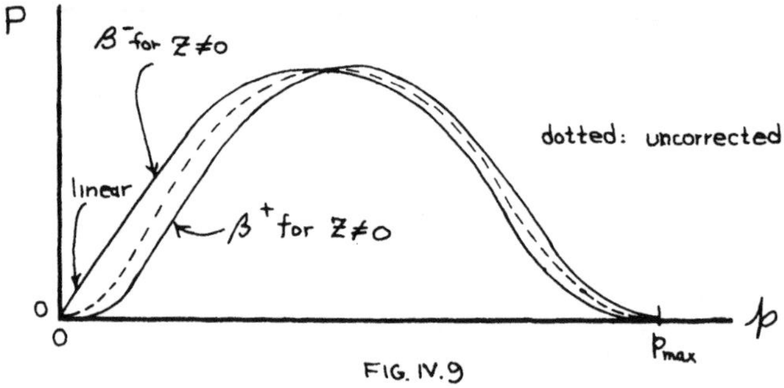


FIG. IV.9

The corresponding plot against energy is given in FIG. IV.10.*

$$\left(\sqrt{m^2c^4+p_{max}^2c^2}-\sqrt{m^2c^4+p^2c^2}\right)^2 p^2 dp \propto (E_\beta^{max}-E)^2\sqrt{E}\,dE \qquad \text{IV.26}$$

For $E \ll mc^2$ (E stands for kinetic energy here)

since $p^2dp = 1/2\ p\ d(p^2) \propto \sqrt{E}\ dE$.

The curves corrected for non-zero nuclear charge are shown. The corrected negative electron curve has the form near E = 0 of $(E_\beta^{max} - E)^2\ dE \sim$ constant x dE, therefore the curve has finite ordinate at E = 0.

## G. Experimental Verification.

There has always been uncertainty in the experimental results for the low energy part of the spectrum. Improvement in exper-

*Some forbidden $\beta$ decays have a different spectrum shape. These forbidden spectra have been experimentally verified.

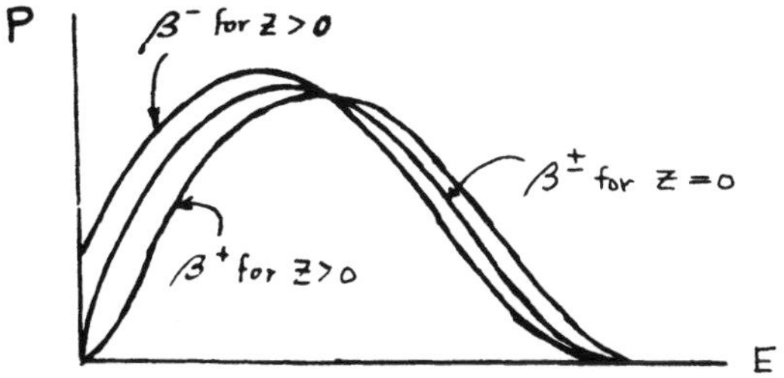


FIG. IV.10

imental technique has so far improved the agreement between experiment and theory.

The theoretical shape of the curve near $E_\beta^{max}$ depends on the mass of the neutrino. For neutrino mass 0, there is second order contact; for neutrino mass $\neq 0$ the curve has a vertical tangent at the point of contact. See Bethe A, p. 191.

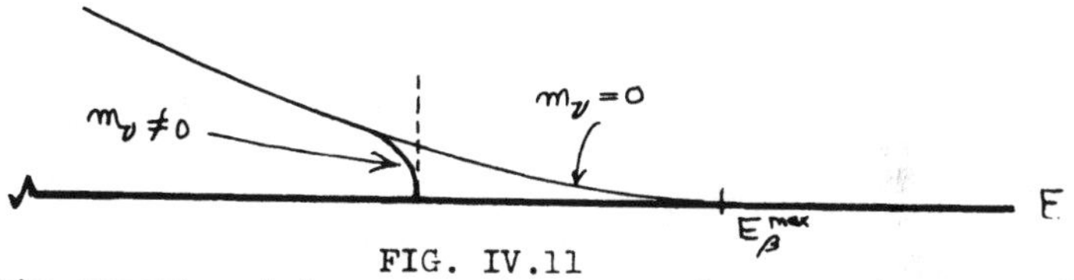


FIG. IV.11

Within experimental error, the curve for neutrino mass 0 is correct. The mass is certainly small, less than 10 Kev, in energy units.

The point at which the curve reaches the horizontal axis is difficult to determine experimentally because the curve is tangent to the axis there. It is therefore difficult to determine $E_\beta^{max}$ directly. More accurate determination of $E_\beta^{max}$ is made possible by the Kurie plot. From equation IV.15, the intensity of emission at p is

$$I(p) = (E_\beta^{max} - E)^2 p^2 C(Z,p) \qquad \text{IV.27}$$

where C(Z,p) includes the constants and also the dependence on nuclear charge. This can be written

$$(E_\beta^{max} - E) = \sqrt{\frac{I(p)}{p^2 C(Z,p)}} \qquad \text{IV.28}$$

Now the plot of the radical against energy should be a straight line whose intercept with the horizontal axis is easy to determine, FIG. IV.12(a).

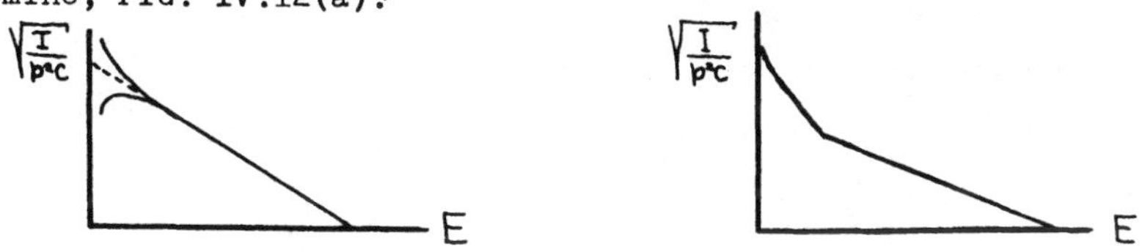


(a) Kurie Plots (b)

FIG. IV.12

Kurie plots deviate more or less from a straight line at low energy. These deviations are being reduced by more refined experimental techniques and by refinements in the theory of the factor $C(Z,p)$.

Occasionally a Kurie plot has the form in FIG.12(b). This is interpreted as the superposition of two β decay phenomena with different values of $E_\beta^{max}$, as exemplified by the diagram:

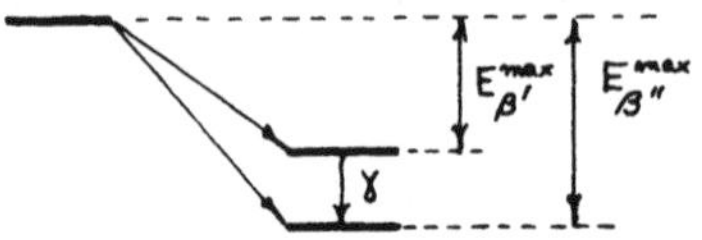


Often the gamma for the decay from the excited state is observed.

Sometimes β spectra have the form given in FIG. IV.13. The line spectrum is not true β decay, but is due to internal conversion, discussed in Ch. V.

> Problem: Design an experiment to study β decay in $Cu^{64}$. Indicate what type of electron spectrograph you would use. Assume data is desired down to energies of about 50 Kev. Specify the thickness of the source. Specify the thickness and type of backing for the source in order not to have too great distortion at low energies. Distortion of the energy spectrum is due to back scattering and absorption.
>
> 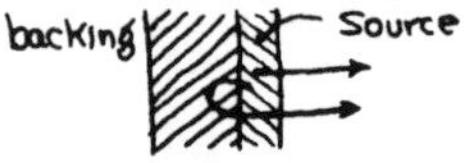
> 

## H. Selection Rules

In optical (atomic) transitions, the complete and exact matrix element is complicated and rarely evaluated. Rather, the integral of the optical retardation factor $\exp(-i\,\underline{n}\cdot\underline{r}\,\bar{\lambda}^{-1})$ is expanded in powers of $R/\bar{\lambda}$, where R is the extension of the atom. The first term gives the rate of dipole radiation, the second, quadrupole, etc. The successive terms of the unsquared matrix element differ by factors of $R/\bar{\lambda}$. The expansion is valid only if $R \ll \bar{\lambda}$, which means that the first nonzero term is dominant. If the first (dipole) term is zero, the transition is said to be "forbidden". The transition may still proceed, but at the much smaller rate given by the higher order terms. In optical emission, $R/\bar{\lambda} \approx 10^{-3}$.

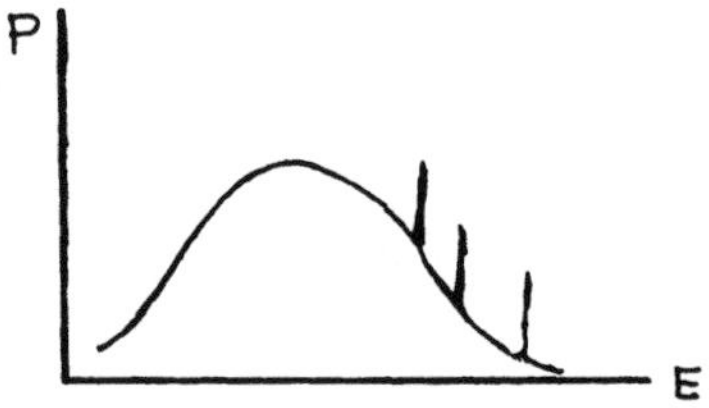


FIG. IV.13

The probability of transition depends on the square of the matrix element, therefore the forbidden optical transitions are about $10^{-6}$ as probable as "allowed" transitions, i.e., transition for nonzero first order terms. The "selection rules" give the necessary conditions on the change of state of the atom for the first term to be nonzero.

A similar situation obtains in the case of β decay. The matrix element is $\int \Psi^*_{final}\,\mathcal{M}\,\Psi_{initial}\,d\tau$, and it is the <u>wave function</u> that is integrated over the finite size of the nucleus. In the calculation of section D it was assumed that the nucleus had zero extension. This amounts to ignoring (in the unsquared matri element) all but the first term in an expansion of the wave function in powers of $R/\bar{\lambda}$, where R is now the extension of the nucle (This is analogous to ignoring higher order terms in the expansion of the retardation factor in atomic emission.)

If the first term is zero, the higher order terms, so far

ignored, must be considered. Since for a nucleus $(R/\lambda\!\!\!^{-})^2 \approx (10^{-12}/10^{-11})^2 = 1/100$ (instead of $10^{-6}$ as in the atomic case), forbidden transitions are relatively more prominent in β decay than in atomic emission.

An additional approximation has been made so far, and is an additional source of correction terms, and therefore of forbidden transitions. Strictly, there is a relativistic correction term of order V/c which should be added to the matrix element. V = velocity of nucleons. Since the matrix element is squared, and since V/c = 1/10, this correction results in terms again about 1/100 as large as for the first term in allowed transitions.

The selection rules for a given order of transition, i.e., for a given term in the expansion, give the necessary conditions on the change of the quantum numbers specifying the state of the nucleus in order that the term of given order in the complete matrix element be <u>nonzero</u>. Clearly there must be <u>different</u> selection rules for different orders in the expansion of the matrix element, otherwise all would be zero simultaneously, and there would be no forbidden transitions. The reason that each order of the expansion has characteristic selection rules is that in the expansion in powers of $R/\lambda\!\!\!^{-}$, the different orders contain different powers of <u>R</u>, a <u>polar</u> vector.**

The quantum "numbers" whose change in a transition are governed by selection rules are <u>total angular momentum</u> and <u>parity</u>. Due essentially to the isotropy of space, these two are characteristics of the state of any isolated system. Spin for a nucleus is denoted by I= (angular momentum)/$\hbar$. Parity refers to the property that the wave function either changes sign or does not change sign (magnitude unchanged in both cases) if the space coordinates are transformed by inversion: $(x,y,z) \longrightarrow (-x,-y,-z)$. If the wave function changes sign, its parity is <u>odd</u>; if not, <u>even</u>. If the matrix element integral is to differ from zero, the total integrand must be even. This imposes a correlation between the parity of $\mathcal{M}$ and the parity of $\psi_f^* \psi_i$ ; in other words, a correlation between the parity of $\mathcal{M}$ and the change in parity of the wave function. In the two cases in the next paragraph, $\mathcal{M}$ is even, and hence there is no change in parity.

Of the several possible forms for $\mathcal{M}$ in beta decay*, two are mentioned here. One is the "scalar" form, which gives a matrix element essentially of the form:

$$\int \psi^*_{final}\, \psi_{initial}\, d\tau \qquad \text{IV.29}$$

This gives the following selection rule for allowed transitions:

$$I = I', \quad \text{"no"} \qquad \text{IV.30}$$

where I is the spin of the initial state, I' the spin of the final state, and "no" means no parity change.

A second form is the "tensor" matrix element, essentially like:

$$\int \psi^*_{final}\, \underline{\sigma}\, \psi_{initial}\, d\tau \qquad \text{IV.31}$$

---

* The various possible forms of interaction, each with different matrix element and selection rules, are given in Konopinski's comprehensive article on β decay, <u>Rev</u>.<u>Mod</u>.<u>Phys</u>. <u>15</u> 209 (1943)

**Readers unfamiliar with the general ideas behind selection rules may first consult Ch. V, section B, p. 96.

Here $\sigma$ is a generalization of the Pauli spin matrices. The selection rule for this matrix element for allowed transitions is

$$I' = \begin{cases} I+1 \\ I \\ I-1 \end{cases} \quad \text{"no"} \qquad \text{(G-T)} \qquad \text{IV.32}$$

$$(I=0 \rightarrow I'=0 \text{ forbidden})$$

This spin change specification is the same as in optical (dipole) radiation and always results when the matrix element is a vector. "No" parity change depends on the fact that the vector is an axial vector, that is, inversion of space does not change the sign of the vector.* (In optical transitions, parity changes because the vector (the dipole moment) is a "polar" vector, that is, its sign changes upon inversion). The second selection rule, IV.32 is the Gamow-Teller (G-T) selection rule, and is favored by the scanty experimental evidence accumulated so far.

The distinction between allowed and forbidden transitions is blurred by uncertainty in the value of the matrix element for allowed transitions. The matrix element may give a factor 1/10, and therefore a factor 1/100 in the transition probability, and thus the transition will appear first order forbidden, but, in principle, be allowed. Situations like this make it difficult to know the selection rules from experimental data.

Any influence that increases the curvature of the wave function at the nucleus will increase the size of the second order terms relative to first order in the expansion of the exact matrix element. One such influence, particularly in heavy elements, is the Coulomb field of the nucleus.

## J. $F\tau$ Tables **

From equation IV.25

$$\frac{1}{\tau} = (\text{universal constant})\,|m|^2\, F(Z,\eta_0)$$

Therefore

$$F\tau = \frac{\text{constant}}{|m|^2} \qquad \text{IV.33}$$

Evidently the product $F\tau$ is a measure of the forbiddenness of a transition. The lowest $F\tau$ values correspond to transitions permitted by the selection rules discussed here. The very lowest $F\tau$ values, 1000-5000 represent superallowed transitions, and are allowed transitions between nuclei having similar nuclear wave functions (see next paragraph). $F\tau$ values of $10^4$ to $10^5$ represent allowed transitions between nuclei not having very similar nuclear wave functions. If the successive orders of forbiddenness differed in transition probability by factors of 1/100, as is predicted by the simple considerations of section H, we would find clusters of points on the $F\tau$ plot at $F\tau = 10^6$ to $10^8$, representing first forbidden transitions, at $F\tau = 10^8$ to $10^{10}$, representing second forbidden transitions, and so forth. Actually, the points do not cluster nearly so nicely. In fact, the clustering is barely discernible with imagination. Rigid classification of the empirical $F\tau$ values seems hopeless.

---

* An example of an axial vector is ordinary angular momentum $\underline{r} \times \underline{p}$.

**Nordheim discusses $F\tau$ values and their relation to other selection rules and to the shell model in P.R. 78, 294 (1950).

The size of the matrix element for a transition depends essentially on the extent to which the initial wave function of the nucleus overlaps the final wave function. In general, the nature of the wave functions is not known, and not much can be said about the overlap, except by observing β decay rates. But in the case of the Wigner or "mirror" pairs of elements, there is reason for assuming that the wave functions for both members of a pair are very much alike.* (The results of this assumption accord with experiment and serve to confirm the assumption.) Wigner pairs are elements such that one results from the other by interchange of neutrons and protons. The simplest pair is of course, the neutron and the proton. The next simplest pair is $He^3$, $H^3$. In general, the nucleus $(Z)^{2Z+1}$ is the mirror of the nucleus $(Z+1)^{2Z+1}$. β transitions between mirror elements are superallowed, because in the matrix element integral, $\Psi_{initial}$ differs little from $\Psi_{final}$. $F\tau \approx$ 1000 to 5000.

---

* The following is a crude argument to show why the wave function of one of a mirror pair does not change much when the nucleus changes to its "image", Assume that except for the presence or absence of one unit of + charge, all nucleons are the same. We might ascribe to each nucleon two dichotomic variables, one for spin orientation and one for charge. The latter's two values correspond to charge +1 or 0. Assume that the nucleus is in its lowest energy state, and that this corresponds to having the lowest space states of the system filled. To each space state there may be four nucleons, without violating the exclusion principle. The four correspond to the four possible combinations of spin and charge variables. That is, each space state may be occupied by two neutrons with opposing spins and by two protons with spins opposed. Now consider two examples of β transformation. In the first example, the number of neutrons is much greater than the number of protons. This means that probably some space states are not completely filled with the full quota of four nucleons:

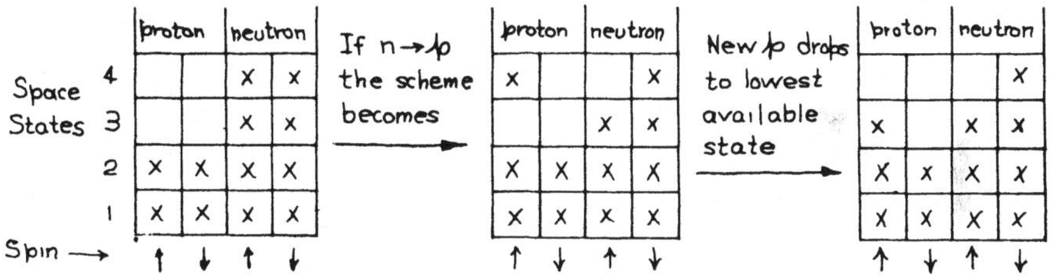


Now if one nucleon changes from p to n, or vice versa, it can also change its space state in order to occupy the lowest available to it. Therefore the wave functions before and after will differ in their space dependence. In the second example, the number of neutrons almost equals the number of protons, as in a mirror pair. Now if a nucleon changes its charge, it probably will not also change its space state:

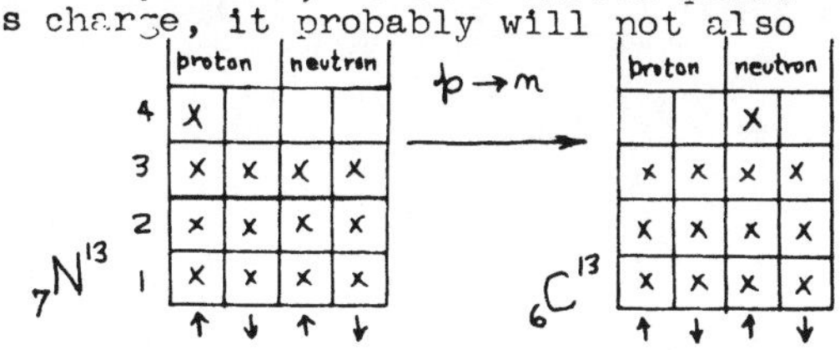

## K. Remarks on K-Capture

As described in section B, K-capture follows the scheme

(Atom (Z) with usual quota of 2 K-shell electrons) ⟶
(Neutral atom (Z-1) with one K-shell orbit empty, plus a neutrino

No observable particle comes out of the atom, but the excited final atom emits an X-ray when a bound electron drops into the empty K-shell state. Thus K-capture is observed by means of the X-rays emitted. The X-rays are soft, and difficult to observe; this makes K-capture hard to observe. $_{29}Cu^{64}$ can undergo either K-capture, positron emission, or $\beta^-$ emission to $_{30}Zn^{64}$. Referring to Fig. IV.3, the energy level of $Cu^{64}$ occupies a position of type C with respect to $_{28}Ni^{64}$, but of type A with respect to $_{30}Zn^{64}$.

K-capture becomes more important at the heavy end of the periodic table. The reason is that there the K orbits are small and the probability for the electron to be at the nucleus is large. On the other hand, the positron wave function is small at the positively charged nucleus, and the probability for positron emission is correspondingly small.

Problem. Discuss methods for observing K-capture.

## L. Remarks on the Neutrino Hypothesis

A promising line of research on the neutrino hypothesis is the study of the recoil of the nucleus in β decay, particularly in K-capture. If the nucleus and the β particle account for all the momentum, then of course the nuclear recoil momentum is equal and opposite to the momentum of the β. Experiments to determine the angular correlation between the nuclear recoil and the emitted β particle are at the limit of available technique. They show that recoil is not opposite to the direction of the β particle*.

In K-capture only the neutrino is emitted. If K-capture occurs, the nucleus will recoil with momentum equal and opposite to that of the emitted neutrino. This experiment is best performed with a light nucleus, such as $Be^7$, in order that the recoil energy be as large as possible, and therefore easy to detect. Observed recoil of the nucleus in K-capture does not rule out the hypothesis that energy is not conserved in β decay (no neutrino) provided one also postulates non-conservation of momentum.

Such experiments can discriminate between a one-neutrino theory and a two- or more neutrino theory. In a one-neutrino theory, in the case of K-capture, the nucleus always recoils with the same energy. In a theory in which more than one neutrino are emitted, the nuclear recoil energy is not always the same. A detailed study of the features of recoil would show which is the true situation.

Assuming one neutrino, one should find that

$$E_\nu^2 = m_\nu^2 c^4 + p^2 c^2$$

connects the energy and the momentum attributed to the neutrino. Experimental determination of E and p, by recoil experiments, would then determine the mass of the neutrino in a way independent of the β decay theory.

Assuming that experiments above would yield the result that $E = cp$, i.e., mass = 0, then a detailed knowledge of the angular

---

* Such experiments have been performed by Allen, Paneth and Morrish, Phys.Rev. 75 570 (1949); Sherwin, Phys.Rev. 73 216 (1948); and by others. A comprehensive article on the search for the neutrino is by H.R.Crane, Rev.Mod.Phys. 20 278 (1948).

correlation between emitted $\beta$ and $\nu$ would discriminate among the forms of the $\beta$ theory, that is, show what interaction operator should form the integrand of the matrix element integral. The shape of the spectrum does not discriminate. The angular distributions to be expected from the various forms of interaction are given by Hamilton, Phys.Rev. 71 456 (1947).

A more direct method to verify the existence of the neutrino is to observe it by collision processes. This has been attempted, using the intense neutrino flux that presumably is given off by a chain reacting pile. A very well shielded counter was placed near a pile. There were no collisions observed. The experiment sets an upper limit on the cross section for detectable collisions, within the counter, of $10^{-31}$ or $10^{-32}$ cm$^2$.

Another possibility for direct detection of the neutrino is by an inverse $\beta$ process. The existence of the inverse process is assured by a principle of detailed balancing, stemming from the basic concepts of quantum mechanics.* Assuming that quantum mechanics (in some form retaining this principle) is applicable to $\beta$ decay, inverse $\beta$ processes are possible, such as

$$\nu + e + (Z+1)^A \longrightarrow (Z)^A$$

USUAL REACTION ($\longleftarrow$)

Such a reaction might take place if a neutrino impinges on a nucleus, and the nucleus simultaneously takes an electron from the bound orbits, or from the positron sea. The cross section for this process is

$$\sigma \approx \frac{2g^2}{\pi \hbar^4 c} |\mathcal{M}|^2 \frac{p^2}{V} \quad \left(\approx 10^{-44} \text{ for neutrino energy} \sim \text{few Mev}\right)$$

For favorable cases, $|\mathcal{M}|^2 \sim 1$. p and V are the momentum and velocity of the electron. Such a small cross section permits a neutrino to cross the sun with little probability of being absorbed in an inverse process. Perhaps the most conclusive proof for the existence of the neutrino, and the most remote of attainment, would be to observe $\beta$ decay with recoil of nucleus and momentum of electron known so as to give the direction of the neutrino, and then on the path of the neutrino to detect almost simultaneously an inverse $\beta$ reaction whose energy relations agree with the energy of the neutrino emitted in the first reaction.

The greatest source of neutrino flux at the earth is the sun. About 10% of the sun's energy is spent in $\beta$ emission, and roughly 50% of this (5% of the total) energy escapes in the form of neutrinos. This corresponds to roughly $10^{11}$ neutrinos per second per cm$^2$ at the earth's surface.

## M. Neutrinos and Anti-neutrinos

The theory of the neutrino may be formulated in a way like that for the electron, that is, postulating negative energy states which are almost always filled, FIG. IV.14. If the rest mass of the neutrino is 0, the spectrum of filled negative energy states joins that for the unfilled positive energy states at energy = 0. Holes or vacancies in the negative energy neutrino sea are called antineutrinos, $\nu^*$. The postulation of of antineutrinos permits formulation of all $\beta$ processes in terms of one, namely,

$$N \longrightarrow P + \beta^- + \nu \qquad \text{IV.34}$$

The reverse reaction is $\beta^- + \nu + P \longrightarrow N$. Now since the $\beta^-$ taken on the left side leaves a hole in the electron negative energy sea, we may write it as the production of a positron on the right side (the 2mc$^2$ threshold energy, section C, appears clearly as the

* Detailed balance is discussed in greater detail in Ch. VIII, C.

energy needed to lift the electron from the positron sea. If it takes an orbital electron as in K-capture, no hole is produced and hence no positron appears). Exactly in the same manner, we may write an antineutrino, $\nu^*$, on the right side, to represent the hole left by the neutrino absorbed, and the equation takes the form:

$$P \rightarrow N + \beta^+ + \nu^* \qquad \text{IV.35}$$

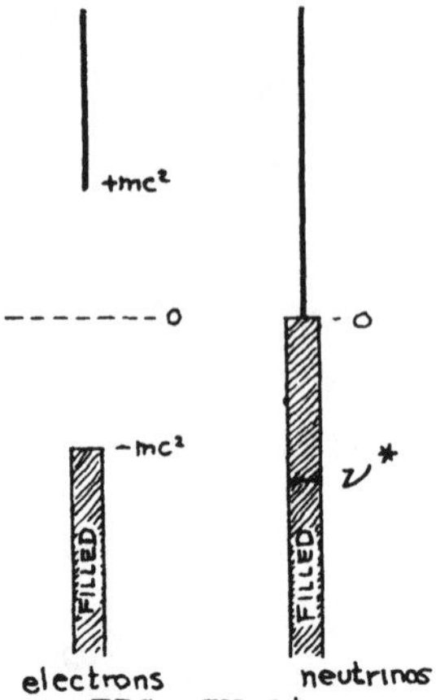

FIG. IV.14

Equations IV.34 and 35 can be taken as defining the antineutrino. $\nu$ and $\nu^*$ are equally difficult to detect. They are, in principle, different particles. The difference cannot be detected by electric charge, as it can for electrons and positrons. According to theory, $\nu$ has spin 1/2. If it also has magnetic moment $\mu$ (very small, to accord with very slight interaction between it and matter), then the ratio $\mu$/(angular momentum) would be the same for $\nu$ and $\nu^*$, but with opposite sign. If $\nu$ and $\nu^*$ have no electromagnetic properties, and cannot be distinguished by them, we can in principle only determine the source of a particular $\nu$, and if it comes from a reaction like IV.34, it is a $\nu$, if from a reaction IV.35, a $\nu^*$.

<u>Double β decay</u> provides a means of finding which form of the β decay theory is correct, the form postulating $\nu$ and $\nu^*$ or the (Majorana) form in which all $\nu$ are the same. Consider the isobaric triplet $Sn^{124}$, $Sb^{124}$, $Te^{124}$, whose energy levels may be as shown:

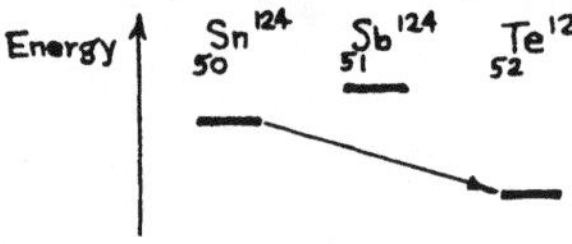


Direct transformation from Sn to Te with the emission of two $\beta^-$'s is conceivable. Both forms of the β decay theory say this is possible, but in second order approximation. Since the constant $g$ is very small, transition probabilities arising from second order transitions are extremely weak. The probability of transition depends acutely on which form of the theory is used.

For the form in which IV.34 is the fundamental reaction, and $\nu$ is different from $\nu^*$, the double process is simply the sum of two processes, and four particles, two electrons and two neutrinos, come out:

$$\begin{array}{l} N \rightarrow P + \beta^- + \nu \\ N \rightarrow P + \beta^- + \nu \\ \hline Sn^{124} \rightarrow Te^{124} + 2\beta^- + 2\nu \end{array} \qquad \text{(either form)}$$

The mean life for this reaction is $\tau = 10^{24}$ years, by this theory.

For the Majorana form, in which all $\nu$'s are the same, the process may go as above, that is, two separate β reactions with four emitted particles. But in this theory, the emission of <u>either</u> $\beta^-$ or $\beta^+$ is accompanied by <u>either</u> the emission <u>or</u> <u>absorption</u> of a neutrino. Therefore another mode is possible, namely one in which the neutrino produced by the first reaction is absorbed in the second:

$$\begin{array}{l} N \rightarrow P + \beta^- + \nu \\ \nu + N \rightarrow P + \beta^- \\ \hline Sn^{124} \rightarrow Te^{124} + 2\beta^- \end{array} \qquad \text{(Majorana form only)}$$

This neutrino is "virtual"; no neutrino comes out. This mode has much greater probability because the virtual neutrino has a much larger region of phase space accessible to it than an emitted neutrino. The mean life under this theory is $10^{16}$ - $10^{17}$ years.

Experimentally, a decay rate with $\tau = 10^{24}$ years is undetectable. For $10^{16}$ or $10^{17}$ years, it may be just within detection. The double β decay from $Sn^{124}$ has been reported as observed.* The experiment needs further checking. In particular, it can be seen from the above that in the last mode. (virtual neutrino), in which no neutrinos are emitted, the sum of the energies of the β particles is a constant. This crucial point should be checked.

(The theory of double β decay and summary of the differences between the two forms is given in Furry, Phys.Rev. 56 1184 (1939). Also, M. Goeppert-Mayer, Phys.Rev. 48 512 (1935)).

* Note in Phys.Rev. 75 323 (1949) by Fireman. See also Inghram and Reynolds, Phys. Rev. 76 1265 (1949).

> Problem. Plan an experiment to find the angular correlation between nuclear recoil and emitted β, in β decay by $He^6$. The articles by Allen, et.al, by Sherwin, and by Crane contain information pertinent to this problem.

Note: The neutrino and antineutrino are sometimes defined

$$N \rightarrow P + \beta^- + \nu^*$$

$$P \rightarrow N + \beta^+ + \nu$$

Note: Inghram and Reynolds get half-life of 1.4 x $10^{21}$ years for double beta decay of $Te^{130}$. P.R., 78, 822 (1950).

# CHAPTER V. GAMMA RADIATION

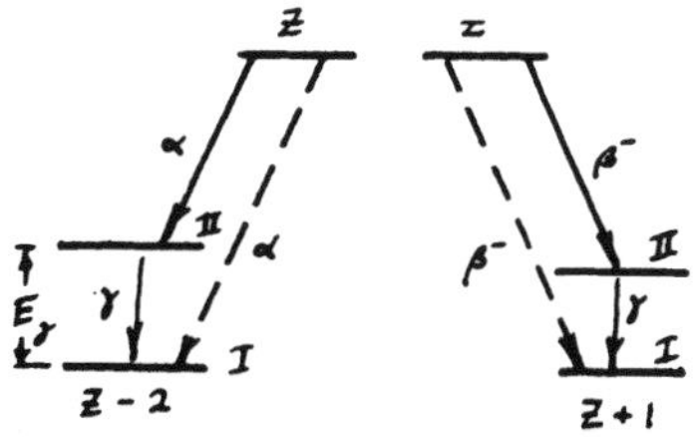


FIG. V.1 Post-α and post-β gammas. Gamma radiation will follow the transitions indicated by solid lines.

Gamma radiation and the ejection from the atom of internal conversion electrons are two of the ways that an excited nucleus may lose energy. The excitation may have arisen from bombardment, or the decay of some other nucleus to an excited state of the nucleus under consideration (illustrated in FIG. V.1), or by the absorption of a photon.

If the $\gamma$-emitting nucleus is formed by α-decay, it is improbable that $E_\gamma$ will be $> \frac{1}{2}$Mev. This is because of the extreme energy dependence of the Gamow barrier penetration probability (Ch. III, p. 57).

On the other hand, if the nucleus is formed by β-decay, then subsequent $\gamma$'s are commonly observed with energies around 2 Mev. We can explain this fact by remembering that, for β-decay (IV.24, p. 77)

$$F(\eta_0)\tau \longrightarrow \frac{1}{30}\eta_0^5\tau \propto E^5\tau = \text{constant} \qquad \begin{aligned} E &>> mc^2 \\ \eta &\equiv p/mc \end{aligned}$$

This dependence of the transition rate, $1/\tau$, upon the fifth power of E is less sensitive **than** the exponential dependence for α-emission.

In this chapter we shall discuss the different ways in which an excited nucleus can decay to a state of lower energy. We shall determine the most probable sort of transition between a given pair of states, and its transition probability per unit time. At the end we shall present only the briefest application of the theory to isomerism. Some of the ideas developed will, however, be used in the chapter on nuclear reactions.

For a more complete treatment, not only of radiation, but also of isomerism, both theoretical and experimental (with a discussion of selected cases), see the review article by Segre and Helmholz, "Nuclear Isomerism," Rev. Mod. Phys. 21, 271, ('49).

## A. SPONTANEOUS EMISSION

The object of this section is to present eqn. V.5 (p. 94), which gives the transition probability per unit time for multipole gamma emission, and eqn. V.9 (p. 95 ) for the transition probability per unit time for dipole emission. We shall need V.5 to discuss the relative importance of dipole, quadrupole, etc., radiation by nuclei. We shall use V.9 for discussion of selection rules and of the internal conversion coefficient.

The reader who is familiar with these equations or who would like to save time by merely accepting them, should skip the intervening review. In any case we shall not prove these equations, but shall discuss them and indicate how one might guess them from the classical (non-quantum mechanical, non-relativistic*) equations of radiation. They are also illustrated with problems at the end of the chapter. For a more rigorous treatment of radiation, see Schiff, chapters X and XIV.

We use the following notation:

$\underline{J} = \rho \underline{V}$ (not $\rho \underline{V}/c$) is the current density.

$\underline{S} = \frac{c}{4\pi} \underline{\mathcal{E}} \times \underline{\mathcal{H}}$ is the instantaneous value of the Poynting vector (electromagnetic intensity).

$\overline{\underline{S}}$ is its time average

$\int \underline{S} \cdot d\underline{A}$ = instantaneous total power radiated.

A single particle of charge e and no inherent magnetic moment radiates according to**

$$\int \underline{S} \cdot d\underline{A} = \frac{2}{3} \frac{e^2}{c^3} \left( \ddot{\underline{r}} \right)^2 \qquad \dot{r} \ll c$$

Because of interference we cannot, in general, use simple addition to find the power radiated by two particles moving close to one another***. However, if the particles are describing periodic motion, each with a different frequency, there is no time average interference. The first step then, in solving a radiation problem, is to analyze the current distribution into its frequency components, each of which can then be considered separately.

We shall define the Fourier components, $\underline{j}(\underline{r})$, as follows:

$$\underline{J}(\underline{r}, t) = \sum_{\omega} \underline{J}_{\omega}(\underline{r}, t); \qquad \omega > 0$$

$$\underline{J}_{\omega}(\underline{r}, t) = \underline{j}_{\omega}(\underline{r})\, e^{-i\omega t} + \text{complex conj.} \qquad \text{V.1}$$

From now on we shall deal with only one Fourier component, and omit the subscript $\omega$.

We can follow the conventional treatment of radiation and arrive at a suitable formula whence to start our review. In the "wave zone" ($r \gg r'$; $r\lambda \gg r'^2$),

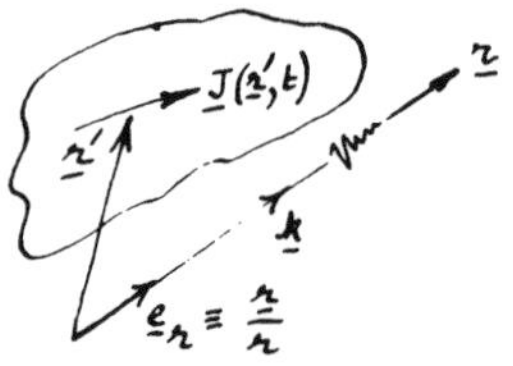


FIG. V.2

$$\overline{\underline{S}} = \frac{c}{4\pi} \overline{\underline{\mathcal{E}} \times \underline{\mathcal{H}}}$$

$$= \frac{\underline{k}k}{2\pi r^2 c} \left| \int \underline{j}_{\perp}(\underline{r}') e^{-i\underline{k}\cdot\underline{r}'} d\tau' \right|^2 \qquad \text{V. 2****}$$

where $\underline{k} = \frac{\omega}{c} \underline{e}_r$ (see sketch) and $\underline{j}_{\perp}$ is such a vector that $\underline{j} = \underline{j}_{\perp} + \underline{j}_{\parallel}$, the subscripts "perpendicular" and "parallel" referring to $\underline{k}$; i.e.

$$\underline{j}_{\perp} = \underline{j} - (\underline{j}\cdot\underline{e}_r)\underline{e}_r$$

---

*Nuclear radiation may be treated non-relativistically, since $v/c \sim 0.1$ for the particles in a nucleus.

**Abraham and Becker, "Classical Electricity and Magnetism," Ch. X, sect. 11.

***See, for example, problem 2, p. 108.

****Schiff, (36.11) p. 251. Some students will expect an 8 rather than a 2 in the denominator. This difference of a factor of $2^2$ arises as follows.

(footnote continued on p. 91)

For a quantum mechanical treatment we may consider the current density $\underline{J}(\underline{r}',t)$ not as continuous, but as due to the motion of particles of charge e, mass M, density n, and momentum $\underline{P}$. We can then write

$$\underline{J}(\underline{r}',t) = \frac{e}{M}\underline{P}n(\underline{r}',t)$$

where $\underline{P}n$ is the momentum per unit volume.

A time-independent momentum density, $\underline{p}n(\underline{r}')$, may be defined analogously to $\underline{j}(\underline{r}')$:

$$\underline{P}\, n(\underline{r}',t) = \underline{p}\, n(\underline{r}')\, e^{-i\omega t} + \text{c.c.}$$

So

$$\underline{j}(\underline{r}') = \frac{e}{M}\, \underline{p}\, n(\underline{r}')$$

As a first step in the transition to quantum mechanics, let the momentum vector, $\underline{p}_\perp(\underline{r}')$, take on the meaning of the momentum operator, $\underline{p}'_{\perp\,op} = \frac{\hbar}{i}\nabla'_\perp$, where $\nabla'_\perp$ is defined analogously to $\underline{j}_\perp$, i.e. it is the vector component of the gradient perpendicular to $\underline{k}$.

Quantum mechanics attributes the radiation not to a steady-state particle density, $|\psi|^2$, but to a transition from the initial state, $\psi_i$, to a final state, $\psi_f$. These new interpretations are indicated schematically as follows.

$$\overline{\underline{S}} = \frac{\underline{k}k}{2\pi r^2 c}\left|\int \frac{e}{M}\, \underline{p}'_\perp\, e^{-i\underline{k}\cdot\underline{r}'}\, |\psi(\underline{r}')|^2\, d\tau'\right|^2 \qquad \text{First guess}$$

$$\rightarrow \frac{\underline{k}k}{2\pi r^2 c}\left|\int \psi^*(\underline{r}')\, e^{-i\underline{k}\cdot\underline{r}'}\, \frac{e}{M}\, \underline{p}'_{\perp\,op}\, \psi(\underline{r}')\, d\tau'\right|^2 \qquad \text{Better}$$

$$\rightarrow \frac{\underline{k}k}{2\pi r^2 c}\left|\int \psi_f^*\, e^{-i\underline{k}\cdot\underline{r}'}\, \frac{e}{M}\, \underline{p}'_{\perp\,op}\, \psi_i\, d\tau'\right|^2 \qquad \text{Quantum Mechanics}^*$$

We generalize to the case where $\psi$ describes N particles

$$\overline{\underline{S}} = \frac{\underline{k}k}{2\pi r^2 c}\left|\sum_{s=1}^{N} \int \psi_f^*\, e^{-i\underline{k}\cdot\underline{r}'_s}\, \frac{e_s}{M_s}\left(\underline{p}'_{\perp\,op}\right)_s \psi_i\, d\tau'\right|^2$$

Finally, we assume that $\dot{Q}(\theta,\phi)$, the number of photons emitted per sec. per unit solid angle, of frequency $\omega/2\pi$, is given by.

$$\dot{Q}(\theta,\phi) = \frac{r^2\, \overline{S}}{\hbar\omega}$$

---

(footnote con. from p. 90) Frequently V.1 is written differently:

$$\underline{J}_\omega(\underline{r},t) = \underline{J}_\omega(\underline{r}) \cos(\omega t + \delta) = \text{Re}\left[\underline{J}(\underline{r})\, e^{-i\omega t}\right]$$

A comparison of the two definitions shows that

$$\underline{j}(\underline{r}) = \tfrac{1}{2}\, \underline{J}(\underline{r}).$$

We are going to convert V.2 to a quantum mechanical equation; for this purpose it is more convenient to work with quantities that vary as exp $(-i\omega t)$ rather than $\cos\omega t$.

*The frequency of the radiation $\omega = \frac{E_f - E_i}{\hbar} = kc$ no longer corresponds to the classical motion.

So

$$\dot{Q}(\theta,\phi) = \frac{\omega}{2\pi\hbar c^3}\left|\sum_{s=1}^{N}\int \psi_f^* e^{-i\underline{k}\cdot\underline{r}'_s}\frac{e_s}{M_s}\underline{p}'_{\perp s}\psi_i\, d\tau'\right|^2 \qquad \text{V.3}$$

For atomic and nuclear radiation problems, the dimensions of the radiator are small compared to the wave length of the radiation, so that we may break the above integral into a sum of terms using the expansion

$$e^{-i\underline{k}\cdot\underline{r}} = 1 - i\underline{k}\cdot\underline{r} - \tfrac{1}{2}(\underline{k}\cdot\underline{r})^2 + \cdots$$

and then consider only the first non-zero integral.

$$\dot{Q}_l(\theta,\phi) = \frac{\omega}{2\pi\hbar c^3}\left|\sum_{s=1}^{N}\int \psi_f^* \frac{(-i\underline{k}\cdot\underline{r}'_s)^{l-1}}{(l-1)!}\frac{e_s}{M_s}\underline{p}'_{\perp s}\psi_i\, d\tau'\right|^2 \qquad \text{V.4}$$

If the size of the current distribution is denoted by R, then the contribution from successive non-zero terms decreases roughly as $[(kR)/(l-1)]^2$. As an example, let us assume a wavelength corresponding to a $\frac{1}{2}$ Mev photon, and set R = 7 x $10^{-13}$ cm, the radius of a nucleus with A = 100. Then kR is about 1/60, and the power ratio, $[(kR)/(l-1)]^2 = 3 \times 10^{-4}/(l-1)^2$.

The first term, $\dot{Q}_1$, is said to contribute electric dipole radiation. The second term represents electric quadrupole and magnetic dipole radiation. We shall now consider qualitatively the significance of this terminology and at the same time point out that V.4 must be amended to take into account the radiation arising from the intrinsic magnetic moments of the nucleons.

In order to be able to discuss classical models, let us consider not V.4, but its macroscopic classical analog. Dropping primes:

$$\dot{Q}_l(\theta,\phi) \propto \left|\int (\underline{k}\cdot\underline{r})^{l-1}\, \underline{j}_\perp(\underline{r})\, d\tau\right|^2 \qquad \text{V.4C}$$

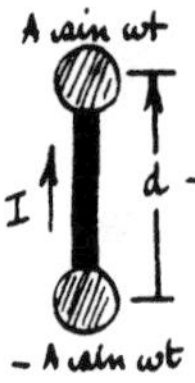


FIG. V.3 Electric Dipole.

First consider an electric dipole, whose strength varies sinusoidally in time. Physically we could represent this with two spheres (FIG. V.3), the charge on the top one being A sin ωt, the charge on the lower one being -A sin ωt. If the spheres are separated by a distance d, then the dipole moment would be $\underline{M} = A\underline{d}$ sin ωt, and the current between the two spheres would be I(t)=ωAcos ωt, upwards(thru a plane separating the spheres) if we assume conservation of charge.

The integral in $Q_1$, $\int j_\perp d\tau$, will be non-zero; in fact it will be $\frac{1}{2}\omega A d \sin\theta$. For later comparison with the quadrupole case, notice that the amplitude of the radiated electric and magnetic fields will vary as sin Θ, where Θ is the angle between $\underline{d}$ and the observer.

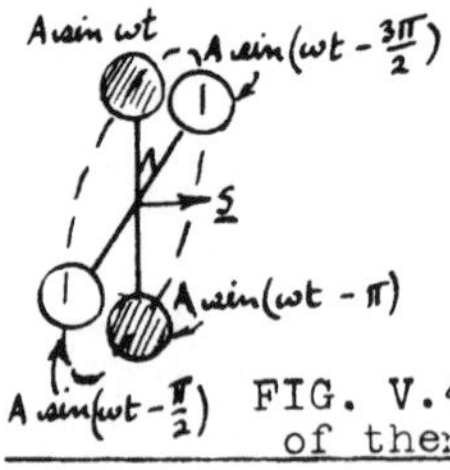


Instead of letting the charges vary in time, we could produce another type of dipole radiation by keeping a charge +A on one ball, -A on the other, and rotating the whole dumbbell about the vector labelled $\underline{s}$ in Fig. V.3. The fields thus generated could be duplicated (FIG. V.4) by superimposing the fields of the original fixed dipole and those of another

FIG. V.4 Two static dipoles equivalent to removing one of them and rotating the other.

similar dipole whose time variation is $\pi/2$ out of phase with it and pointed in the direction $\underline{d} \times \underline{s}$.

Next we examine the radiation of a magnetic dipole. Again, it could stem from the time variation of a dipole fixed in space or from the physical rotation of the whole dipole. The directions of $\mathcal{E}$ and $\mathcal{H}$ due to the magnetic dipole will be interchanged with those of $\mathcal{E}$ and $\mathcal{H}$ of the electric dipole, but this will be the only difference between the two cases. We have not mentioned any electric current, $\underline{J}(\underline{r},t)$; consequently the radiation originating from the magnetic dipole is not covered by eq. V.3. Since nucleons do have a permanent magnetic moment that can radiate as it rotates, we shall have to incorporate a spin operator into V.3 (see page 94).

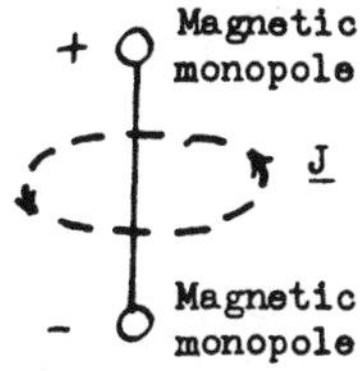

FIG. V.5 Magnetic Dipole

So long as an observer is far from a magnetic dipole, he cannot tell whether the radiation is caused by an inherent dipole or by an electric current loop (see FIG. V.4). Notice that, in the second case,

$$\int \underline{j}\, d\tau = 0$$

so that magnetic dipole radiation generated by currents will be covered only by the second term of V.4C, not by the first term, which accounts for the electric radiation of a dipole.

In addition to a magnetic dipole, an electric quadrupole would also contribute to the term $Q_2$ of V.4C. Examples of electric quadrupoles are shown in FIGS. V.6 and V.7.

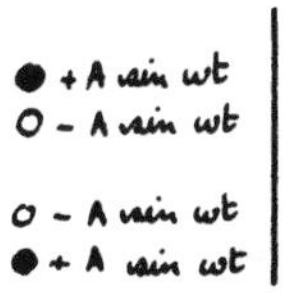

Quadrupoles
FIG. V.6 FIG. V.7

For both figures $\int \underline{j}\, d\tau = 0$, but when the integrand is multiplied by $\underline{k}\cdot\underline{r}$, we get a non-zero integral. Although $Q_2$ accounts for both electric quadrupoles and magnetic dipoles, the radiation from the two is quite different. The fields of the quadrupole vary as $\sin\theta \cos\theta$ whereas those of the dipole of Fig. V.5 vary as $\sin\theta$. The frequency dependence of the two sorts of field amplitudes is also different*

---

*The quadrupole operator of V.4C is proportional to

$$\underline{k}\cdot\underline{r}\;\underline{j} = \frac{e}{M}\underline{k}\cdot\underline{r}\;\underline{p}$$

Using $\underline{L} = \underline{r}\times\underline{p}$ we can get the vector identity

$$\underline{k}\times\underline{L} = \underline{k}\cdot(\underline{p}\,\underline{r} + \underline{r}\,\underline{p}) - 2\underline{k}\cdot\underline{r}\,\underline{p}$$

Using $\underline{p} = M\dot{\underline{r}}$ we can then write

$$\underline{k}\cdot\underline{r}\;\underline{j} = -\frac{e}{2M}\underline{k}\times\underline{L} + \frac{e}{2}\underline{k}\cdot(\dot{\underline{r}}\,\underline{r} + \underline{r}\,\dot{\underline{r}})$$

$$= -\frac{e}{2M}\underline{k}\times\underline{L} + \frac{e}{2}\underline{k}\frac{d}{dt}(\underline{r}\,\underline{r})$$

Now $\frac{e}{2Mc}\underline{L}$ is the magnetic dipole moment associated with an orbital current whirl, and $e\underline{r}\,\underline{r}$ is an electrostatic quadrupole moment. So $\underline{k}\cdot\underline{r}\;\underline{j}$ is proportional to a magnetic dipole plus the time derivative of an electric quadrupole.

Footnote continued on next page

$Q_3$ of V.4C will describe the radiation of electric octupoles and magnetic quadrupoles. In general $Q_\ell$ will describe the radiation of electric $2^\ell$-poles and magnetic $2^{\ell-1}$-poles.

Now let us see how we can amend V.3 so that it takes into account the radiation of <u>intrinsic</u> magnetic transitions. $(e/M)\underline{p} = \underline{j} = e\dot{\underline{r}}$ is the time derivative of an electric dipole moment $\underline{d}$. It is plausible that the complete operator should be $\dot{\underline{d}} + \dot{\underline{\mu}}$ where $\underline{\mu}$ is the dipole moment of the intrinsic magnetic dipole. It turns out that if we write $\dot{\underline{\mu}} = \omega\mu\underline{\sigma}$, where $\underline{\sigma}$ is the Pauli spin operator, we get the complete expression for V.3

$$\dot{Q}(\theta,\phi) = \frac{\omega}{2\pi\hbar c^3}\left|\sum_{s=1}^{N}\int \psi_f^* e^{-i\underline{k}\cdot\underline{r}'_s}\left\{\frac{e}{M}\underline{p}'_\perp + \omega\mu\underline{\sigma}_\perp\right\}_s \psi_i\, d\tau'\right|^2 \qquad \text{V.5}$$

<u>Spontaneous Electric Dipole Emission</u>

If we integrate V.5 over solid angle, we shall get the transition probability per unit time for the system. For simplicity we shall consider only the electric dipole case, where $\exp(-i\underline{k}\cdot\underline{r}')$ is set equal to one and the integral reduces, for some particular value of the subscript s, to the matrix element $\frac{e}{M}(\underline{p}'_\perp)_{fi}$. From a classical point of view it seems logical to write*

$$\frac{e}{M}\underline{p}'_\perp = e\underline{V}'_\perp = e\dot{\underline{r}}'_\perp = -ie\omega\underline{r}'_\perp$$

so (for one particle) V.5 reduces to

$$\dot{Q}_1(\theta,\phi) = \frac{\omega^3 e^2}{2\pi\hbar c^3}\left|(\underline{r}'_\perp)_{fi}\right|^2 .$$

Since $\underline{k}$ disappears in the dipole approximation, $|(\underline{r}'_\perp)_{fi}|^2$ now varies simply as $|\underline{r}'|^2 \sin^2\theta$, where $\theta$ is the angle between the axis of the dipole and the observer. The average value of $\sin^2\theta$ over the surface of a sphere is 2/3, so, instead of integrating the equation above, we shall just multiply by $4\pi \times 2/3$.

$$\int_{4\pi}\dot{Q}_1(\theta,\phi)\,d\omega = \frac{1}{\tau} = \frac{4}{3}\,\frac{\omega^3 e^2}{\hbar c^3}\left|\underline{r}'_{fi}\right|^2 \qquad \text{V.6}$$

$1/\tau$ should be called the transition probability per <u>unit time</u>; however, often it is simply called the transition probability.

The various assumptions made above are justified by quantum electrodynamics. Furthermore the Planck radiation law follows

Footnote continued from p. 93

If we are interested in an order-of-magnitude expression only, the "quadrupole" term of V.4 can be rewritten

$$\dot{Q}_2 \sim \frac{\omega}{\hbar c^3}\left|\int \psi_f^*\left(\omega\mu_{Nuclear} + e\frac{\omega^2}{c}rr'\right)\psi_i\, d\tau'\right|^2$$

In dropping the summation over the N particles we have assumed that the states of only one, or at most a few nucleons, have changed during the transition. This assumption is in the spirit of the nuclear shell model, p. 167 ff.

If we assume that the integral of $|\psi|^2 d\tau$ over the nucleus yields simply the nuclear dimension R, times a factor of order of magnitude one, the

$$\dot{Q}_2 \sim \frac{\omega^3}{\hbar c^3}\mu^2_{Nuclear} + \frac{\omega^5}{\hbar c^5}e^2 R^4 \qquad \text{(A)}$$

from V.6 together with the rather similar equation for induced dipole emission and absorption (Schiff (35.23) page 247).

$$\frac{1}{\tau} = \pi^2 \frac{4}{3} \frac{e^2}{\hbar^2 c} I(\omega) \left| \underline{r}_{fi} \right|^2 \qquad \text{V.8}$$

where I is the electromagnetic intensity per unit $\Delta\omega$.

## Spontaneous Emission by Intrinsic Magnetic Dipoles

From V.5, when considering only the first term in the expan- of exp $(-i\underline{k}\cdot\underline{r})$, we get, in addition to the electric dipole term V.6, also the magnetic dipole term

$$\frac{1}{\tau} = \frac{4}{3} \frac{\omega^3}{\hbar c^3} \left| \mu \underline{\sigma}_{fi} \right|^2$$

We can generalize V.6 to read

$$\boxed{\frac{1}{\tau} = \frac{4}{3} \frac{\omega^3}{\hbar c^3} \left| \underline{M}_{fi} \right|^2} \qquad \text{V.9}$$

For electric dipoles: $\underline{M} = e\underline{r}$
For spin dipoles: $\underline{M} = \mu\underline{\sigma}$

## Transition Rates for Electric and Magnetic Multipoles

The rate of transition resulting from an intrinsic magnetic dipole (V.9) and from an orbital magnetic dipole (eq. A, footnote, p. 94) are both of the order of magnitude

$$\frac{1}{\tau} = \frac{\omega^3}{\hbar c^3} \mu_{Nuclear}$$

which is smaller than the rate of transition due to an electric dipole by the factor $\left(\frac{\mu_{Nuclear}}{eR}\right)^2 = 5 \times 10^{-3} A^{-2/3}$
where $R = 1.5 \times 10^{-13} A^{1/3}$.

Inserting $A = 100$, this ratio becomes $2.4 \times 10^{-4}$, close to the ratio $3 \times 10^{-4}$ for quadrupole to dipole rates at ½ Mev (p. 92) It thus becomes convenient to think of electric quadrupole and both sorts of magnetic dipole transitions as having a mean life on the order of 1000 to 10,000 times that of an electric dipole. Similarly we group as one order of magnitude the mean lives of electric octupole and magnetic quadrupole radiation, and so on. This grouping is invalid for photon energies much different from ½ Mev, as seen in FIG. V.8.

We can now summarize: comparing V.9 with V.4 and assuming* that numerical factors in the integration are of order of magnitude one:

For electric $2^\ell$-poles, (dipoles excluded)

$$\frac{1}{\tau_{Elect.\,\ell}} = \frac{e^2}{\hbar c}\,\omega \left(\frac{\omega}{c}\right)^{2\ell} \frac{R^{2\ell}}{[(\ell-1)!]^2} \qquad \text{V.10}$$

For magnetic $2^\ell$-poles,

$$\frac{1}{\tau_{Mag\,\ell}} = \frac{1}{\tau_{Elect.\,\ell}} \times 5 \times 10^{-3} A^{-2/3}$$

$R = 1.5 \times 10^{-13} A^{1/3}$.

*We shall show later that this assumption is unacceptable if $\ell = 1$.

In FIG. V.8 Moszkowski* shows how the experimental data fit V.10. The integrations in V.4 have been done correctly, however, and the $\ell$-dependence turns out to be not $[(\ell-1)!]^{-2}$ but rather $\dfrac{2\ell+1}{\ell(1\cdot 3\cdot 5\cdots 2\ell+1)^2}$ for the case plotted, a change of nuclear spin I from $\ell+\frac{1}{2}$ to $\ell-\frac{1}{2}$.

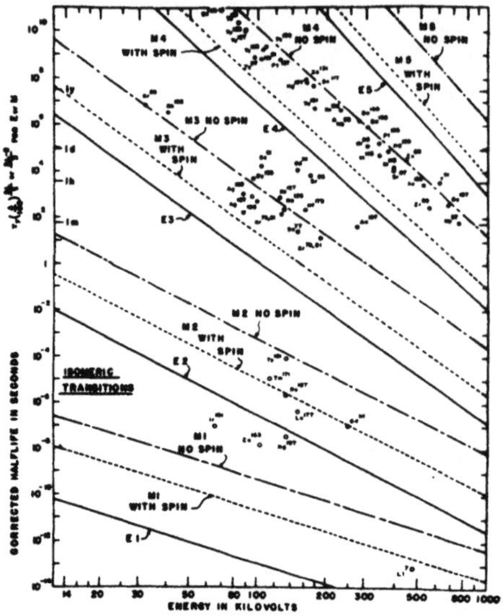


Corrected halflife of Isomeric Transitions as function of energy. The points show experimental values of halflives corrected for internal conversion (using theoretical values), and to A = 100.

FIG. V.8 *

## B. SELECTION RULES

On page 92 we showed that we could predict the transition rate of a radiation process by expanding V.3 in powers of $\underline{k}\cdot\underline{r}'$ and then considering the first non-zero term. In this section we shall show that it may frequently be necessary to go several terms down the expansion before finding a non-zero term. The reason is that more than half of all conceivable radiation processes are forbidden because of conservation of angular momentum or because of parity considerations. These selection rules apply to both atomic and nuclear radiation. Then there is an additional **limitation** on nuclear dipole radiation- because of the homogeneous distribution of nuclear charge, this process is only about 1/1000 as probable as would otherwise be expected.

### 1. Angular Momentum

Heitler** shows that a photon resulting from a $2^\ell$-pole electric or magnetic transition carries away an angular momentum of $\ell\hbar$ with respect to the origin to which the multipole is referred. As illustrated in the sketch, we can then use conservation of angular momentum to write

$\ell < I$

$$|I - I'| \leq \ell \leq |I + I'| \qquad \text{V.11}$$

which is a compact way of writing that, for a given $\ell <$ I:

$|\underline{I}'| = |\underline{I} - \underline{\ell}|$ can take on $2\ell+1$ values: $I+\ell,\ I+\ell-1,\ I+\ell-2 \ldots I-\ell$

and for a given $\ell >$ I:

$|\underline{I}'| = |\underline{I} - \underline{\ell}|$ can take on 2I+1 values: $\ell+I,\ \ell+I-1,\ \ell+I-2 \ldots \ell-I$

*"Interpretations of Isomeric Transitions," Phys. Rev. to be published. See also Ch. XII of a forthcoming book on nuclear physics by Weisskopf and Blatt, John Wiley, 1951.

**Proc. Camb. Phil. Soc. 32, 112 ('36)

Eq. V.11 is not the only restriction; others will soon be developed.

Illustration with spherical harmonics. V.11 could be derived without explicitly mentioning conservation of angular momentum, by expressing in terms of spherical harmonics all the factors in the matrix elements of V.5

$$\int \psi_f^* (\underline{k}\cdot\underline{r}')^{l-1} \, \underline{p}_\perp \, \psi_i \, d\tau' \qquad \text{Current radiation} \qquad \text{V.12}$$

$$\int \psi_f^* (\underline{k}\cdot\underline{r}')^{l-1} \, \underline{\sigma}_\perp \, \psi_i \, d\tau' \qquad \text{Spin radiation} \qquad \text{V.13}$$

Then, using product and orthogonality relations of spherical harmonics, we can prove not only V.11, but also the parity rules to be discussed in the next section.

We shall illustrate this statement with an electric dipole transition in a one-body system. We shall ignore inherent spin (non-orbital angular momentum), so that $\underline{I}$ becomes $\underline{L}$*. $\psi$ may be written as the product of a function of r, f(r), and a spherical harmonic. As shown in the derivation of V.6, V.12 reduces to

$$\int r^2 dr \, f_f^*(r) \, f_i(r) \int Y_{L'}^{m'*} \, \underline{r} \, Y_L^m \, d\omega$$

where $d\omega$ is an element of solid angle, and we have used primes to indicate the final spherical harmonic.

The vector $\underline{r}$ has the properties of a spherical harmonic of order one; its components, x,y, and z, may be expressed as follows: $x = r \sin\theta \cos\phi$, $y = r \sin\theta \sin\phi$, $z = r\cos\theta$ (It will not be necessary for this simple example, but these may then be expressed as follows:

$$x = \frac{r}{2}\sin\theta \, e^{i\phi} + \frac{r}{2}\sin\theta \, e^{-i\phi} = \frac{r}{2}\sqrt{\frac{8\pi}{3}}\left[Y_1^1(\theta,\phi) + Y_1^{-1}(\theta,\phi)\right]$$

Similarly $$y = \frac{r}{2i}\sqrt{\frac{8\pi}{3}}\left[Y_1^1 - Y_1^{-1}\right], \qquad z = r\sqrt{\frac{4\pi}{3}}\,Y_1^0$$

Here we shall use the simpler sin θ and cos θ expressions.)

Equations such as those given in Margenau and Murphy "The Mathematics of Chemistry and Physics," (3-48) p. 103:

$$\cos\theta \, P_L^m(\cos\theta) = c_1 P_{L-1}^m + c_2 P_{L+1}^m$$

$$\sin\theta \, P_L^m = c_3 P_{L-1}^{m+1} + c_4 P_{L+1}^{m+1}$$

show that the product $\underline{r}P_L$ will give only terms in $P_{L\pm1}$, so that unless L' = L ± 1 the integration over θ ($d\omega = -d(\cos\theta)\,d\phi$) will give zero. Since V.11 implies that I' can equal I (L' = L for

*It is conventional to write the orbital angular momentum in a one-body problem ($\ell$) rather than (L). Here this would lead to confusion between the angular momentum and the degree of the multipole. Spherical harmonics will be written

$$Y_L^m(\theta,\phi) = \left[\frac{(2L+1)(L-|m|)!}{4\pi \; (L+|m|)!}\right]^{1/2} P_L^m \, e^{im\phi},$$

where $P_L^m(\cos\theta)$ is an associated Legendre function.

this problem) one might have expected the product $\underline{r}P_L$ to give a term in $P_L$, but this is forbidden by parity considerations. The spin operator, $\underline{\sigma}$ , does not operate upon space coordinates. Thus for spin dipoles we shall get a non-zero integral when L' = L.

## 2. Parity.

Parity is a rather fundamental quantum-mechanical concept, with which many readers will be familiar. All that we shall prove is that the integrand of matrix elements like V.12 and V.13 must have even parity. If the reader already knows this, he should skip to the last paragraph of this numbered part.

The concept of parity arises from the fact that neither the Hamiltonian nor the energy of an isolated system will change if we invert simultaneously thru the origin the coordinates of all the particles of the system.

To express this mathematically, take a system of many particles. Set up the Hamiltonian; call it $H(\underline{r}_1, \underline{r}_2, \underline{r}_3, \ldots \underline{r}_n)$ or $H(\underline{r}_i)$ for short. Now change each $\underline{r}_i$ to $-\underline{r}_i$. The Hamiltonian will not change. That is

$$H(\underline{r}_i) = H(-\underline{r}_i) \qquad \text{V.14}$$

We shall next show that, because of V.14, any physically important eigenfunction of a time-independent Hamiltonian has the form of one of the following alternatives:

$$\psi(-\underline{r}_i) = +\psi(\underline{r}_i) \qquad \text{Even Parity}$$
$$\psi(-\underline{r}_i) = -\psi(\underline{r}_i) \qquad \text{Odd Parity} \qquad \text{V.15}$$

Parity is a quantum-mechanical concept; we do not know of any classical analog.

Parity of Eigenfunctions Write Schrödinger's equation

$$H(\underline{r}_i)\psi(\underline{r}_i) = E\psi(\underline{r}_i)$$

Now consider the same equation where the variable is $-\underline{r}_i$. Using V.14

$$H(\underline{r}_i)\psi(-\underline{r}_i) = E\psi(-\underline{r}_i)$$

Thus $\psi(\underline{r}_i)$ and $\psi(-\underline{r}_i)$ satisfy the same equation. If there is no degeneracy,* $\psi(\underline{r}_i)$ must be proportional to $\psi(-\underline{r}_i)$.

$$\psi(\underline{r}_i) = K\psi(-\underline{r}_i)$$

If, now, we change $\underline{r}_i$ to $-\underline{r}_i$, this last equation becomes

$$\psi(-\underline{r}_i) = K\psi(\underline{r}_i)$$

The only way in which these two equations can hold simultaneously is for k to be ±1. This is what we set out to prove.

Parity of the Radiation Operators, $\underline{p}$, $\underline{\sigma}$, and $\underline{k}\cdot\underline{r}$

Having taken care of $\psi$, we shall next examine the parity of all the factors in the radiation integrals V.12 and V.13.

*If there is degeneracy, it can always be removed with a small perturbation. The present argument will then hold. At the end of the proof we can let the perturbation approach zero. The eigenfunction cannot change suddenly, so it will retain its property of parity. This is the sort of eigenfunction that we called "physically important" above.

The electric operator, $\underline{p} = \frac{\hbar}{i}\underline{\nabla}$, has odd parity. This is evident, because $\frac{\partial}{\partial x} = -\frac{\partial}{\partial(-x)}$, etc. Vectors of odd parity, such as $\underline{p}$, $\underline{\nabla}$, $\underline{J}$, $\underline{r}$, $\underline{F}$, are called polar vectors. The spin operator, $\underline{\sigma}$, does not even operate on the space coordinates, which are the only coordinates inverted in the parity operation. $\underline{\sigma}$ must have even parity. Vectors of even parity (all angular momenta) are called axial vectors.

Notice that the product of two functions of odd parity gives one of even parity, and that, in general, the <u>parity</u> of a product of odd and even functions is given by the same rules as the <u>sign</u> of a product of negative and positive quantities.

The only function in the radiation matrices that we have not yet considered is $(\underline{k}\cdot\underline{r})^{\ell-1} = (kr\cos\theta)^{\ell-1}$. Using our analogy, we say that this has parity $(\text{odd})^{\ell-1}$; where $(\text{odd})^0 = (\text{odd})^2 = \ldots = (\text{even})$.

<u>Parity of Integrands</u> We are now in a position to perform the integration of the matrix elements like V.12 and V.13. For a nucleus of mass number A, $d\tau$ is really the product of many differentials:

$$d\tau = r_1^2\,dr_1\,d\omega_1\;\; r_2^2\,dr_2\,d\omega_2\;\ldots\;r_A^2\,dr_A\,d\omega_A$$

Let the integrand be $f(\underline{r}_i)$, and let us integrate over solid angle first:

$$I \equiv \int_\omega f(\underline{r}_i)\,d\omega_1\,d\omega_2\ldots d\omega_A$$

Evidently, this may be written as the sum of two integrations, each over a hemisphere:

$$I = \int_{\omega/2}\left[f(\underline{r}_i) + f(-\underline{r}_i)\right]d\omega_1\,d\omega_2\ldots d\omega_A.$$

<u>If $f(\underline{r}_i)$ has odd parity, I vanishes.</u>

When discussing electric and magnetic radiation, the rule that the whole integrand must have even parity is usually expressed in terms of the necessity for a change in the parity of the state function. Thus, for electric dipole radiation, $\psi_f^*$ and $\psi_i$ must have different parities; that is to say, the answer to the question "Must there be a change in parity?" is "Yes." For electric quadrupole radiation, however, because of the extra factor $\underline{k}\cdot\underline{r}$ in the integrand, the answer is "No." For magnetic dipole radiation (either by current or spin) the answer is, again, "No."

## <u>3. Nuclear Dipole Radiation is Improbable.</u>

Let us consider a very crude model of the nucleus, namely, a <u>uniformly</u> charged sphere, and ask what motions of the sphere give the various electric and magnetic multipole radiations.

A glance at the radiation integrals V.12 and V.13 shows the following:

Electric Dipole radiation could be caused only by a translational oscillation of the whole drop. In other words, the center of mass of the drop would have to oscillate at the frequency of the radiation emitted. A nucleus would not behave this way. In other words, we would not expect any dipole radiation at all <u>if</u> a nucleus behaved like a rather loosely bound collection of neutron-proton pairs, the neutron and the proton in each pair sticking very closely together. Bethe B treats this problem quantitatively on p. 222.

Experimentally, nuclei are observed to emit some electric dipole radiation, but it is several thousand times less probable than what would be expected from V.10, p. 96. For this reason no curve corresponding to $l = 1$ is included in Fig. V.8. Apparently, then, nuclei do act somewhat like homogeneously charged bodies.

Electric Quadrupole radiation could arise from ellipsoidal oscillations of the body. This sort of motion seems plausible, and V.10 accords with experiment.

Electric Multipole radiation corresponds to more complicated wave motions on the surface of the nucleus.

Magnetic Dipole radiation can be explained if the different constituents of the nucleus have different gyromagnetic ratios, so that the total magnetic moment does not lie in the direction of the total angular momentmm.

In summary, V.10 describes adequately all modes of radiother than electric dipole.

## 4. Selection Rules, Summary

Table V.I summarizes the selection rules for $\gamma$-radiation. Entries between the same vertical lines are observed experimentally to have mean lives of the same order of magnitude for $\hbar\omega \sim 1$ Mev, as shown in FIG. V.8.

| Half life, T, sec. $\hbar\omega$ = 1 Mev | $10^{-12}$ | | $10^{-7.5}$ | $10^{-3}$ |
|---|---|---|---|---|
| Electric Radiation | Dipole | Quadr. | Octu. | $2^4$-pole |
| $\lvert I + I' \rvert \geq$ $\lvert I - I' \rvert \leq$ | 1 | 2 | 3 | 4 |
| Change in Parity | Yes | No | Yes | No |
| Magnetic Radiation | | Dipole | Quadr. | $2^3$-pole |
| $\lvert I + I' \rvert \geq$ $\lvert I - I' \rvert \leq$ | | 1 | 2 | 3 |
| Change in Parity | | No | Yes | No |

TABLE V.I Selection Rules

## 5. Gamma Absorption at High Energies

At energies greater than 10 Mev, nuclei may be less reluctant to absorb and emit dipole radiation. At these energies "resonance" absorption has been observed in many elements.*

*Diven and Almy, Phys. Rev. 80, 408 ('50), Bethe and Levinger, Phys. Rev. 78, 115 ('50).

The length of the discussion of spontaneous emission should not be taken to imply that it is the only important process by which a nucleus may lose energy. A process which competes strongly with emission, at least for heavy nuclei, is internal conversion. This is the general name given to K-conversion, L-conversion, etc., meaning the ejection of an atomic K-, L-, etc., electron into a free state.

The ejected electron will have a kinetic energy less than the $\gamma$ energy by the binding energy of the K-, L-, etc. orbit.

The name internal conversion may not always be a good name for the process, since in some cases (when electrons penetrate the nucleus) the electrons may be raised to an excited level even though the $\gamma$ selection rules forbid the emission of a quantum. Besides, if both $\gamma$ rays and internal conversion electrons are emitted, $\tau_{partial}$ for the $\gamma$ ray is the same for an atom and a stripped nucleus. (This was first pointed out by Mott and Taylor) Thus it may not be correct to think of a $\gamma$ first leaving the nucleus and subsequently exciting the atom as it is absorbed.

The ratio between the number of K-electrons and photons emitted is called the K-orbit partial internal conversion coefficient $\alpha_K \equiv N_K/N_\gamma$; similarly $\alpha_L \equiv N_L/N_\gamma$. etc.
Then the internal conversion coefficient is defined by

$$\alpha = \alpha_K + \alpha_L + \cdots = N_e/N_\gamma$$

where $N_e$ is the total number of electrons emitted. Sometimes $\alpha$ is defined differently, namely as $\alpha = N_e/(N_\gamma + N_e)$; but our definition follows more recent custom.

After internal conversion there is an empty electronic state, the atom will emit x-radiation and possibly Auger electrons. Auger emission and internal conversion are similar processes. In the former process the energy comes from an atomic transition, in the latter, the energy comes from a nuclear transition, but in both cases there may be, alternatively, radiation, or the emission of an electron.

For nuclear processes other than internal conversion, any x-radiation or Auger electrons will be characteristic of the daughter atom. In the case of internal conversion, of course, the parent and daughter atom are one and the same. The characteristic x-radiation is convenient for identifying the nucleus undergoing transition. Segré and Helmholz discuss the detection of internal conversion electrons on page 274

## 1. Theory of Internal Conversion

Calculation of the internal conversion coefficient is by no means simple, but it is essentially an atomic, rather than a nuclear problem. For this reason, $\alpha$ may be computed much more accurately than the mean life for $\gamma$ transitions.

Segré and Helmholz give a complete list of references. Rasetti, p. 134, gives the non-relativistic treatment for the case of K-conversion of electric dipole radiation. We shall sketch how the problem is set up and quote his result.

Assume the existence in the nucleus of an electric dipole pointed along the z-axis, whose moment is*

$$\underline{M} = \underline{k} M (e^{-i\omega t} + e^{+i\omega t})$$

Near the nucleus the dominant term in the potential will be

$$\phi = -\underline{M}\cdot\nabla \frac{1}{r} = M \frac{\cos\theta}{r^2}(e^{-i\omega t} + e^{+i\omega t}) \qquad \text{V.16}$$

which decreases as $r^{-2}$ and has odd parity. (On the other hand the "near" potential of a quadrupole decreases as $r^{-3}$ and has even parity -- we shall refer to this difference later).** If we apply the arguments of page 97, we see that the dipole potential will excite a 1s electron only to states with $l=1$ and $m = 0$. Though improbable, it is conceivable that $E_\gamma$ would be just sufficient to raise a K electron to one of the empty optical states. Usually, however, the electron is raised to a state of the continuum. Thus, for large r, the final state function is (neglecting the long-range coulomb potential)

$$\psi_f = N_f \frac{1}{kr} \sin(kr + \delta) \cos\theta \qquad kr \gg 1$$

If we follow the usual procedure of arbitrarily quantizing the system by surrounding it with a large sphere of radius R, we find that the normalizing factor, $N_f = \left(\frac{2}{R}\,\frac{3}{4\pi}\right)^{1/2}$

The matrix element $H_{fi}$ may then be calculated:

$$H_{fi}(t) = \int \psi_f^* \, e\phi \, \psi_i \, d\tau$$

$$= N_f \, 2\pi e M \int_0^\pi \cos^2\theta \sin\theta \, d\theta \int_0^\infty N_i \frac{u_f^*(r)}{r} \frac{u_i(r)}{r} r^2 dr \left[e^{-i\omega t} + e^{+i\omega t}\right]$$

Where $\psi(r,\theta) \equiv N \frac{u(r)}{r} P_L(\cos\theta)$; $\psi_i$ is the Hydrogenic 1s function.

The integration over $\Theta$ gives 2/3. Call the radial integral (including $N_i$) $I(E_\gamma, Z)$. Then

$$H_{fi}(t) = \frac{4\pi}{3} e M I(E_\gamma, Z)\left[e^{-i\omega t} + e^{+i\omega t}\right] N_f \qquad \text{V.16A}$$

$$\equiv H_{fi}\left[e^{-i\omega t} + e^{+i\omega t}\right]$$

When $H_{fi}(t)$ is written in this way, $H_{fi}$ may be substituted directly into the expression for the transition probability per unit time (Golden Rule No. 2***)

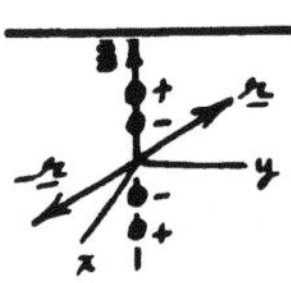

$$\frac{1}{\tau_K} = \frac{2\pi}{\hbar} |H_{fi}|^2 \frac{dn}{dE}$$

*Quantum mechanically $\underline{M}$ would be proportional to the expression V.12, p. 97, plus its complex conjugate, where $l=1$, and the $\psi$'s are nuclear state functions.

**Illustration of this statement: sketched right above this note is a typical quadrupole. There is no preferred direction along the z axis, no preferred direction in the x-y plane. Consequently $\phi(\underline{r}) = \phi(-\underline{r})$.

***On page 193 of Schiff's "Quantum Mechanics" this expression is derived for perturbations that are constant except for being switched on and off.

One might at first be tempted to write

$$\frac{dn}{dE} = \frac{4\pi p^2}{h^3 V} \times (\text{Vol. of large sphere})$$

This would be wrong because it includes all the states of the continuum that lie in the energy interval dE. Because the angular dependence of the final state has already been determined by the perturbing potential, most of these states are not available to the electron.

For dn/dE we want to write only the number of different radial states in the energy interval, where the radial factor $u_f(r)/r$ is determined by the equation

$$\left[\frac{d^2}{dr^2} + \frac{2M}{\hbar^2}(E_f - U) - \frac{l(l+1)}{r^2}\right] u_f(r) = 0$$

Since $\psi$ must be zero at the distant surface of our quantizing sphere,

$$R = n\frac{\lambda}{2} = n\frac{h}{2p} = n\frac{\pi\hbar}{p}$$

from which

$$\boxed{dn = \frac{R}{\pi\hbar}dp = \frac{R}{\pi\hbar V}dE} \quad *$$

Putting dn/dE and V.16A into Golden rule No. 2,

$$\frac{1}{\tau_K} = \frac{2\pi}{\hbar}\left|\frac{4\pi}{3}eMI(E_\gamma, Z)\right|^2 \frac{R}{\pi\hbar V} N_f^2$$

But, $N_f^2 = \frac{3}{2\pi R}$, so

$$\frac{1}{\tau_K} = \frac{16\pi}{3\hbar^2}\frac{e^2M^2}{V} I^2(E_\gamma, Z) \qquad \textbf{V.17}$$

The internal conversion coefficient is a ratio of mean lives, so we need also V.9, the transition probability for dipole $\gamma$ emission

$$\frac{1}{\tau_\gamma} = \frac{4}{3}\frac{\omega^3}{\hbar c^3}\left|\underline{M}_{fi}\right|^2 \qquad \textbf{V.9}$$

If we set $\underline{M}_{fi}$ = M and remember that there are 2 electrons in the K shell

$$\alpha_K = \frac{2\tau_\gamma}{\tau_K} = \frac{8\pi e^2 c^3 I^2(E_\gamma, Z)}{\hbar\omega^3 V}$$

Rasetti and Bethe B (formula 735) give $I(E_\gamma, Z)$ assuming that that $E_\gamma >>$ binding energy of the K shell. In this case $\alpha_K << 1$. The expression is (note the similarity with II.33, p. 39)

$$\alpha_K = 4\pi\alpha^5\frac{Z^4}{\nu^4}\frac{e^{-4n\cot^{-1}n}}{e^{2\pi n} - 1} \qquad \textbf{V.18}$$

where $\alpha = \frac{e^2}{\hbar c}$ (finestructure constant)

$$\nu = \frac{E_\gamma}{mc^2}; \quad n = \frac{Z\alpha}{\sqrt{2\nu - Z^2\alpha^2}}$$

Unfortunately it has become customary to use both $\alpha$ and $\alpha_K$ in these equations. The function, $\alpha_K$, should not be confused with the constant, $\alpha$.

Segrè and Helmholz ** add the condition that the velocity of the K electron be small compared to the velocity of the conversion electron, and get

$$\alpha_K = 2^{\frac{5}{2}} Z^3 \alpha^4 \nu^{-\frac{7}{2}} \qquad \textbf{V.19}$$

Note the rapid increase with Z. This indicates that as we proceed to the heavy end of the periodic table, internal conversion becomes of increasing importance.

*This expression will not be referred to again in this book, but it is boxed because of its frequent occurence in photoelectric calculations.

**Rev. Mod. Phys. 21, 282 ('49).

Values of $\alpha_K$ calculated from the formulas above may be smaller than the experimental values by as much as a factor of 5. The reason is mainly that a non-relativistic treatment is inadequate because the velocity of the electron is large both in the bound state and in the ionized state. Taylor and Mott have made calculations using the Dirac relativistic wave equation. The result is presented in FIG. V.9. Segrè and Helmholz give more exact equations and references. An idea of the general behavior is given by Segré and Helmholz' generalization of V.19 for any order multipole

$$\alpha_K^l = Z^3 \alpha^4 \frac{l}{l+1} \left(\frac{2}{\nu}\right)^{l+5/2} \qquad V.20$$

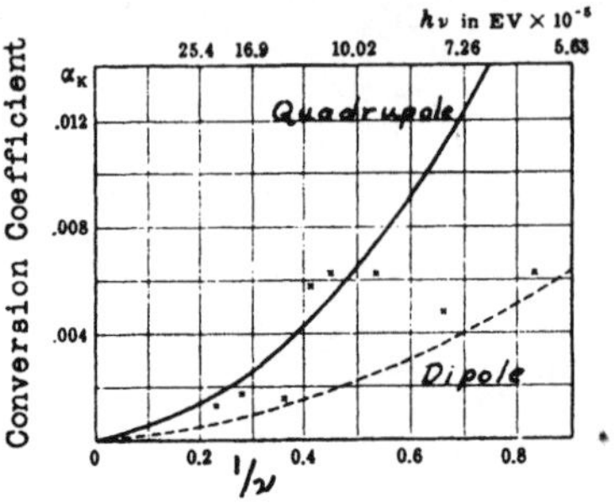


FIG. V.9 Conversion Coefficient for the K shell. Z = 84. Theoretical curve by Mott and Hulme. Experimental data (x) by Ellis and Aston. From Rasetti, p. 132.

For curves of half lives, see FIG. V.8, p. 96

As long as any $\gamma$ radiation which we are investigating has a measurable conversion coefficient, we can make use of the dependence of this coefficient upon $l$ to differentiate, for example, between electric quadrupole and magnetic dipole radiation.

The reason for this dependence of $\alpha_K$ upon $l$ is the variation with $l$ of the multipole potential of the nucleus as seen by the K electron. This was mentioned in the discussion below V.16.

## 2. Selection Rules.

Whenever a nucleus is capable of emitting photons, it will also be able to excite orbital electrons even though these do not penetrate the nucleus. The $\gamma$ selection rules, Table V.I, apply.

Penetration of the Nucleus However, an electron which penetrates the nucleus may be raised to an excited state even though radiation is forbidden. As a simple model, consider a charged spherical shell whose radius oscillates in time. $\phi$ outside the maximum radius of the shell is always pure coulomb, so there is no radiation. Inside the maximum radius the fields vary with time, so an electron in this region can receive energy.

When RaC' decays from its third excited state (1.414 Mev) to the ground state, K electrons, but no $\gamma$'s at all, are observed. Both the states mentioned have I = 0. According to table V.I, transitions from I = 0 to I = 0 are strictly forbidden (forbidden to all multipoles). This is evident, because there is no order of multipole radiation for which the photons carry off zero angular momentum. Thus the 1s electron must penetrate the nucleus in order to receive energy. Calculations give a mean life of the order of $10^{-10}$ sec. for this penetration process. The conversion coefficient is infinity, even though, on an absolute basis, $\tau$ is fairly long compared with the $\tau$ for the ejection of an electron that can acquire energy without entering the nucleus.

For I = 0 to I = 0 transitions, even though the electron penetrates the nucleus, it can be shown that there is still a selection rule that the parity of the final and initial states be the same ("No").

3. Other Processes.

If $E_\gamma > 2mc^2$ a nuclear multipole moment may eject an electron from the negative energy sea rather than a K electron. In this case we would observe an electron-positron pair leaving the atom.

Much rarer processes, of importance only when both radiation and internal conversion are forbidden (transitions between two nuclear states, both with I = 0, but with different parity), are simultaneous emission of two electrons or of two photons.

4. Experimental Determination of Conversion Coefficient, $\alpha$.

$\alpha$ is most easily determined when $\gamma$ radiation is followed or preceeded by $\alpha$- or $\beta$-decay. Suppose, for instance, the sequence of transitions is as pictured in the sketch. If the $\beta$ transition

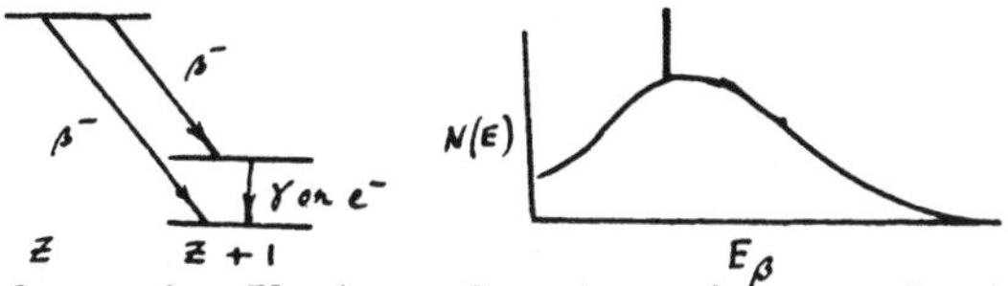


Conversion Electrons Superimposed upon $\beta$ Spectrum

to the ground state is rare, we can simply state $\alpha = \frac{N_e}{N_\gamma} = \frac{N_e}{N_\beta}$. If both sorts of $\beta$ rays are present, we can then make use of $\beta$-$\beta$ and $\beta$-$\gamma$ coincidence techniques.

In any case, it is simpler to investigate $\gamma$'s by examining their conversion electrons in a $\beta$ spectrometer than it is to observe the $\gamma$'s directly.

When the "spin" (strictly, I) of an excited state is quite different from that of the ground state, particularly if its energy is not too high, we see from Fig. V.8, p. 96, that it may be years before the nucleus radiates. Such metastable states are said to be isomeric with the ground state. About 100 isomers have $\tau$'s of the order of minutes or greater.

Indium $^{115}$ is an example of a typical isomer. The excited state (0.35 Mev) has a spin of $\frac{1}{2}$, but the ground state has a spin of 9/2. 16-pole radiation is required for the transition. $\tau$ = 4.5 h.

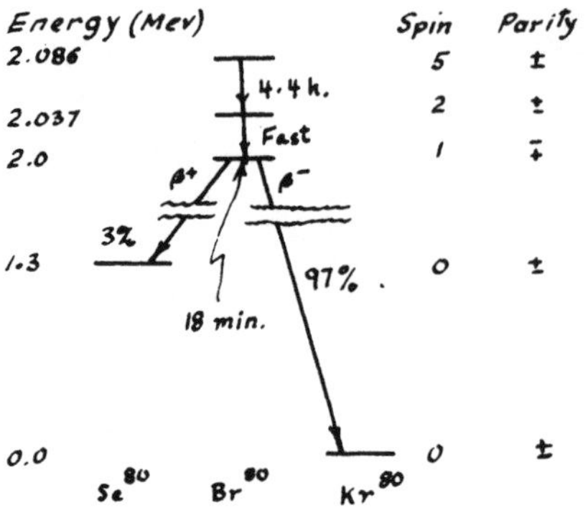


A more complicated case is $Br^{80}$. Segrè and Helmholz give the energy level diagram reproduced below on page 291 of their review article, Rev. Mod. Phys. 21, 271 ('49). They discuss this and other selected cases of isomerism and give a complete table of isomers.

REFERENCE. An additional review is "Spectroscopy of some Artificially Radioactive Nuclei," Mitchell, Rev. Mod. Phys. 22, 36 ('50).

## E. PROBLEMS

1. A classical particle, with charge e, moves so that

$$\underline{r} = 2x_0 \cos \omega t \, \underline{e}_x = x_0 e^{-i\omega t} \underline{e}_x + c.c.,$$

where $x_0 \ll c/\omega$. Calculate its radiation using (A) the formula

$$\int \underline{S} \cdot d\underline{A} = \frac{2}{3}\frac{e^2}{c^3}(\ddot{\underline{r}})^2$$

and (B) V. 2, p. 90. (C): Solve the equivalent one-dimensional quantum-mechanical harmonic oscillator problem, and show that the radiation predicted by V.9, p. 95, agrees with both of the calculations (A) and (B).

Solution. (A)

$$\ddot{\underline{r}} = -\omega^2 2x_0 \cos \omega t \, \underline{e}_x$$

$$(\ddot{\underline{r}})^2 = 4\omega^4 x_0^2 \cos^2 \omega t$$

$$\overline{(\ddot{r})^2} = 2\omega^4 x_0^2$$

So

$$\int \overline{\underline{S}} \cdot d\underline{A} = \frac{4}{3}\frac{e^2}{c^3}\omega^4 x_0^2 \qquad \text{V.21}$$

For comparison with part (C), we attribute the oscillation of the particle to its motion in a harmonic potential well, $V = \frac{1}{2}M\omega^2 x^2$. The energy of the system is then

$$E = \frac{1}{2}M\omega^2(2x_0)^2$$

so

$$\int \overline{\underline{S}} \cdot d\underline{A} = \frac{2}{3}\frac{e^2}{c^3}\omega^2 \frac{E}{M} \qquad \text{V.22}$$

Part B. To get the dipole term of V.2, p. 90, set $e^{-i\underline{k}\cdot\underline{r}'} = 1$, so that $\underline{j}_\perp$ goes to $\underline{j}$ sin θ. Integrating this expression over the surface of a sphere with area $4\pi r^2$, and dropping the primes

$$\int \underline{\overline{S}}\cdot d\underline{A} = \frac{2}{3} 4\pi \frac{\omega^2}{2\pi c^3} \left| \int \underline{j}(\underline{r})\, d\tau \right|^2$$

$$= \frac{4}{3} \frac{\omega^2}{c^3} \left| \int \underline{j}(\underline{r})\, d\tau \right|^2 \qquad \text{V.23}$$

where, as previously defined, $|\underline{j}(\underline{r})|$ is one half the space part of a current distribution that is assumed to vary as cos $(\omega t + \delta)$.

For this simple problem, $\int \underline{j}(\underline{r})\, d\tau$ may be written down by inspection as equal to $-i\omega x_0 e \underline{e}_x$. Nevertheless, we shall verify this by the use of δ-functions. The reader who is unfamiliar with this sort of treatment may find this exercise helpful for understanding problem 2.

First we express the charge distribution

$$\rho(\underline{r}) = e\,\delta(x - r)\,\delta(y)\,\delta(z)$$

This statement is easily checked, because $\int \rho\, d\tau$ gives a charge e, located at the point r,0,0,

The velocity of the particle is

$$v(t) = -\omega 2 x_0 \sin \omega t\ \underline{e}_x$$

So $$\underline{J}(\underline{r}, t) = -2\omega x_0 e\,\delta(x-r)\,\delta(y)\,\delta(z) \sin \omega t\ \underline{e}_x \equiv \sum_\omega \underline{j}_\omega(r)\, e^{-i\omega t} + \text{c.c.}$$

If we now multiply this equation by $e^{+i\omega t}$ and integrate over a complete period, we find

$$\underline{j}_\omega(\underline{r}) = \frac{1}{T}\int_0^T \underline{J}(\underline{r}, t)\, e^{+i\omega t}\, dt$$

Do not perform the integration, but put this integral into the integral that we need to compute the radiation.

$$\int \underline{j}_\omega(\underline{r})\, d\tau = \frac{1}{T}\int d\tau \int_0^T \underline{J}(\underline{r}, t)\, e^{+i\omega t}\, dt$$

Now integrate over volume, and, once this is done, perform the Fourier analysis, i.e., the integration over time.

To summarize, the simplest way to work a classical problem like this one is to find $\int \underline{J}(\underline{r}, t)\, d\tau$ and then Fourier analyze. Most often the Fourier analysis may be carried out by inspection.

Thus, for this problem

$$\int \underline{J}(\underline{r}, t)\, d\tau = -2\omega x_0 e \sin \omega t\ \underline{e}_x$$

and, by inspection, $$\int \underline{j}_\omega(\underline{r})\, d\tau = -i\omega x_0 e\, \underline{e}_x$$ *as stated above.*

Substituting into V.23 yields

$$\int \underline{\overline{S}}\cdot d\underline{A} = \frac{4}{3} \frac{\omega^2}{c^3} \left| \omega^2 x_0^2 e^2 \right| = \frac{4}{3} \frac{e^2}{c^3} \omega^4 x_0^2$$

in agreement with V.21

Part C, Quantum-mechanical treatment. In order to have correspondence with the classical case, we shall assume that the harmonic oscillator is in a highly excited state (n>>1).

To get the radiated power, instead of the transition rate, V.9, p. 95, must be multiplied by $\hbar\omega$.

$$\int \bar{\underline{S}} \cdot d\underline{A} = \frac{4}{3} \frac{\omega^4 e^2}{c^3} \left| \underline{r}_{fi} \right|^2$$

For this one-dimensional problem, $|r|^2 = |x|^2 + |y|^2 + |z|^2$ reduces to

$|x_{fi}|^2 = \frac{n\hbar}{2M\omega}$ , using Schiff's equation (13.18). Substituting $n\hbar\omega = E$,

$$\int \bar{\underline{S}} \cdot d\underline{A} = \frac{2}{3} \frac{\omega^2 e^2}{c^3} \frac{E}{M}$$

in agreement with V.22, obtained by classical arguments in parts (A) and (B).

******

PROBLEM 2 Compute the radiation of two electrons traversing a circle at opposite ends of a diameter.

Solution

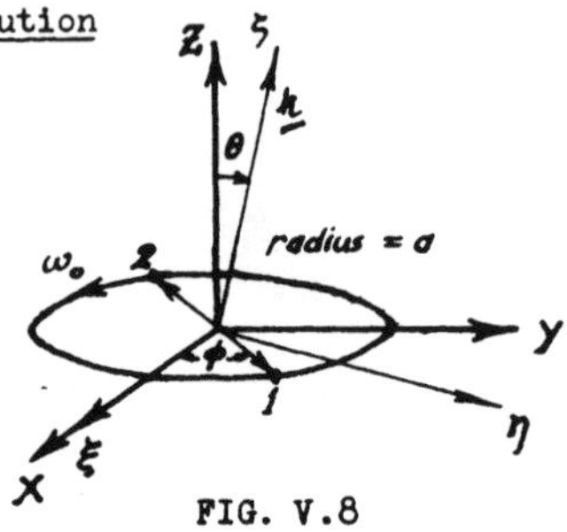


FIG. V.8

It can be seen by inspection that the dipole radiation (defined by V.4C, p. 92) will be zero, so we concentrate on the quadrupole term. Dropping the primes in V.2, p. 90, this is

$$\bar{S}(\theta,\phi) = \frac{\omega^2}{r^2_{obs}\, 2\pi c^3} \left| \int \underline{j}_\perp(\underline{r})\, \underline{k}\cdot\underline{r}\, d\tau \right|^2$$

Assume that the electrons are revolving in the x-y plane (see FIG. V.8). The unit vectors in this plane will be denoted by $\underline{e}_x$ and $\underline{e}_y$. Let $\underline{e}_x$ and $\underline{k}$ define a second cartesian system, as illustrated.

We shall follow the procedure discussed in part B of prob. 1.

Let $$I(t) \equiv \int I_\perp(\underline{r},t)\, \underline{k}\cdot\underline{r}\, d\tau = \int \rho(\underline{r}) \underline{V}_\perp(t)\, \underline{k}\cdot\underline{r}\, d\tau \qquad \text{V.24}$$

If we substitute the Fourier components of I(t) into V.2, p. 90, we shall get the power radiated into the element of solid angle $2\pi \sin\theta\, d\theta$. We can then integrate over $\theta$ to get the total power radiated.

First we must find expressions for the three factors of V.24

$$\rho(\underline{r}) = e\delta(\underline{r}-\underline{r}_1) + e\delta(\underline{r}-\underline{r}_2); \quad \text{where } \underline{r}_1 = -\underline{r}_2 = \underline{e}_x\, a\cos\omega_0 t + \underline{e}_y\, a \sin\omega_0 t$$

$$\underline{V}_{\perp 1} = \underline{V}_{\xi 1} + \underline{V}_{\eta 1} = \underline{V}_{x1} + V_{y1}\cos\theta\, \underline{e}_\eta$$

$$= \underline{e}_x(-\omega_0 a \sin\omega_0 t) + \cos\theta\, \underline{e}_\eta(\omega_0 a\cos\omega_0 t); \quad V_{\perp 1} = -V_{\perp 2}$$

$$\underline{k}\cdot\underline{r} = k\zeta = \frac{\omega}{c}\zeta = \frac{n\omega_0}{c}\zeta \qquad \text{for the } n^{th} \text{ Fourier component}$$

$$= \frac{n\omega_0}{c}\, y \sin\theta$$

$$= \frac{n\omega_0}{c}\, a \sin\phi \sin\theta \qquad \text{where } \rho(\underline{r}) \neq 0.$$

Now substitute these last three expressions into V.24:

$$I(t) = \frac{nea^2\omega_0^2}{c} \int \left[\delta(\underline{r}-\underline{r}_1) - \delta(\underline{r}-\underline{r}_2)\right]\left[\underline{e}_x(-\sin\omega_0 t) + \underline{e}_\eta \cos\omega_0 t \cos\theta\right] \sin\theta \sin\phi\, d\tau$$

$$= \frac{nea^2\omega_0^2}{c} \sin\theta \left[\underline{e}_x(-2\sin\omega_0 t) + \underline{e}_\eta\, 2\cos\omega_0 t\cos\theta\right] \sin\omega_0 t$$

$$= \frac{nea^2\omega_0^2}{c} \sin\theta \left[\underline{e}_x(\cos 2\omega_0 t - 1) + \underline{e}_\eta \cos\theta\left[\sin 2\omega_0 t\right]\right]$$

We can then write, by inspection

$$\int \underline{j}_n(\underline{r})\,d\tau = 0 \quad \text{for } n \neq 2$$

$$\int \underline{j}_{2\xi}(\underline{r})\,d\tau = \frac{e a^2 \omega_0^2}{c} \sin\theta \; \underline{e}_\xi$$

$$\int \underline{j}_{2\eta}(\underline{r})\,d\tau = -\frac{e a^2 \omega_0^2}{ic} \sin\theta \cos\theta \; \underline{e}_\eta$$

Since all the harmonics except the second are zero, this means that the frequency of the radiation will be twice the frequency of revolution of the electrons.

Then, using V.2

$$\overline{\underline{S}}(\theta) = \frac{1}{r^2_{\text{observer}}} \frac{(2\omega_0)^2}{2\pi c^3} \left| \frac{e a^2 \omega_0^2}{c} \sin\theta \left(\underline{e}_\xi + i\cos\theta \; \underline{e}_\eta\right)\right|^2 \underline{e}_k$$

$$= \frac{2 e^2 a^4 \omega_0^6}{\pi r^2_{obs} c^5} \sin^2\theta \left(1 + \cos^2\theta\right) \underline{e}_k$$

$$\int \overline{\underline{S}} \cdot d\underline{A} = \frac{2 e^2 a^4 \omega_0^6}{\pi c^5} \int_0^\pi \sin^2\theta \,(1 + \cos^2\theta)\, 2\pi \sin\theta \, d\theta$$

$$= \frac{2 e^2 a^4 \omega_0^6}{\pi c^5} \; \frac{16\pi}{5} = \frac{32}{5} \frac{e^2 a^4 \omega_0^6}{c^5}$$

********

PROBLEM 3 Consider a model of a nucleus in which a single proton moves in a square well of depth 20 Mev, radius $6 \times 10^{-13}$ cm. The lowest state will be one with zero angular momentum; call it 1s. The next lowest state that can decay to 1s by electric dipole radiation is $1p_1$. Find the energy levels of these two states and the mean life for radiation.

Solution The computation is tedious, since one has to match the values and slopes of the inside and outside functions. The expressions are given in Schiff, Sec. 15. We shall give only the results.

| | |
|---|---|
| For the 1s state: | $E = -15.96$ Mev |
| For the 1p state: | $E = -11.81$ Mev |
| The mean life, | $\tau = 7.5 \times 10^{-17}$ sec. |

**********

PROBLEM 4 Place a proton, with magnetic moment $\mu = 2.79$ nuclear magnetons in a magnetic field, $\mathcal{H}_z$. Classically the proton will precess with 2 x 2.79 times the Larmor Frequency, and emit magnetic dipole radiation. Quantum mechanically, there will be two eigenstates, and a proton in the higher will eventually decay to the lower. Find the transition rate from $\binom{0}{1}$, the state antiparallel to $\mathcal{H}_z$, to $\binom{1}{0}$, the parallel state. $\mathcal{H}_z = 10{,}000$ gauss.

Solution If we call the unit cartesian vectors $\underline{i}$, $\underline{j}$, $\underline{k}$, the matrix element of V.9, p. 95, becomes

$$\underline{M}_{fi} = \mu \begin{pmatrix}1 & 0\end{pmatrix}\left[\sigma_x \underline{i} + \sigma_y \underline{j} + \sigma_z \underline{k}\right]\begin{pmatrix}0\\1\end{pmatrix}$$

$$= \mu \begin{pmatrix}1 & 0\end{pmatrix}\left[\begin{pmatrix}0 & 1\\1 & 0\end{pmatrix}\underline{i} + \begin{pmatrix}0 & -i\\i & 0\end{pmatrix}\underline{j} + \begin{pmatrix}1 & 0\\0 & -1\end{pmatrix}\underline{k}\right]\begin{pmatrix}0\\1\end{pmatrix}$$

$$\underline{M}_{fi} = \mu(\underline{i} - i\underline{j})$$

$$|\underline{M}_{fi}|^2 = 2\mu^2$$

The difference in energy between $\binom{1}{0}$ and $\binom{0}{1}$ is $2\mu\mathcal{H}_z$.

Equation V.9 then becomes

$$\frac{1}{\tau} = \frac{4}{3}\frac{(2\mu\mathcal{H}_z)^3}{\hbar^4 c^3} \times 2\mu^2$$

$$\tau = 2.7 \times 10^{+28} \text{ sec.}$$

# CHAPTER VI NUCLEAR FORCES

## A. INTRODUCTION

The structure of nuclei and of (P,P), (N,P), and (N,N) interactions will be analysed using the concept of strong, short-range nuclear forces. The range of these forces will be on the order of the size of the nucleons or $3 \times 10^{-13}$cm. The use of this nuclear force concept with the help of quantum mechanics (already shown applicable to some extent to the nucleus in the case of alpha decay) will prove useful in understanding and predicting some nuclear phenomena, altho there are some doubts as to whether the usual concepts of geometry hold for such small regions of space.

### 1. Meson Theory

At present the meson theories of nuclear forces are the main guides and give valuable qualitative results. However, there are serious difficulties in these theories which lead people to believe that the answer to the problem cannot be found by their further development. In the section on exchange forces (next page) more is said about the meson theory of nuclear forces (see FIG. VI.2). In addition the next chapter will be devoted entirely to a discussion of mesons.

### 2. Saturation of Nuclear Forces

One of the requirements to be fulfilled by a theory of nuclear forces is the saturation property. By saturation of nuclear forces, it is meant that the binding energy of the nucleus is directly proportional to the size of the nucleus and that the densities of all nuclei are the same. This is the case with liquids and solids where the force between the particles is usually made up of a strong, short-range repulsive force and a longer range attractive force. However, the assumption of a short range repulsive nuclear force cannot be made to fit all of the existing experimental data. This is shown in a paper by G. Parzen and L. Schiff, Phys. Rev. 74, 1564 (1948).

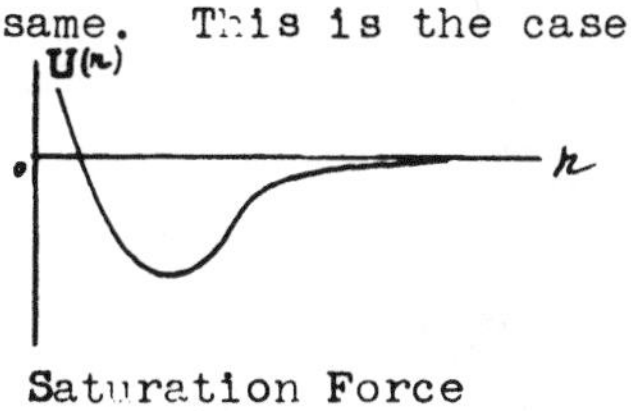


Saturation Force
FIG. VI.1

The problem is to explain why the nucleus doesn't collapse if only attractive forces can be used. At first thought one might think that the Pauli exclusion principle could prevent collapse, since not more than four nucleons (2 protons and 2 neutrons) can be in the same space state. However, it can be shown that then the most stable nucleus is one which has collapsed to a diameter equal to the range of the nuclear force.*

This difficulty can be removed by inventing a nuclear force which can give interaction only for nucleons which have close to the same state functions. Then the total interaction energy could be proportional to **A** rather than $A^2$. In the footnote on the next page it will be made plausible that an exchange type of force could meet these requirements.

3. <u>Exchange Forces</u>

If the potential is a function of the space and spin coordinates only, there are three possible types of exchange for two particles:

$$\psi(\underline{r}_1,\sigma_1,\underline{r}_2,\sigma_2) \rightarrow \begin{cases} \psi(\underline{r}_2,\sigma_1,\underline{r}_1,\sigma_2) & \text{Majorana} \\ \psi(\underline{r}_1,\sigma_2,\underline{r}_2,\sigma_1) & \text{Bartlett} \\ \psi(\underline{r}_2,\sigma_2,\underline{r}_1,\sigma_1) & \text{Heisenberg} \end{cases}$$

The potential energy operator for the Majorana type exchange force can be written in the form $U(\underline{r})\, P_{12}$ where the operator $P_{12}$ interchanges the space coordinates of particles 1 and 2.

This can be pictured physically in the case of (N,P) by considering that a charged meson ($\pi^+$ or $\pi^-$) travels back and forth between the neutron and proton.

$N = P + \pi^-$ $P = N + \pi^+$

In FIG. 2, the neutron and proton change positions because of the $\pi^-$ travelling from N to P. These virtual mesons extend a short range from the "parent" nucleon and are considered as quanta of the nuclear force field. See Ch. VII for a more complete discussion of this. Wigner gives an argument

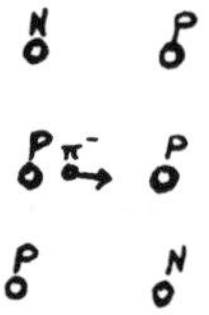

FIG. VI.2

---

* This is easily seen by finding the maximum binding energy as a function of the diameter of the nucleus. The binding energy is defined by BE = -(T + U). The total potential energy is negative and proportional to the number of interacting pairs. As the diameter of the nucleus is reduced to the range of the nuclear force, this number of interacting pairs reaches its maximum, $A(A - 1)/2$. Since the nucleus is to be in the ground state, Fermi statistics show that the average kinetic energy per particle is proportional to the 2/3 power of the density of particles. Thus the total kinetic energy is proportional to $A \times A^{2/3}$ or $A^{5/3}$. However, it has just been shown that the total potential energy can go as a higher power (as $A^2$) with **A**. Thus for large **A** the potential energy term can exceed the kinetic energy. Then the BE is maximum when the potential energy reaches its maximum which occurs when the diameter of the nucleus is approximately equal to the range of the nuclear forces.

that the Majorana and Heisenberg type exchange forces can give saturation.*

The nature of nuclear forces is given by the form of the interaction potential between (P,P), (N,P), and (N,N). This information is obtained from (N,P) and (P,P) scattering experiments and from the study of the deuteron, which in nuclear physics plays a role analogous to that of the hydrogen atom in atomic physics. The fact that nuclei with the same number of protons and neutrons are the most stable (ignoring the electrostatic energy contribution; see eq. I.7) suggests that (P,P) and (N,N) forces (except for electrostatic forces) are similar.

## B. THE DEUTERON

### 1. Non-central and Spin-dependent Forces

The spin of the deuteron is I=1. The magnetic moment is

---

* Proc. Nat. Acad. Sci. 22, 662 (1936)
This can be made plausible by the following argument:

In the nucleus assume particle no. 1 in a state $a(\underline{r}_1)$ and particle no. 2 in a state $b(\underline{r}_2)$. Then the interaction energy of these two particles due to an exchange force is

$$W = \iint \{a^*(\underline{r}_1)b^*(\underline{r}_2)\}\, U(r_{12})\, P_{12}\{a(\underline{r}_1)b(\underline{r}_2)\}\, d^3r_1\, d^3r_2$$
$$= \int\left\{\int U(r_{12})\, a^*(\underline{r}_1)b(\underline{r}_1)\, d^3r_1\right\} a(\underline{r}_2)b^*(\underline{r}_2)\, d^3r_2$$

Now if $a(\underline{r}_1)$ and $b(\underline{r}_1)$ have considerably different wave lengths and if $U(r_{12})$ is slowly varying compared to the product $a^*(\underline{r}_1)b(\underline{r}_1)$,

then $\int U(r_{12})a^*(\underline{r}_1)b(\underline{r}_1)\, d^3r_1$ is close to zero. If one considers the nuclear model where the nucleons are plane waves confined to a box of the nuclear volume and obeying Fermi statistics, then for any given nucleon, only those nucleons in neighboring momentum states will give any appreciable W. Thus the total W is proportional to A.

In the case of non-exchange forces, the contribution will be large since

$$W = \iint U(r_{12})\, |a(\underline{r}_1)|^2\, |b(\underline{r}_2)|^2\, d^3r_1\, d^3r_2 \quad \text{for non-exchange forces.}$$

In this case all pairs of particles will interact, giving a total W proportional to $A^2$.

Actually the situation is more complicated than implied here. Since the total wavefunction must be anti-symmetric with respect to the exchange of any two protons or two neutrons, both types of forces give both types of integrals. However the mixture of the different kinds of forces may be adjusted so that the energy terms proportional to powers of A greater than one drop out.

$\mu = 0.85647$. If the deuteron consisted of a proton and neutron with spins parallel and with zero orbital angular momentum ($^3S_1$ state), the spin would be one and the magnetic moment would be

$$\mu_p + \mu_n = 2.7896 - 1.9103 = 0.8793 \pm .0015$$

which differs by 2.6% from the observed value. This discrepancy does not fall within the limits of experimental error. However a mixture of 96% of $^3S_1$ and 4% of $^3D_1$ for the ground state of the deuteron would give the correct magnetic moment. Also it would give the correct quadrupole moment! There is no P state contribution in this mixture since $\ell = 1$ has different parity from $\ell = 0$ and $\ell = 2$. It will be remembered that parity is a good quantum number since the Hamiltonian is invariant to an inversion of all the coordinates. See page 98.

A purely central force cannot give a mixture for the ground state. This is because $u_{n,\ell}(r)\, Y_{\ell,m}(\theta,\phi)$ is the general exact eigenfunction and $E_{n,\ell}$ the eigenvalue for any central force field. However a non-central force will give a solution $\Psi(r,\theta,\phi)$ which can be expanded in spherical harmonics, and thus is a mixture of various states. Thus the interaction potential of (N,P) must contain a term or terms which are non-central. Since the force is invarient with respect to displacement, rotation, and inversion of the observer's coordinate system, it can be shown that the most general form of two particle potential is narrowed down to three terms.* If exchange is to be included, three more terms are necessary.

$$U = U_a(r) + U_b(r)\, \underline{\sigma}_1\cdot\underline{\sigma}_2 + U_c(r)\, S + P_{12}\left\{U_d(r) + U_e(r)\, \underline{\sigma}_1\cdot\underline{\sigma}_2 + U_f(r)\, S\right\} \qquad \text{VI.1}$$

where $S \equiv \frac{(\underline{\sigma}_1\cdot\underline{r})(\underline{\sigma}_2\cdot\underline{r})}{r^2} - \frac{1}{3}\underline{\sigma}_1\cdot\underline{\sigma}_2$ is the non-central tensor force term.

The problem is to determine the form of VI.1 for (N,P). At first the non-central components will be neglected, since it is known that they are small (deuteron is 96% pure S state). Then the wave function will not be a mixture of different states and

$$P_{12}\, \Psi_\ell(\underline{r},\sigma_N,\sigma_P) = \Psi_\ell(-\underline{r},\sigma_N,\sigma_P)$$

$$P_{12}\, \Psi_\ell(\underline{r},\sigma_N,\sigma_P) = (-1)^\ell\, \Psi_\ell(\underline{r},\sigma_N,\sigma_P) \qquad \text{VI.2}$$

We already know that $\Psi_0$ (S state) is the best description of the deuteron, so for the following discussion $\ell = 0$ and $P_{12} = 1$. Let $f_1(r) \equiv U_a + U_d$ and $f_2(r) \equiv U_b + U_e$

then VI.1 gives $U = f_1(r) + f_2(r)\, \underline{\sigma}_1\cdot\underline{\sigma}_2$

Now $\underline{\sigma}_1\cdot\underline{\sigma}_2 = 1$ for spins parallel
$= -3$ for spins antiparallel VI.3

This property of the spin operators can most quickly be seen from the vector model approach. Classically,

$$I^2 = s_1^2 + s_2^2 + 2\underline{s}_1\cdot\underline{s}_2$$

$$\underline{s}_1\cdot\underline{s}_2 = \frac{I^2 - (s_1^2 + s_2^2)}{2}$$

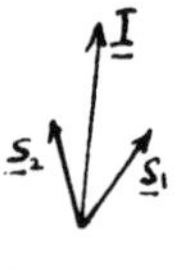


FIG. VI.3

* A justification of this is given in Bethe D, page 73.

Quantum mechanically,

$$\left(\tfrac{1}{2}\sigma_1\right)\cdot\left(\tfrac{1}{2}\sigma_2\right) = \frac{I(I+1) - s_1(s_1+1) - s_2(s_2+1)}{2} \qquad \text{since } s_1 = s_2 = \frac{\sigma}{2}$$

$$\sigma_1\cdot\sigma_2 = 2\left[I(I+1) - 2s(s+1)\right]$$

for the triplet state I = 1 and

$$\sigma_1\cdot\sigma_2 = 2\left[2 - 2\times\tfrac{3}{4}\right] = 1$$

for the singlet state I = 0 and

$$\sigma_1\cdot\sigma_2 = 2\left[0 - 2\times\tfrac{3}{4}\right] = -3$$

Let $U_3(r)$ now stand for the potential for the triplet interaction and $U_1(r)$ for the singlet.

Then $U_3(r) = f_1 + f_2$

$U_1(r) = f_1 - 3f_2$

Let $u(r) = r\Psi(r)$, then Schroedinger's equation becomes

$$\frac{d^2u(r)}{dr^2} + \frac{2\mu}{\hbar^2}\left[E - U(r)\right]u(r) = 0$$

For (N,P), $\mu \approx M/2$ where M is the average mass of neutron and proton.

## 2. Ground State of the Deuteron

The wave equation will first be examined for the triplet interaction ( $U_3$). If the negative of the binding energy of the deuteron is put in the wave equation for E and $U_3$ for U, then the equation will closely represent that for the ground state of the deuteron. The BE of the deuteron can be determined independently from the threshold energy for photodisintegration of the deuteron and from mass spectrograph data. The experimental result is

$$\text{BE of } H^2 \equiv W = 3.58 \times 10^{-6} \text{ ergs} = 2.23 \text{ Mev}$$

The wave equation becomes

$$\frac{d^2u}{dr^2} - \frac{M}{\hbar^2}(W + U_3)u = 0$$

Let $r_0$ be the range of $U_3(r)$. For $r > r_0$ ,

$$u \propto e^{-\sqrt{\frac{MW}{\hbar^2}}\,r}$$

Let $\alpha \equiv \sqrt{\frac{MW}{\hbar^2}} = 2.32 \times 10^{12}\ \text{cm}^{-1}$

Then $1/\alpha = 4.31 \times 10^{-13}$ cm can be taken as a measure of the size of the deuteron. The exact shape of $U_3(r)$ is not critical. Only high energy scattering experiments can give some information about the shape. The square well is the easiest to work with and will be used here. Potentials of the form $e^{-r}$ and $e^{-r^2}$ are worked out in Bethe **A**, p. 110, and $\frac{1}{r}e^{-kr}$ is treated in Phys. Rev. 53, 991.

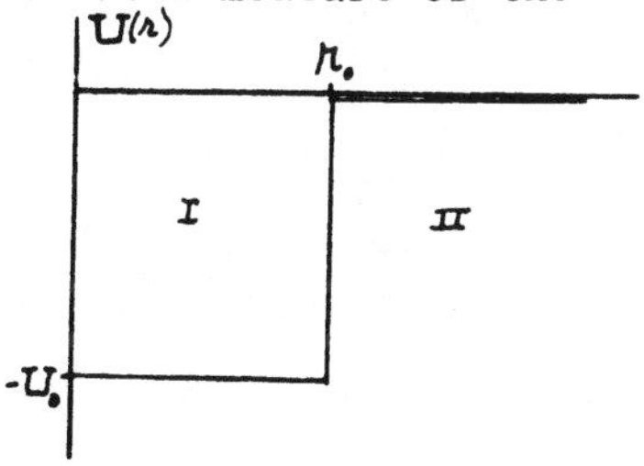


FIG. VI.4

The value $r_0 = 2.82 \times 10^{-13}$cm. will be used. By accident this value also happens to be the radius of the electron $= \frac{e^2}{mc^2}$. This value is determined within 5% by fitting the energy dependance of (P,P) scattering to a rectangular potential well. See page 128.

In region I: $\frac{d^2u_I}{dr^2} + \frac{M}{\hbar^2}(U_0 - W)u_I = 0$

where $U_0$ is the absolute value of the well depth.

$$u_I = A\sin\left(\sqrt{\frac{M}{\hbar^2}(U_0-W)}\; r\right)$$

$$u_{II} = Be^{-\alpha r}$$

At the boundary:

$$u_I'(r_0) = u_{II}'(r_0)$$

$$u_I(r_0) = u_{II}(r_0)$$

$$\left.\frac{u_I'}{u_I}\right|_{r_0} = \left.\frac{u_{II}'}{u_{II}}\right|_{r_0}$$

gives $\sqrt{\frac{M}{\hbar^2}(U_0-W)}\,\cot\left(\sqrt{\frac{M}{\hbar^2}(U_0-W)}\; r_0\right) = -\sqrt{\frac{MW}{\hbar^2}} = -\alpha = -2.32 \times 10^{12}$ cm$^{-1}$

$$U_0 = 21 \text{ Mev}$$

$$U_3 = \begin{cases} -21 \text{ Mev} & r < r_0 \\ 0 & r > r_0 \end{cases}$$

$$\sqrt{\frac{M}{\hbar^2}(U_0-W)} = 6.73 \times 10^{12} \text{ cm}^{-1}$$

$$\sqrt{\frac{M}{\hbar^2}(U_0-W)}\; r_0 = 108°$$

This potential gives a wave function as shown in FIG. 5. This function gives an expectation of 54% that the deuteron is outside the range of the nuclear force.

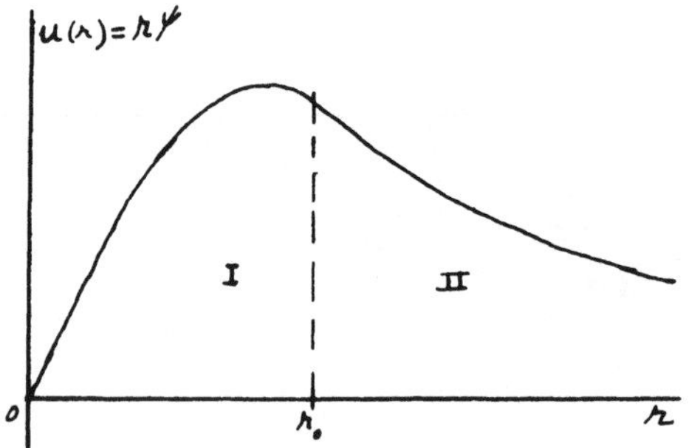


Deuteron Radial Wave Function
FIG. VI.5

**Problem:**

Determine the depth $U_0$ for the potential of the shape

$$U_3 = \begin{cases} \frac{U_0}{r_0} r - U_0 & r < r_0 \\ 0 & r > r_0 \end{cases}$$

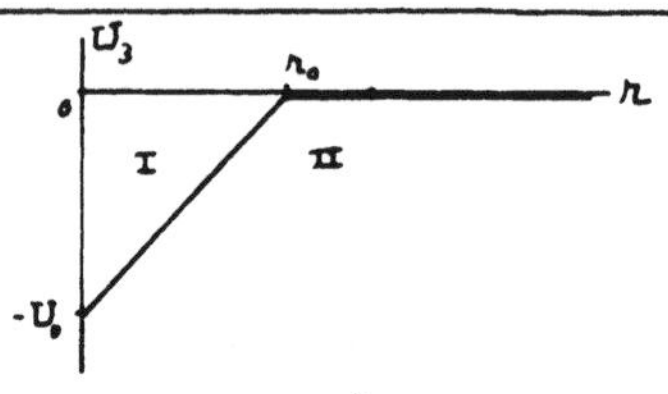


FIG. VI.6

Making the transformation $x = \frac{r}{r_0}$ gives

$$\frac{d^2u_I}{dx^2} = r_0^2\alpha^2[g x u_I - (g-1)u_I] \qquad \text{VI.4}$$

where $g = \frac{U_0}{W}$ and $\alpha = 2.32 \times 10^{12}$ as before.

The boundary conditions are

$u_I(0) = 0$

$u_I'(0) =$ any convenient number

$$\left.\frac{u_I'}{u_I}\right|_{x=1} = \left.\frac{u_{II}'}{u_{II}}\right|_{x=1} = -r_0\alpha = -0.65$$

A guess is made for the value $g$ and VI.4 is numerically integrated* from x = 0 to x = 1. Then $\left.\frac{u_I'}{u_I}\right|_{x=1}$ is compared with $-r_0\alpha = -0.65$. Three such numerical integrations should be sufficient to interpolate $g = 27.3 \pm 0.2$ Thus $U_0 = 61$ Mev.

A machine which automatically plots the wave function for any given potential is described in a paper by R.L.Garwin, R.S.I. 21, 411 (1950)

## C. NEUTRON - PROTON SCATTERING

### 1. Method of Partial Waves

First it will be necessary to state some of the pertinent results of quantum mechanics. The wave equation will be set up for a spherically symmetric potential, which acts like a scattering center at r = 0. The two-body problem can be transformed into this one-body form by using reduced mass and relative coordinates. It is shown on page 125 that this potential U(r) must drop off faster than 1/r. The wave equation is

$$\nabla^2\psi + \frac{2\mu}{\hbar^2}[E - U(r)]\,\psi(r,\theta) = 0. \qquad \text{VI.5}$$

$$\psi(r,\theta) \rightarrow e^{i\frac{p}{\hbar}z} + f(\theta)\frac{1}{r}e^{i\frac{p}{\hbar}r} \qquad \text{VI.6}$$

is an asymptotic solution of the above equation and is in the convenient form of an incoming wave and a scattered wave:

$$\psi_{scatt} \equiv f(\theta)\frac{1}{r}e^{ikr}$$

f(θ) can be determined using the method of partial waves.** In this method f(θ) is expanded in Legendre polynomials:

$$\psi_{scatt.} = \sum_{l=0}^{\infty} C_l P_l(\cos\theta)\frac{1}{r}e^{i\frac{p}{\hbar}r} \qquad \text{VI.7}$$

The problem is to determine the form of the scattered wave, or the $C_l$. This is done by first breaking up the complete $\Psi$ into its partial waves:

$$\psi(r,\theta) = \sum_{l=0}^{\infty}\frac{1}{r}\mathcal{N}_l(r)P_l(\cos\theta)$$

which, when put into VI.5, gives

$$\mathcal{N}_l'' + \left[\frac{2\mu E}{\hbar^2} - \frac{2\mu U(r)}{\hbar^2} - \frac{l(l+1)}{r^2}\right]\mathcal{N}_l = 0 \qquad \text{VI.8}$$

This equation shows that $\mathcal{N}_l$ behaves as a sine wave for large r and $\mathcal{N}_l \sim r^{l+1}$ for small r. This equation can easily be solved in the case of no potential (plane wave) with the result that

*Margenau and Murphy, Mathematics of Physics and Chemistry, p.468.

**Mott and Massey, The Theory of Atomic Collisions, Ch. II, Schiff, page 103.

$v_l = A_l \sqrt{r}\, J_{l+\frac{1}{2}}\left(\frac{p}{\hbar} r\right)$ where $J_{l+\frac{1}{2}}$ is the $l+\frac{1}{2}$ order Bessel fn.

It can be shown that these Bessel functions behave asymptotically such that

$$v_l \to \cos\left(\frac{p}{\hbar} r - \frac{l+1}{2}\pi\right) \qquad \frac{pr}{\hbar} >> 1$$

With the potential, the asymptotic form becomes

$$v_l \to \cos\left(\frac{p}{\hbar} r - \frac{l+1}{2}\pi + \beta_l\right)$$

Mott and Massey (also Schiff, section 19) show that

$$C_l = \frac{\hbar(2l+1)}{p} e^{i\beta_l} \sin\beta_l \qquad \text{VI.9}$$

Using VI.6, the flux scattered into the solid angle $d\omega$ is

$$|\psi_{scatt}|^2 v r^2 d\omega = \left| e^{i\frac{p}{\hbar}z} \right|^2 v\, \sigma(\theta)\, d\omega$$

where $\sigma(\theta)$ is the differential cross section by definition.

Thus $\sigma(\theta) = \left| \sum_{l=0}^{\infty} C_l P_l(\cos\theta) \right|^2$ VI.10

The total scattering cross section is obtained by integrating $\sigma(\theta)$ over the entire solid angle.

$$\sigma = \int \sigma(\theta)\, d\omega = \sum_{l=0}^{\infty} \sum_{l'=0}^{\infty} C_l C_{l'}^* \iint P_l P_{l'} \sin\theta\, d\theta\, d\phi$$

$$\sigma = 2\pi \sum_{l=0}^{\infty} \sum_{l'=0}^{\infty} C_l C_{l'}^* \int_{-1}^{1} P_l(x) P_{l'}(x)\, dx$$

$$\int_{-1}^{1} P_l(x) P_{l'}(x)\, dx = \frac{2}{2l+1} \delta_{l,l'}$$

Using VI.9,

$$|C_l|^2 = \frac{\hbar^2}{p^2} (2l+1)^2 \sin^2\beta_l$$

$$\therefore \quad \boxed{\sigma = 4\pi \bar{\lambda}^2 \sum_{l=0}^{\infty} (2l+1) \sin^2\beta_l} \qquad \text{VI.11}$$

If the potential could be switched on and off, $\beta_l$ would be the phase difference at large r between the two solutions. In FIG. 7 the solid line represents the solution with the potential on, and the dotted line is the solution, $\sqrt{r}\, J_{l+\frac{1}{2}}\left(\frac{p}{\hbar} r\right)$, with the potential off.

It can be shown for energies such that $\bar{\lambda} = \frac{\hbar}{p} >> r_0$, that $\beta_l \simeq 0$ for $l \neq 0$.

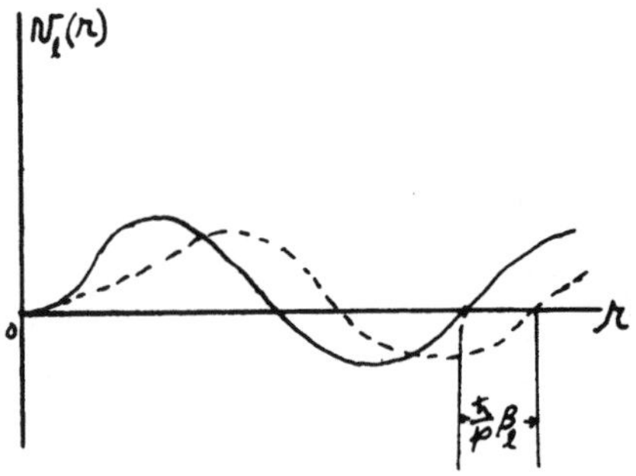


FIG. VI.7

This is essentially because $v_l$ goes as $(kr)^{l+1}$ for small r. Thus $\int_0^\infty v_l^2 U(r)\,dr$ is small for small p. For small $\beta_l$, this integral is proportional to $\beta_l$. See Schiff p.165. Thus the $C_l$ 's would be almost zero except for $C_0$.

This may be seen more easily using semi-classical concepts. Let b be the classical distance of closest approach. Then the angular momentum = bp = $l\hbar$. If b > $r_0$, the particle doesn't enter the region of short range force. Thus $\beta_l = 0$ for

$$b > r_0$$

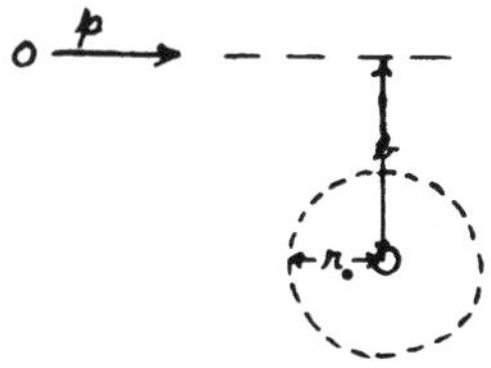


FIG. VI.9

Since $b = \frac{l\hbar}{p}$, we have

$$\beta_l = 0 \quad \text{for } l > \frac{p}{\hbar} r_0 \qquad \text{VI.12}$$

For slow neutrons ( ~1 ev), $\frac{\hbar}{p} \sim 10^{-9}$ and $r_0 \sim 10^{-13}$. Then $\beta_l = 0$ for $l > 10^{-4}$. VI.9 shows that in this case only $C_0$ is non-zero. Thus for slow neutrons only S-wave or isotropic scattering occurs.

2. Solution for σ for Low Energy Scattering

An approximate expression for $\beta_0$ in terms of the depth of the potential well will now be developed. This will give the cross section as a function of the well size, since

$$\sigma = 4\pi \bar{\lambda}^2 \sin^2\beta_0$$

The equation for the zeroth wave in the region of the potential is

$$v_0'' - \frac{2\mu U}{\hbar^2} v_0 \approx 0 \qquad \text{for } E \ll U$$

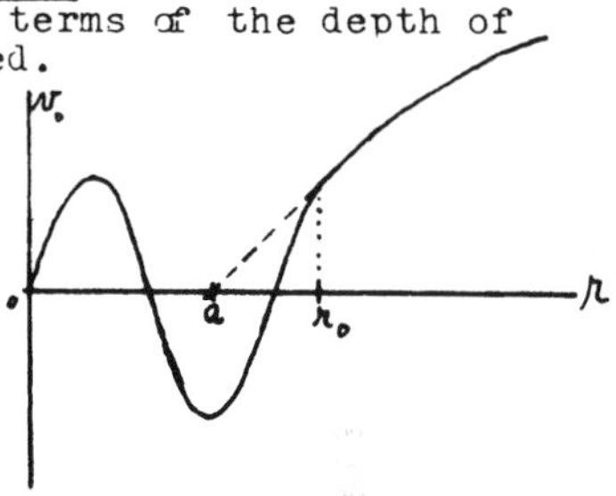


$v_0$ for Arbitrary Well & Low E
FIG. VI.10

Outside $r_0$, the wave length is much longer and this slow sine wave starts out like the straight line

$$v_0 = c(r - a)$$

where c is the slope and a is the intercept on the r axis.

Then
$$A \sin\left(\frac{pr}{\hbar} + \beta_0\right) \approx A\left(\frac{pr}{\hbar} + \beta_0\right) = c(r - a)$$

$$\begin{cases} A\frac{p}{\hbar} = c \\ A\beta_0 = -ca \end{cases}$$

$$\beta_0 = -\frac{1}{\bar{\lambda}} a$$

$$\boxed{\sigma = 4\pi a^2}$$

For the case of a square well,

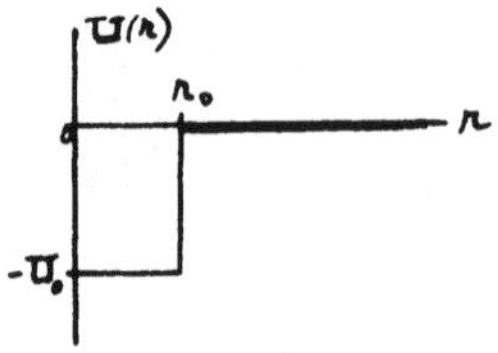


$$v_0'' + \frac{2\mu U_0}{\hbar^2} v_0 = 0 \quad \text{for } r < r_0$$

$$v_0 = A \sin k_3 r \quad \text{where } k_3 = \sqrt{\frac{2\mu U_0}{\hbar^2}}$$

$$\frac{d v_0}{dr} = A k_3 \cos k_3 r$$

Using FIG. 10, the distance cut off on the r axis from $r_0$ by the tangent at $r_0$ is

$$\frac{v_0(r_0)}{\text{slope}} = \frac{A \sin k_3 r_0}{A k_3 \cos k_3 r_0}$$

a = $r_0$ minus this distance

$$a = r_0 - (1/k_3) \tan k_3 r_0$$

$$\boxed{\sigma = 4\pi r_0^2 \left(1 - \frac{1}{k_3 r_0} \tan k_3 r_0\right)^2} \qquad \text{VI.14}$$

3. Virtual State of the Deuteron

The experimental scattering cross section for slow (N,P) is shown in FIG. 11. $\sigma$ is seen to be 20.3 barns for low energies. (The increase of $\sigma$ at very low energies is due to molecular binding of the hydrogen atoms and thermal agitation. See Ch IX p. 194 ). However VI.14 gives $\sigma = 4.48$ barns rather than the known value of 20.3 barns for $r_0 = 2.32 \times 10^{-13}$ and $U_3 = -21.3$ Mev. This discrepancy is due to $U_1 \neq U_3$. $U_1$ can now be calculated from the experimental value for $\sigma$. Since the statistical weights of the triplet and singlet states are 3 and 1 respectively, for random orientations,

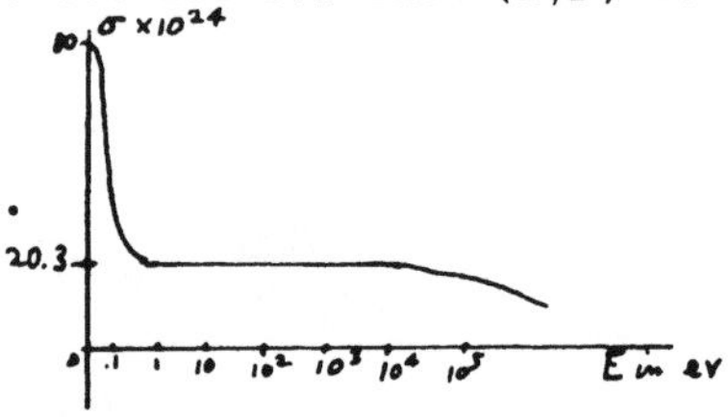


σ for Neutrons against H
FIG. VI.11

$$\sigma = \tfrac{3}{4}\sigma_3 + \tfrac{1}{4}\sigma_1 \qquad \text{VI.15}$$

$$20.3 = \tfrac{3}{4}(4.48) + \tfrac{1}{4}\sigma_1 \qquad \sigma_1 = 68 = 4\pi a_1^2$$

$$a_1 = 2.32 \times 10^{-12} \text{ cm.}$$

$$k_1 = 5.28 \times 10^{12} \quad \text{where } k_1 = \sqrt{\frac{2\mu U_1}{\hbar^2}}$$

$$U_1 = -11.5 \text{ Mev}$$

From the results of neutron scattering on ortho- and para-hydrogen the sign of $a_1$ is found to be opposite to that of $a_3$. See Ch. IX p. 199. For slow neutrons a negative $a_1$ gives a wave function as shown in FIG. 12.

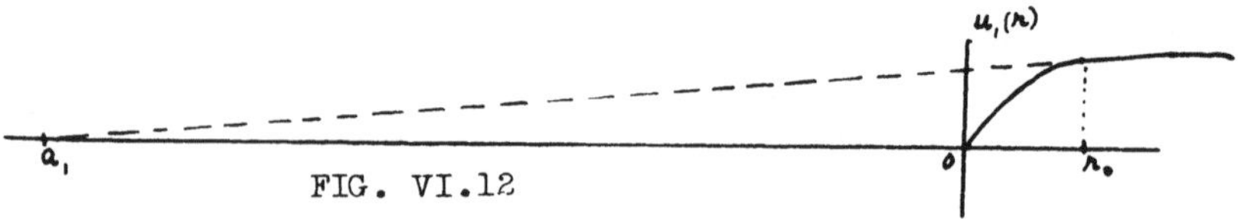


FIG. VI.12

Since the r intercept is so far to the left, this state is nearly bound. If the slope of the dotted line were slightly negative rather than positive, a bound singlet state of the deuteron would be possible. The singlet deuteron is said to be a virtual state.

## 4. Evidence for Exchange Forces

To explain the saturation properties of nuclear forces, it was found advantageous to postulate the existence of some type of exchange force (see p. 112). The recent (1946) availability of high energy ( ~90 Mev) neutron beams (see Ch. VIII p. 177) has given conclusive proof of the existence of exchange phenomena. The older "low" energy scattering experiments cannot tell much about exchange forces since only the S partial wave is significant ( $C_l \approx 0$ for $l > 0$ as explained on p.119 ). For higher energies the P scattering appears. If the potential is of the form:

$$U(r) = U_1(r) + U_2(r)P_{12} \qquad \text{see VI.1}$$

$$U = U_1 + U_2 \qquad \text{for even } l$$

$$U = U_1 - U_2 \qquad \text{for odd } l \quad \text{see VI.3}$$

In case $U_2$ is a bigger well than $U_1$, odd $l$ would give a repulsive force, while even $l$ would give an attractive force. The theoretical results for an ordinary force (no exchange) are such that the greater the energy of the neutron beam, the more will it be scattered in the forward direction. This can be seen classically using two simple considerations of collisions (in the lab system) between particles of equal mass:

1. The angle between the two final velocity vectors is $90^o$. This is independent of the mechanism of collision. In the lab system let $V_P'$ be the velocity picked up by the proton. Let $V_N'$ be the final neutron velocity and $V_N$ its initial velocity. By the conservation of energy,

$$\tfrac{1}{2}MV_N^2 = \tfrac{1}{2}MV_N'^2 + \tfrac{1}{2}MV_P'^2 \qquad V_N^2 = V_N'^2 + V_P'^2$$

Thus these three velocity vectors make a right triangle and $V_P'$ must be perpendicular to $V_N'$. (By conservation of momentum $\underline{V}_N = \underline{V}_N' + \underline{V}_P'$ .)

2. If the interaction is a square well of depth $U_0$, then, independent of the energy of the incident particle, the target proton cannot acquire an energy of order of magnitude greater than $U_0$.

In the case of exchange forces only, the identity of the particles would be exchanged, and the neutrons would appear mostly scattered at $90^o$ in the lab system.

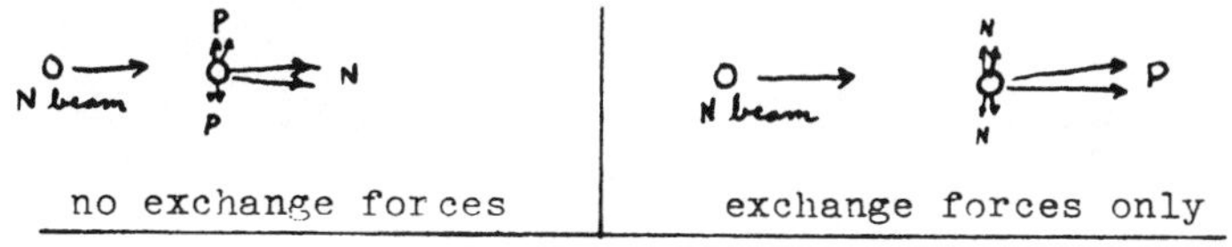


Scattering of High Energy Neutron Beam in Lab System
FIG. VI.13

The above conclusions can also be reached using a simple application of the Born approximation:

$$\sigma(\theta) = \frac{\mu^2}{4\pi^2\hbar^4}\left|\int e^{-i\underline{k}_f\cdot\underline{r}}\, U\, e^{i\underline{k}_i\cdot\underline{r}}\, d\tau\right|^2 \qquad \text{(in c-m system)}$$

where $\underline{k}_i$ is the initial relative propagation constant and $\underline{k}_f$ the final. The $P_{12}$ part of U changes $e^{i\underline{k}_i\cdot\underline{r}}$ to $e^{-i\underline{k}_i\cdot\underline{r}}$

thus
$$\sigma(\theta) \propto \left|\int U_1\, e^{-i(\underline{k}_f-\underline{k}_i)\cdot\underline{r}}\, d\tau + \int U_2\, e^{-i(\underline{k}_f+\underline{k}_i)\cdot\underline{r}}\, d\tau\right|^2$$

For $(\underline{k}_f-\underline{k}_i)\cdot\underline{r}_0 \gg 1$ the exponential in the first integral oscillates rapidly over the region of integration. Thus its contribution is large only for $\theta \approx 0$. The reverse is true for the second integral which is large for scattering when $\theta \approx 180^\circ$.

First Integral (non-exchange) Second Integral (exchange)

The experimental results for 90 Mev neutrons are given in the right hand curve of FIG. 14.*

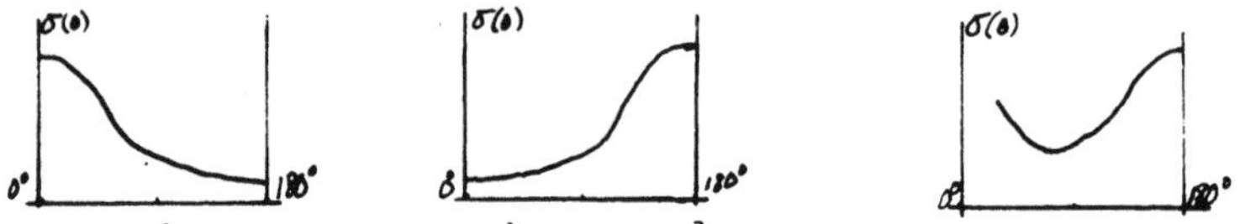


no exchange exchange only

Theoretical Results Experimental Results

90 Mev Neutron Scattering Curves in C-M System

FIG. VI.14

The data for scattering near zero degrees can't be obtained because of interference with the main beam. The experimental results imply that the exchange forces predominate slightly.

There is later evidence from the Berkeley cyclotron for the existence of exchange forces. This is the inverse experiment, the scattering of a ~400 Mev proton beam by neutrons in the nuclei of any target. The experimental results are that neutrons of an average energy ~350 Mev are observed in a narrow beam in the forward direction. These neutrons must originally have been protons of the cyclotron beam. In addition there are lower energy (non-exchange) neutrons knocked out at right angles. Also there is what appears to be an isotropic "low" energy neutron background due to neutrons evaporating from excited nuclei.

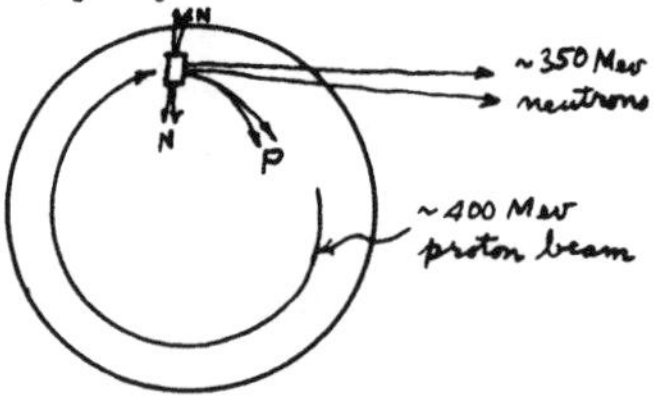


The factor $\frac{1+P_{12}}{2} = \begin{cases} 1 & \text{for even } l \\ 0 & \text{for odd } l \end{cases}$ is used in most of the "recipes" given. This would give a symmetric curve for FIG. 14. Serber recommends for the (N,P) potential:*

$$-U(r) = g^2\frac{1}{r}e^{-kr}\,\frac{1+P_{12}}{2}$$

* Phys. Rev. 75, 351 (1949); Phys. Rev. 79, 96 (1950).

where $\frac{g^2}{\hbar c} = \begin{cases} .405 & \text{for triplet} \\ .280 & \text{for singlet} \end{cases}$

and $\frac{1}{k} = 1.2 \times 10^{-13}\,\text{cm}.$ (see Ch. VII, p. 135)

An older (1941) "recipe" which fitted the data then, but gives wild results for the new 90 Mev scattering data is:

$$U(r) = J(r)\{1 - \tfrac{1}{2}g + \tfrac{1}{2}g(\sigma_1 \cdot \sigma_2) + \gamma S\}$$

where $J(r) = \begin{cases} -13.9 \text{ Mev} & r < r_0 = 2.8 \times 10^{-13}\text{cm.} \\ 0 & r > r_0 \end{cases}$

$g = 0.0715$

$\gamma = 1.725$

The derivation is given in Phys. Rev. 59, 436 (1941) by Rarita and Schwinger. This potential explains the quadrupole and magnetic moments by using a non-central force $f(r)S$ which gives a mixture of 96% S and 4% D states. However, there is a big difficulty in reconciling this non-central force with 90 Mev scattering results.

## D. PROTON - PROTON FORCES

### 1. Pauli Principle Complications

(P,P) interactions are more complicated than (N,P) interactions in two respects. First, there is the coulomb scattering "superimposed" on the nuclear force scattering. Secondly, since two protons are identical particles of spin ½, they must obey the Pauli principle, and consequently have an anti-symmetric wave function. If the scattered wave is of the form $\psi_{scatt} = \frac{1}{r} f(\theta)\, e^{ikr}$ then the expression for the differential cross section $\sigma(\theta)$ that either the scattered particle or the scattering particle come off at an angle $\theta$ in the c-m system is

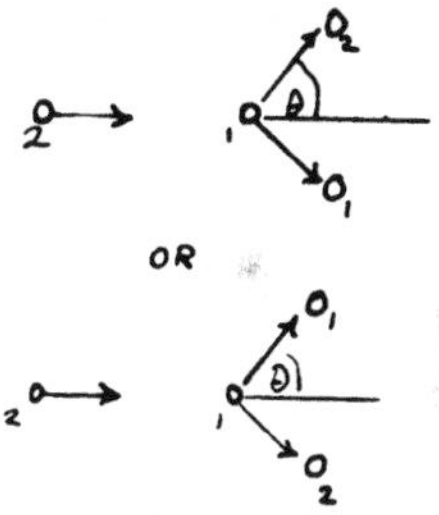


FIG. VI.15

$$\sigma(\theta) = |f(\theta)|^2 + |f(\pi - \theta)|^2$$
(for distinguishable particles)

However, for identical particles the probability is the square of the sum of the amplitudes rather than the sum of the squares.

$$\sigma(\theta) = |f(\theta) \pm f(\pi - \theta)|^2 \quad \text{(for identical particles)}$$

The plus or minus sign depends on whether the space part of the wave function is symmetric or anti-symmetric. This in turn depends on the symmetry of the spin function. In order to obtain a final result, the properties of spin must now be reviewed.

### 2. Spin Functions

Since any kind of interaction energy of the (P,P) magnetic moments is very weak compared to other interactions, $\psi(r_1,s_1,r_2,s_2)$

may have its space ($\underline{r}$) and spin ($\mathfrak{s}$) coordinates separated:

$$\psi = u(\underline{r}_1, \underline{r}_2)\, v(\mathfrak{s}_1, \mathfrak{s}_2)$$

For spin of $\frac{1}{2}$ the possible primary $v(\mathfrak{s}_1, \mathfrak{s}_2)$ functions are $(\frac{1}{2},\frac{1}{2})$; $(\frac{1}{2},-\frac{1}{2})$; $(-\frac{1}{2},\frac{1}{2})$; and $(-\frac{1}{2},-\frac{1}{2})$. These are already eigen-functions of $S_z = S_{1z} + S_{2z}$ where $s_z = \frac{1}{2}\sigma_z$ . However, linear combinations must be chosen such that they are also eigen-functions of $S^2$. This is worked out in Schiff, p. 227 with the four resulting spin functions:

| $S^2$ eigen value: | 2 | 2 | 2 | 0 |
|---|---|---|---|---|
| $S_z$ eigen value: | 1 | 0 | -1 | 0 |
| $(\frac{1}{2},\frac{1}{2})$ | 1 | 0 | 0 | 0 |
| $(\frac{1}{2},-\frac{1}{2})$ | 0 | $\frac{1}{\sqrt{2}}$ | 0 | $\frac{1}{\sqrt{2}}$ |
| $(-\frac{1}{2},\frac{1}{2})$ | 0 | $\frac{1}{\sqrt{2}}$ | 0 | $-\frac{1}{\sqrt{2}}$ |
| $(-\frac{1}{2},-\frac{1}{2})$ | 0 | 0 | 1 | 0 |
| | triplet functions | | | singlet function |

The four eigen-functions are the above four columns where the numbers give the coefficients of the four primary functions to the left.

Problem: For two identical particles each of spin 3/2, the resultant spin can be either 3,2,1, or 0. Find the degeneracy of each of these and their symmetry properties

Solution: There are the following 16 linearly independent primary eigen-functions of $S_Z$:

$(\frac{3}{2},\frac{3}{2})$ $(\frac{3}{2},\frac{1}{2})$ $(\frac{3}{2},-\frac{1}{2})$ $(\frac{3}{2},-\frac{3}{2})$

$(\frac{1}{2},\frac{3}{2})$ $(\frac{1}{2},\frac{1}{2})$ $(\frac{1}{2},-\frac{1}{2})$ $(\frac{1}{2},-\frac{3}{2})$

$(-\frac{1}{2},\frac{3}{2})$ $(-\frac{1}{2},\frac{1}{2})$ $(-\frac{1}{2},-\frac{1}{2})$ $(-\frac{1}{2},-\frac{3}{2})$

$(-\frac{3}{2},\frac{3}{2})$ $(-\frac{3}{2},\frac{1}{2})$ $(-\frac{3}{2},-\frac{1}{2})$ $(-\frac{3}{2},-\frac{3}{2})$

These can be made into 10 primary symmetric functions and 6 primary anti-symmetric functions by pairing the 12 off-diagonal functions into 6 symmetric and 6 anti-symmetric. For example the two functions $(\frac{3}{2}, \frac{1}{2})$ and $(\frac{1}{2}, \frac{3}{2})$ give

$(\frac{3}{2},\frac{1}{2}) + (\frac{1}{2},\frac{3}{2})$ symmetric

$(\frac{3}{2},\frac{1}{2}) - (\frac{1}{2},\frac{3}{2})$ anti-symmetric

Since the 4 diagonal functions are already symmetric, this gives 4 + 6 possible symmetric functions and 6 possible anti-symmetric functions.

Using the vector model approach, a spin of S has a degeneracy of 2S + 1 corresponding to its possible projections on a given (the z) axis. This is shown in the table to the right. We already know for S = 3 that the spin function

| TOTAL SPIN | DEGENERACY |
|---|---|
| 3 | 7 |
| 2 | 5 |
| 1 | 3 |
| 0 | 1 |

must be symmetric since it represents spins parallel. Thus 7 of the possible 10 symmetric functions must be assigned to S = 3. 3 symmetric functions are left to be assigned to the remaining groups of 5, 3, or 1. Thus S = 1 which has 3 degeneracies is also symmetric and the 6 anti-symmetric functions go with S = 2 and S = 0.

Several students preferred using the exact matrix operator formulation. This gave secular equations from which the 16 correct spin functions were determined in terms of the 16 primary functions. Needless to say, this procedure is rather lengthy and only the results can be given here.

| $S^2 = S(S+1)$ | 12 | 12 | 12 | 12 | 12 | 12 | 12 | 6 | 6 | 6 | 6 | 6 | 2 | 2 | 2 | 0 |
|---|---|---|---|---|---|---|---|---|---|---|---|---|---|---|---|---|
| $S_z$: | 3 | 2 | 1 | 0 | -1 | -2 | -3 | 2 | 1 | 0 | -1 | -2 | 1 | 0 | -1 | 0 |
| SYMMETRY: | + | + | + | + | + | + | + | − | − | − | − | − | + | + | + | − |
| $(\frac{3}{2},\frac{3}{2})$ | 1 | | | | | | | | | | | | | | | |
| $(\frac{3}{2},\frac{1}{2})$ | | 1 | | | | | | 1 | | | | | 1 | | | |
| $(\frac{3}{2},-\frac{1}{2})$ | | | 1 | | | | | | 1 | | | | | 3 | | |
| $(\frac{3}{2},-\frac{3}{2})$ | | | | 1 | | | | | | 1 | | | | | | 1 |
| $(\frac{1}{2},\frac{3}{2})$ | | 1 | | | | | | -1 | | | | | | | | |
| $(\frac{1}{2},\frac{1}{2})$ | | | $\sqrt{3}$ | | | | | | | | | | $-\frac{2}{\sqrt{3}}$ | | | |
| $(\frac{1}{2},-\frac{1}{2})$ | | | | 3 | | | | | | 1 | | | | -1 | | -1 |
| $(\frac{1}{2},-\frac{3}{2})$ | | | | | 1 | | | | | | 1 | | | | 1 | |
| $(-\frac{1}{2},\frac{3}{2})$ | | | 1 | | | | | | -1 | | | | 1 | | | |
| $(-\frac{1}{2},\frac{1}{2})$ | | | | 3 | | | | | | -1 | | | | -1 | | 1 |
| $(-\frac{1}{2},-\frac{1}{2})$ | | | | | $\sqrt{3}$ | | | | | | | | | | $-\frac{2}{\sqrt{3}}$ | |
| $(-\frac{1}{2},-\frac{3}{2})$ | | | | | | 1 | | | | | | 1 | | | | |
| $(-\frac{3}{2},\frac{3}{2})$ | | | | 1 | | | | | | -1 | | | | 3 | | -1 |
| $(-\frac{3}{2},\frac{1}{2})$ | | | | | 1 | | | | | | -1 | | | | 1 | |
| $(-\frac{3}{2},-\frac{1}{2})$ | | | | | | 1 | | | | | | -1 | | | | |
| $(-\frac{3}{2},-\frac{3}{2})$ | | | | | | | 1 | | | | | | | | | |

Each column represents the eigen-function (un-normalized) corresponding to the eigen-values given at the top.

## 3. Coulomb Scattering

This will first be worked out for distinguishable particles. Unfortunately the method of partial waves cannot be applied to any potential of the asymptotic form $U \to 1/r^n$ where $n \leq 1$. This can readily be seen by a rough application of the W.K.B. method to the equation of the zeroth partial wave (VI.8).

$$v_0'' + \frac{2\mu}{\hbar^2}[E - U(r)]v_0 = 0 \qquad U(r) = \frac{e^2 zZ}{r^n}$$

$$v_0 \to \begin{matrix}\sin\\ \cos\end{matrix}\Big\} \int \sqrt{k^2 - \frac{2\mu e^2 zZ}{\hbar^2 r^n}}\, dr$$

$$\sqrt{k^2\left(1-\frac{2\mu e^2 \mathfrak{z} Z}{\hbar^2 k^2 r^n}\right)} = k\left(1-\frac{\mu e^2 \mathfrak{z} Z}{\hbar^2 k^2 r^n} + \cdots\cdots\right)$$

Integrating,

$$\int\sqrt{k^2-\frac{2\mu}{\hbar^2}U}\,dr \approx kr - \frac{\mu e^2 \mathfrak{z} Z}{\hbar^2 k}\int\frac{dr}{r^n}$$

The second term is divergent for $n \leq 1$. For $n > 1$, it will contribute an asymptotic phase angle $\beta_0$. Thus for a coulomb potential ($n = 1$), the asymptotic solution is of the form:

$$v \rightarrow e^{i(kr-\alpha \ln r)} \qquad \text{where } \alpha \equiv \frac{\mathfrak{z} Z e^2}{\hbar v} \qquad (v = \text{velocity})$$

It is seen that the term ln( r) gives a slowly but ever varying phase angle for large r. Thus the partial wave equations must be forgotten and we must return to the complete wave equation. Fortunately the coulomb potential gives a differential equation which can be solved. See Mott and Massey Ch. III or Schiff p. 114. The complete wave equation is

$$\nabla^2\psi + \left(k^2 - \frac{2\mu \mathfrak{z} Z e^2}{\hbar^2 r}\right)\psi = 0$$

It is shown that $\psi = e^{ikz}F(-i\alpha, 1, ik(r-z))$ is a solution

where F is related to the confluent hypergeometric function. It is also shown that this has the asymptotic form which separates the incident wave from the scattered wave:

$$\psi(\underline{r}) = I + Sf(\theta) \qquad \text{VI.16}$$

The incident wave $= I = \left[1 - \frac{\alpha^2}{ik(r-z)}\right]e^{ikz + i\alpha \ln k(r-z)}$

This is not a plane wave due to the distortion by the long range coulomb potential.

$$S = \frac{1}{r}e^{ikr - i\alpha \ln 2kr}$$

$$f(\theta) = \frac{\mathfrak{z} Z e^2}{2\mu v^2}\frac{1}{\sin^2\frac{\theta}{2}}e^{-2i\alpha \ln \sin\frac{\theta}{2} + i\pi + 2i\eta_0} \qquad \text{VI.17}$$

where $e^{i\eta_0} = \dfrac{\Gamma(1+i\alpha)}{|\Gamma(1+i\alpha)|}$

$\sigma(\theta) = \dfrac{r^2|Sf(\theta)|^2}{|I|^2}$ is the differential cross section in the c-m system.

$\sigma(\theta) = \left(\dfrac{\mathfrak{z} Z e^2}{2\mu v^2}\right)^2\dfrac{1}{\sin^4\frac{\theta}{2}}$ which happens to be the classical Rutherford scattering formula for the c-m system.

It is interesting to note that all three methods: the exact quantum mechanical (this one), the Born approximation, and the classical all give exactly the same result for coulomb scattering.

Now the solution VI.16 will be adjusted to meet the situation of two identical particles of spin $\frac{1}{2}$. As in VI.15,

$$\sigma = \tfrac{3}{4}\sigma_3 + \tfrac{1}{4}\sigma_1$$

where the subscript 3 denotes triplet state and 1 denotes singlet state. For the spatial parts of the wave functions in these two cases,

$\psi_3(\underline{r}) = -\psi_3(-\underline{r})$ since the spin part is symmetric

$\psi_1(\underline{r}) = +\psi_1(-\underline{r})$ since the spin part is anti-symmetric

$$\left.\begin{aligned}\psi_3(\underline{r}) &= \psi(\underline{r}) - \psi(-\underline{r})\\ \psi_1(\underline{r}) &= \psi(\underline{r}) + \psi(-\underline{r})\end{aligned}\right\}$$ are both good solutions where $\psi(\underline{r})$ is defined by VI.16.

$$\psi_1(\underline{r}) = I(\underline{r}) + I(-\underline{r}) + S(r)\,[f(\theta) + f(\pi-\theta)]$$

$\therefore\ \sigma_1(\theta) = |f(\theta) + f(\pi-\theta)|^2$ see discussion on p. 123.

$$\sigma_1(\theta) = |f(\theta)|^2 + |f(\pi-\theta)|^2 + 2\,\mathrm{Re}\{f(\theta) f^*(\pi-\theta)\}$$

Likewise,

$$\sigma_3(\theta) = |f(\theta)|^2 + |f(\pi-\theta)|^2 - 2\,\mathrm{Re}\{f(\theta) f^*(\pi-\theta)\}$$

$$\therefore\ \sigma(\theta) = |f(\theta)|^2 + |f(\pi-\theta)|^2 - \mathrm{Re}\{f(\theta) f^*(\pi-\theta)\}$$

$$\sigma(\theta) = \frac{e^4}{M^2V^4}\left\{\frac{1}{\sin^4\frac{\theta}{2}} + \frac{1}{\cos^4\frac{\theta}{2}} - \frac{\cos\left(2\alpha \ln\tan\frac{\theta}{2}\right)}{\sin^2\frac{\theta}{2}\cos^2\frac{\theta}{2}}\right\} \qquad \text{VI.18}$$

using VI.16; $\mu = M/2$ and $\theta$ is still in the center of mass system.

Experimental results check with this relation for low energies. Deviations begin to appear above a few hundred kev. This is the first indication of the effect of the short range nuclear forces on the scattering. The calculation of how the short range nuclear forces modify VI.18 will not be done here (see Bethe D, p. 66-69). The partial wave approach is used again. Let $\beta_0$ be the phase shift between the asymptotic partial wave solution with both the coulomb and nuclear potentials and the asymptotic S component of the expansion of the solution with the coulomb potential only. Then if $\beta_\ell = 0$ for $\ell > 0$, the c-m scattering cross section is

$$\sigma(\theta) = \frac{e^4}{M^2V^4}\left[\frac{1}{\sin^4\frac{\theta}{2}} + \frac{1}{\cos^4\frac{\theta}{2}} - \frac{1}{\sin^2\frac{\theta}{2}\cos^2\frac{\theta}{2}} - \frac{2\hbar V}{e^2}\,\frac{\sin\beta_0\cos\beta_0}{\sin^2\frac{\theta}{2}\cos^2\frac{\theta}{2}} + \left(\frac{2\hbar V}{e^2}\right)^2\sin^2\beta_0\right]$$

except for $\frac{\theta}{2} \approx 0°$ or $90°$

Determination of $\beta_\ell$ for even $\ell$ can give information only of the spin anti-parallel or singlet interaction. The reason is that the spatial part of the wave function is symm. for even $\ell$ ($P_\ell(\cos\theta)$ is symmetric for even $\ell$). Thus the spin part is forced to be anti-symmetric. To obtain any knowledge of the triplet interaction, $\beta_1$ must be determined. High energy scattering data is needed for this. So far the only significant experimental results give $\beta_0$, or information of the singlet interaction only. *

In fact, if a rectangular well is assumed, the data is sufficient to specify the well depth within ± 1% and $r_0$ within ± 5%. This is done by determining which potential well best fits the cross section as a function of the energy. However, these results do not give information on the shape of the potential. The results for the square well are shown in FIG. 16.

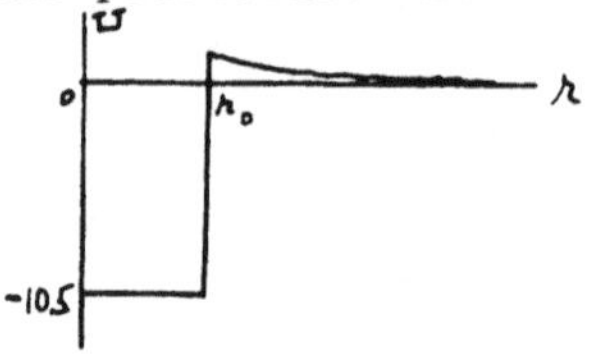

Singlet (P,P) Potential
FIG. VI.16

On page 120 the potential well for (N,P) singlet state (the virtual state of the deuteron) was found to be -11.5 Mev. This is quite close to the -10.5 Mev for the singlet (P,P) interaction, and if the (P,P) well is considered as superimposed on the coulomb $e^2/r$, the agreement is almost exact. This is further evidence for the identity of (P,P) and (N,P) forces.

The highest energy experiments are for 32 Mev done at Berkeley. These unpublished results do not give any significant potential for the triplet interaction. The S wave phase shift was $\beta_0 = 51.2^\circ$. The theoretical curve could best be made to fit the points by making $\beta_1 = 1^\circ$.

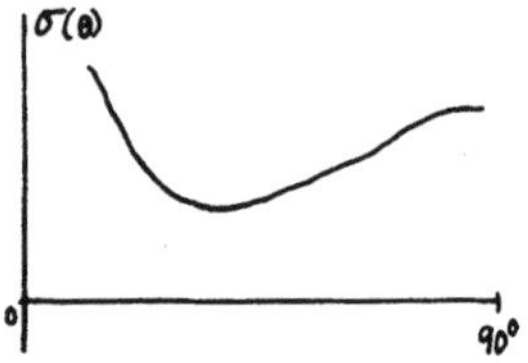

32 Mev (P,P) Scattering in C-M System
FIG. VI.18

Thus our present knowledge of the (P,P) is given by

For $\sigma_1 \cdot \sigma_2 = -3$ $U(r) = \begin{cases} -10.5 & r < r_0 \\ \frac{e^2}{r} & r > r_0 \end{cases}$

For $\sigma_1 \cdot \sigma_2 = 1$ $U(r) = \frac{e^2}{r}$

Problem: Discuss qualitatively how high the energy should be for (P,P) scattering experiments in order for the triplet interaction to show up, if it is to show up at all. The energies must be high enough to give a significant $\beta_1$ without hopelessly complicating the scattering pattern due to higher phase shifts.

The argument given on page 118 shows that $\mu v > l \frac{\hbar}{r_0}$ (see VI.12)

where $\mu$ is the reduced mass and v is the relative velocity after penetrating the coulomb barrier.

$$\frac{M}{2} v > \frac{\hbar}{r_0}$$

$$\frac{M v^2}{2} > \frac{2\hbar^2}{M r_0^2} = 10.5 \text{ Mev}$$

In addition there is the electrostatic barrier to be penetrated which is $\sim \frac{e^2}{r_0} = 0.5 \text{ Mev}$.

Thus $E > 11$ Mev is the final answer.

*Breit, Thaxton, and Eisenbud; Phys. Rev. 55, 1018 (1939), Chamberlain, Segre, and Wiegand, Phys. Rev. 83, 923 (1951).

## E. NEUTRON - NEUTRON FORCES

At present it is impossible to conduct (N,N) scattering experiments due to the low intensities of neutron beams available. The conclusion that (N,N) forces are the same as (N,P) and (P,P) is reached from the symmetrical roles played by the neutron and the proton in the various nuclei. It has been shown on the previous page that (N,P) and (P,P) nuclear forces experimentally are the same for $\ell = 0$. Calculations can be made for the differences in binding energy for mirror image nuclei using $r = 1.5 \times 10^{-13} A^{1/3}$ as the radius to give the electrostatic energy. Assuming the (N,N), (N,P), and (P,P) nuclear interactions to be the same gives agreement with the observed binding energies.

<u>Problem:</u> Design an experimental setup for detecting deviations from pure coulomb scattering of 1 Mev protons.

The article by Breit, Thaxton, and Eisenbud (Phys. Rev. 55, 1018) gives a summary and lists the papers of all such previous work in the 1 Mev range. In the same volume on page 998 is a pertinent paper.

## CHAPTER VII. MESONS

### A. PROPERTIES KNOWN FROM EXPERIMENT

In this section we shall discuss briefly some of the facts known about mesons, and summarize them in a table; except for one of the problems, however, we shall not discuss the experiments behind the facts*.

By mesons we mean unstable particles of mass greater than that of the electron, less than that of the nucleon. The only ones directly observable so far have either a positive or negative fundamental charge.

Mesons were postulated by Yukawa in 1935, and soon thereafter μ-mesons (they will be called "μ's" or muons from here on) were observed as secondary particles in cosmic radiation***. In 1948 π-mesons (π's or pions) were created artificially by bombarding various targets in the Berkeley cyclotron**. During 1949-50, overwhelming evidence has been found for the existence of a neutral pion $\pi^{\circ}$. This is discussed further on p. 237

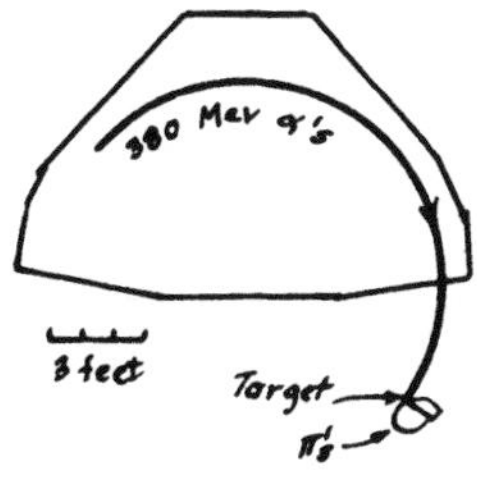


Production of $\pi^{\pm}$'s in a cyclotron.

So far only two sorts of mesons, π and μ, have been identified beyond all doubt, but there are rumors of others.

The names ρ and σ are also used in the literature. This is because the various kinds of meson tracks observed were classified phenomenologically by Powell and his associates according to what was observed at the end of the tracks. This nomenclature is confusing because the number of different kinds of mesons turned out to be less than the number of categories chosen, so that identical mesons may be called by different names.

A ρ meson is one which is observed to stop in the emulsion without producing any observable product. This is a rather time-dependent definition, since more sensitive films are currently being developed. Thus previously unobservable singly-charged relativistic particles (particles travelling at "minimum ionization" -- see Fig. II.4, p. 33) may now be detected.

A σ meson (σ for "star-producing") denotes a meson which produces a nuclear disintegration at the end of its track.

*For nice discussions see "Mesons Old and New" by Keller, Am. Jour. Phys. 17, 356 (Sept. 1949) and a 10-page article by Snyder, Nucleonics 5, 42 (July '49). See also Occhialini and Powell, "Nuclear Physics in Photographs" (1947); and all the references on p. 239 of this book.

**Gardner and Lattes, Phys. Rev. 74, 1236 ('48), Science 107, 270 ('48); Burfening and Lattes, Phys. Rev. 75, 382 ('49).

***Neddermeyer and Anderson, Phys. Rev. 51, 884 ('37), Street and Stevenson, Phys. Rev. 51, 1005 ('37).

The π-Meson (π for "primary" -- for a summary of its properties, see TABLE VII.I):

1. Charged Pions.

The mean life, $\tau \sim 10^{-8}$ sec*, given in the table, applies in the c-m system of the π. Observed in the laboratory system, this time appears dilated by a factor of $\gamma = (1-\beta^2)^{-1/2} = W/M_\pi c^2$. Therefore a $\pi^{\pm}$ formed with an energy of several Bev during a collision of a high energy cosmic ray particle and a nucleus could travel, at a speed approaching that of light, many meters before it decays. In this case it will probably decay at high energy (before it slows down), into a high-energy μ and a neutrino (?).

If a $\pi^{\pm}$ slows down before decaying (or is formed at low energy) then as it slows down to about 10 Mev, its rate of ionization increases slowly to about five times minimum ionization, at which point it becomes visible even in the older nuclear films. The last 10 Mev of its path is about 2500 microns long.

A $\pi^+$, which is repelled by nuclei, simply comes to rest and decays. But a slow $\pi^-$ is attracted****and frequently absorbed by a nucleus, giving up its rest energy and probably boiling off several nucleons. These two sorts of tracks are illustrated in FIG. VII.1.

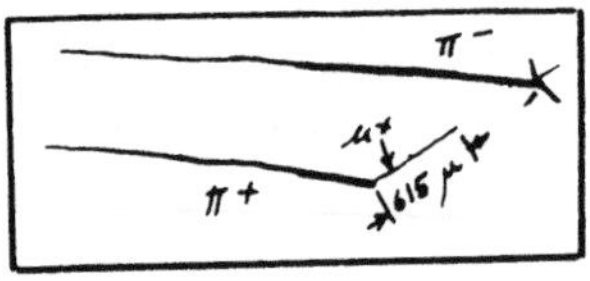


FIG. VII.1 Tracks of π Mesons in Emulsion.

When not captured by a nucleon, a π decays as follows:

$$\pi^{\pm} \xrightarrow{10^{-8}\,sec} \mu^{\pm} + \nu'$$

where $\nu'$ is thought to be a neutrino (we shall refer to it as such). As illustrated in problem 1, p. 138, $M_\nu c^2$ is known to be < 15 Mev.

2. Neutral Pions -- see p. 237.

The μ-Meson (again, see TABLE VII.I for mass, etc.):

If the decay reaction mentioned just above takes place while the π is at rest, the μ has a kinetic energy of 4.1 Mev and travels almost exactly 615 microns in Ilford emulsions. Of course, most of the μ's in cosmic radiation are formed when π's decay at high energy, so their range is >> 615 microns.**

On most film the end of the path looks blank, but with cloud chambers, g-m tubes, or minimum ionization film, it has been determined that, when there are no heavy nuclei around, one of the products of the μ-disintegration is an electron which may have one of several energies and is thought to have a continuous spectrum from 9 to 55 Mev***. No other particles have been detected during the reaction, so that the most logical guess is

$$\mu^{\pm} \xrightarrow{2.15\,\mu sec} e^{\pm} + 2\nu \quad (\textit{i.e. at least } 2\nu) \quad \text{in vacuo}.$$

The electron is so light compared to the μ that, on the average, we can think of the energy as being essentially divided equally among the three particles, all extremely relativistic.

*Richardson, Phys. Rev. 74, 1720 ('48)

**An energy spectrum of cosmic ray μ's is given in FIG. X.5, p.220.

***Steinberger, Phys. Rev. 75, 1136 ('49) and Leighton, Anderson, Seriff, Phys. Rev. 75, 1432 ('49): Current data is inadequate for differentiation between several discreet energies (as for α's) and a continuum.

****See the discussion at the top of p. 133.

The qualification "in vacuo" concerning the mean life of $\mu$'s is needed because, in matter, a positive meson is repelled by nuclei, but a negative meson may fall into a stable Bohr orbit just as an electron does***. The orbits have radius and energy

$$r = \frac{(n\hbar)^2}{MZe^2} \qquad\qquad E = -\frac{M(Ze^2)^2}{2(n\hbar)^2}$$

where M is the reduced mass. The $\mu$ orbit is smaller than the corresponding electronic orbit by a **factor of 216**, and the binding energy is increased by the same factor. For heavy elements, the smallest orbit is only slightly larger than the nucleus itself, so that the $\mu$ spends a large fraction of its time inside the nucleus. If the $\mu$ interacted strongly with the nucleons (as a $\pi$ does) it would be immediately captured by the nucleus, but we find that the interaction is very weak. Ticho * gives a curve showing that $\tau_\mu$-drops from 2.15 $\mu$sec for Z = 1 to 0.7 $\mu$sec for Z = 16, where the capture probability has started to compete seriously with the natural decay. Remember that these mean lives apply to the c-m system of the meson.

TABLE VII.I summarizes the material discussed in this section:

| | Electron Masses | $Mc^2$ | Probable Spin | MEAN Life in Vacuo | Interaction with Nuclei |
|---|---|---|---|---|---|
| $\pi^\pm$ | 277.4±1.1<br>276.1±1.3 | 141 Mev | 0 or 1 | $2.8 \pm .6 \times 10^{-8}$ sec. | Strong, → exchange forces |
| $\mu^\pm$ | 210 ± 4 | 107 Mev | ½ | 2.15 $\mu$sec | Weak, → exchange forces |
| $\pi^0$ | ≈276 - 6<br>see p.237 | 138 Mev | 0 | $<10^{-13}$ sec | Strong, → ordinary forces |

| | Path Length in Emulsion | Decay Products |
|---|---|---|
| $\pi^\pm$ | Non-relativistic ~2500$\mu$. (see text) | $\pi^-$ usually → star in film<br>$\pi^+ \longrightarrow \mu^+$ (4.1 Mev) + $\nu$ |
| $\mu^\pm$ | 615 $\mu$ ($\mu \equiv$ micron) | $e^\pm$ (<55 Mev) + 2$\nu$ (?) |
| $\pi^0$ | not observable | 2 photons |

TABLE VII.I Mesons.**

## B. MESON THEORY

From electrostatics we know that two particles attract or repel one another according to Coulomb's law. For a classical treatment we say that this force arises from the potential field $\phi = e/r$ of one of the particles. However if we wish to take into account the corpuscular nature of light, we can describe this interaction by saying that one particle "emits" a photon which is subsequently absorbed by the other.

Analogously, the interaction of two nucleons can be partially interpreted by the picture of one nucleon "emitting" a quantum which is promptly absorbed by the second nucleon. These quanta are called mesons, and we shall call them $\pi$-mesons in this discussion. The reason for this nomenclature is that we know experimentally that nucleons interact

*Phys. Rev. 74, 1337 ('48)

**This footnote has been expanded and put on p. 237

***Fermi and Teller, Phys. Rev. 72, 399 ('47), J.A. Wheeler, Rev. Mod.Phys. 21, 133 ('49)

more strongly with $\pi$'s than with $\mu$'s. If we are going to attribute nuclear forces to one sort of meson, we might as well call it a $\pi$.

If the $\pi$ is uncharged then we can write the "reaction"

$$N_1 \rightarrow N_1' + \pi^0, \qquad \pi^0 + P_2 \rightarrow P_2'$$

The charge of the individual nucleons (they may be similar or different) undergoes no change during the "reaction", and it turns out that nuclear forces arising from $\pi^0$'s are of the non-exchange, or ordinary type. On the other hand, if the $\pi$ is charged we have either

$$N_1 \rightarrow P_1 + \pi^-, \qquad \pi^- + P_2 \rightarrow N_2 ;$$

or

$$P_1 \rightarrow N_1 + \pi^+, \qquad \pi^+ + N_2 \rightarrow P_2 .$$

In this case it turns out that the $\pi^\pm$ produces an exchange force.

Since there is evidence that nuclear forces are a mixture of both exchange and ordinary forces, an acceptable theory will probably have to involve both charged and neutral mesons.

Yukawa introduced the meson in 1935, and found that he had to assign it a rest mass of 100-200 m in order to fit the experimental data on the range of nuclear forces.

Fields whose quanta have zero rest mass are long-range; those with quanta of finite mass decrease exponentially. We can illustrate this statement classically as follows:

The potential field of a single electric charge <u>fixed</u> at the origin,

$$\rho = e\,\delta(\mathbf{r}) \qquad \text{VII.1}'$$

obeys Laplace's equation,

$$\nabla^2\phi - \frac{1}{c^2}\ddot{\phi} = -4\pi e\,\delta(\mathbf{r}) \qquad \text{VII.2}'$$

and

$$\phi_{static} = \frac{e}{r}. \qquad \text{VII.3}'$$

If there is a second charge e at a distance r, the interaction energy

$$U = e\phi = e^2/r. \qquad \text{VII.4}'$$

A scalar neutral meson field generated by a nucleon of strength g, at the origin

$$\rho = g\,\delta(\mathbf{r}) \qquad \text{VII.1}$$

obeys the Klein-Gordon equation*

$$\nabla^2\phi - K^2\phi - \frac{1}{c^2}\ddot{\phi} = -4\pi g\,\delta(\mathbf{r}) \qquad \text{VII.2}$$

and

$$\phi_{static} = g\frac{e^{-Kr}}{r} \qquad \text{VII.3}$$

---

*The Klein-Gordon equation may be obtained directly by substituting the operators

$$E = -\frac{\hbar}{i}\frac{\partial}{\partial t}, \qquad p = \frac{\hbar}{i}\nabla$$

into the equation for total relativistic energy

$$W^2 = M^2c^4 + p^2c^2$$

where M is the rest mass and p the momentum of the meson.

$$-\hbar^2\frac{\partial^2}{\partial t^2} - M^2c^4 + \hbar^2\nabla^2c^2 = 0$$

If we place a second nucleon at $\underline{r}$, it may be shown that**

$$U = -g\phi = -g^2 \frac{e^{-\kappa r}}{r} \qquad \text{VII.4}$$

The "range" of $\phi$ is $1/\kappa = \lambda\!\!\!^{-}_{\text{compton}} \frac{m}{M} = 3.86 \times 10^{-11} \frac{m}{M}$ cm.***

$\phi$ is the field variable (or one of its components) and must not be confused with the Schrödinger wave function. In the electromagnetic case, for example, the field variable may be a 4-vector (the electromagnetic 4-potential) or two 3-vectors ($\mathcal{E}$ and $\mathcal{H}$), depending upon one's point of view. In the simplest case (the first that one would try for a meson field) $\phi$ is simply a scalar or a pseudo-scalar. A scalar does not change sign on inversion of space; a pseudo-scalar does. When $\phi$ is a $\left\{\begin{matrix}\text{scalar}\\ \text{pseudo-scalar}\end{matrix}\right\}$ then the non-homogeneous right-hand side of VII.2 is also a $\left\{\begin{matrix}\text{scalar}\\ \text{pseudo-scalar}\end{matrix}\right\}$.

The potential of VII.4 serves only as an example and could not adequately explain nuclear forces, since it is not spin dependent. Attempts have been made to employ more complicated interactions and to introduce vector and tensor fields. The various couplings all give fields of the general form

---

Footnote continued from p. 134:
Now introduce a function $\phi(\underline{r},t)$ which has here the significance of a potential and which we shall call the field variable

$$\left(\nabla^2 - \frac{1}{c^2}\frac{\partial^2}{\partial t^2} - \frac{M^2c^2}{\hbar^2}\right)\phi = 0$$

We shall call

$$\frac{\hbar}{Mc} = \kappa^{-1} = \lambda\!\!\!^{-}_{\text{compton}} \frac{m}{M}$$

and then get VII.2, and if $M = 0$ we get VII.2', for $\underline{r} \neq 0$.

Since the nuclear velocity is low, the main features of the problem show up in the time-independent equation

$$(\nabla^2 - \kappa^2)\phi = -4\pi g\,\delta(\underline{r}) \qquad \text{VII.5}$$

Let $\phi = \frac{u}{r}$; then, for $\underline{r} \neq 0$,

$$\left(\frac{d^2}{dr^2} - \kappa^2\right)u = 0$$

$$\phi = \frac{u}{r} = \text{const}\,\frac{e^{\pm\kappa r}}{r}$$

The constant is easily shown to be g by integrating both sides of VII.5 over a small region including the origin, and then equating the results. In this small region $\exp(\kappa r) \to 1$, so we have complete analogy with the electrostatic case. We discard the positive exponential case to get a localized field.

**The analogy is between the meson field $\phi$, and the <u>components</u> of the electromagnetic 4-potential. The $\phi_{em}$ of VII.3' is only a factor in the 4th component $i\phi$ of this 4-vector. Therefore there is a difference of a factor of $i^2 = -1$ in the signs of the potential energies VII.4' and VII.4.

*** $3.86 \times 10^{-11}$ $(m/M_{\pi^\pm}) = 1.40 \times 10^{-13}$ cm.

$$e^{-\kappa r}\left[\frac{1}{r}, \frac{1}{r^2}, \frac{1}{r^3}, \cdots\right]$$

with some directional terms.

Unfortunately it is impossible to find a solution for Schrödinger's equation when the potential diverges faster than $1/r^2$ at the origin. Where $1/r^3$ terms appear, the field must be arbitrarily cut off in a finite volume, but this makes it impossible to formulate the problem in a relativistically invariant way.

Because of these difficulties there are as yet no self-consistent results from meson theory.

In order to point out another important difficulty in meson theory, we must now discuss, exceedingly briefly, the quantum-mechanical formulation of the problem. As an example we shall take one of the "reactions" postulated on p. 134. We have illus-

$$N_1 \rightarrow P_1 + \pi^- \qquad (A \rightarrow C)$$
$$\pi^- + P_2 \rightarrow N_2 \qquad (C \rightarrow B)$$

trated in the sketch at left that the intermediate state (C) is energetically impossible for nucleons at rest, since it "costs" 145 Mev to create a $\pi^-$. In quantum mechanical perturbation theory, however, states with energies above or below that of the system are important as intermediate, or virtual, states. We shall make extensive use of intermediate states (for example in Ch. VIII to derive the Breit-Wigner formula). Since the mean life of the intermediate state is short ($\tau \sim \hbar/\Delta E$ by the uncertainty principle) there is no violation of conservation of energy.

The transition probability and energy perturbation can be calculated with the help of perturbation theory (i.e., there is no better way known). Since the direct matrix element coupling the initial and final states is assumed to be zero, we use "Golden Rule #1" for the second order transition:

$$H'_{BA} = \sum_C \frac{H'_{BC} H'_{CA}}{E_A - E_C}$$

Now we can point out the difficulty. It turns out that only the first non-vanishing matrix element (in this example the second-order one) is finite, but that the higher order elements are sums that are not negligible- in fact they diverge. The divergences in the corresponding terms in the electromagnetic case can be removed relativistically by the recent advances in quantum mechanics, but the way out of the difficulty has not been found in meson theory.

Even if the divergences of the individual higher-order transitions could be removed there is another difficulty. Perturbation theory applied to the electromagnetic case gives an expansion of successive orders of the interaction Hamiltonian in powers of $(e^2/\hbar c) = 1/137$. This parameter is quite small, so that there is hope that the whole series will converge. But meson perturbation theory is an expansion in powers of $(g^2/\hbar c)$. This cannot be made smaller than about 1/5 if the theory is to give the right magnitude of nuclear forces. There is considerably less hope that the entire series will converge, even if the individual terms can be made finite.

Meson Theory and Beta Decay. By writing in sequence the reactions

$$N \rightarrow P + (\text{Meson})^- ; \quad (\text{Meson})^- \rightarrow e^- + n\nu \quad [n = 1, 2, 3 ?]$$

or

$$P \rightarrow N + (\text{Meson})^+ , \quad \text{etc.}$$

Yukawa hoped to explain $\beta$-decay. Now that it is known that there are two sorts of mesons (maybe more), only one of which decays into an electron

$$\mu^\pm \rightarrow e^\pm + 2\nu$$

it is difficult to reconcile $\beta$-decay with the known meson mean lives in a quantitative way.

Summary A great deal of attention has been given to meson theories, from which has come relatively little quantitative results. Qualitatively, however the theory is valuable. Thus physicists predicted the creation of mesons during high-energy collisions before mesons had ever been observed. Meson theory was of considerable weight in the decision to build the large synchro-cyclotrons. Another example of the qualitative application of meson theory is the discussion in Ch. I (p. 14) where we obtain a numerically wrong but qualitatively useful value for the magnetic moment of the deuteron by assuming that part of the time

$$P \rightarrow N + \pi^+$$
$$N \rightarrow P + \pi^-$$

The formalism of meson theory may be greatly modified or abandoned, but the fundamental ideas are likely to survive.

REFERENCES for further reading on meson theory.

Bethe D, Ch.XV
Heisenberg, W., "Cosmic Radiation," 1943. Chap. 10, by C. v. Weizsäcker, reviews the theory of the meson.
Jànossy, L., "Cosmic Rays," 1948
Pauli, L., "Meson Theory of Nuclear Forces," 1948
Rev. Mod. Phys. 13, 203 ('41) "Elementary Field Theory of Elementary Particles."
Rosenfeld, L. "Nuclear Forces," 1947
Primakoff, Nucleonics 4 (2, Jan. '48)
Wentzel, G., "Recent Advances in Meson Theory," Rev. Mod. Phys. 19, 1 ('47)
Yukawa, H., Proc. Math. Phys. Soc. Japan 17, 48 ('35)
"Models and Methods in Meson Theory," Rev. Mod. Phys. 21, 474 ('49)
Wentzel, G., "Quantum Theory of Fields," 1949

PROBLEM 1 How would you determine a range-energy relationship for mesons in emulsion? Outline a method for determining an upper limit for the mass of the neutrino (?) in the reaction

$$\pi^{\pm} \xrightarrow{10^{-8}\ sec} \mu^{\pm} + \nu(?)$$

Solution (Most of this work was done by Lattes, Occhialini, Powell ... , See Nature 160, 453 ('47) and later publications.)

It seems reasonable to assume that the total number of grains in a track is proportional to the energy of the particle producing the track. The constant of proportionality can be determined by counting the grains due to $\alpha$'s or protons of known energy. By utilizing both $\alpha$'s and protons, one can determine whether the constant is actually a constant (it is). Then, by counting grains on $\mu$ tracks of various ranges, one arrives at a range-energy relation.

Once we know the kinetic energy of the $\mu$, all that remains is to determine the mass of the $\pi$ and the $\mu$. This can be done in one of at least two distinct ways:

1. Measure $\mathcal{H}\rho$ in a cyclotron. For motion perpendicular to $\mathcal{H}$ we can write

$$\frac{pc}{e} = \mathcal{H}\rho \quad (\rho = \text{radius of curvature}, \ p = M\gamma V)$$

Knowing the energy (from grain counting) and the momentum of a particle, we can solve for its mass*.

2. Make use of the fact that the ionization depends only upon the velocity, not upon the mass, of a particle:

$$-\frac{dE}{dx} = f_1(V)$$

Then the range, R $= \int_0^R dx = \int_{E_0}^0 - \frac{dE}{f_1(V)}$

$$\text{But } dE = V dp = V d(M\gamma V) = M V d(\gamma V) \equiv -M f_2(V) dV \qquad \text{VII. 6}$$

$$\text{So } R = M \int_{V_0}^0 \frac{f_2(V)}{f_1(V)} dV = M f_3(V_0)$$

Or we can write $V_0 = f_4\left(\frac{R}{M}\right)$.

Integrating VII.6 we get $E_0/M = f_5(V_0)$

Combining the last two statements, we have $\frac{E_0}{M} = f_6\left(\frac{R}{M}\right)$ VII. 7

Suppose that we have a pair of range-energy curves, one for a particle of known mass, one for a particle of unknown mass (and suppose, incidentally, that we were smart enough to plot these curves on log paper in the first place). Can we use VII.7 to determine the ratio of the masses? We can; the reasoning is as follows.

There must be an infinite number of pairs of points, one point on each curve, such that $\frac{R_1}{M_1} = \frac{R_2}{M_2}$. For these pairs $\frac{E_{01}}{M_1} = \frac{E_{02}}{M_2}$. In other words, for these pairs, we have two simultaneous equations:

$$\frac{R_1}{M_1} = \frac{R_2}{M_2}; \qquad \frac{E_{01}}{M_1} = \frac{E_{02}}{M_2}$$

*For more details, see Gardener and Lattes, Phys. Rev. 75, 1468 ('49)

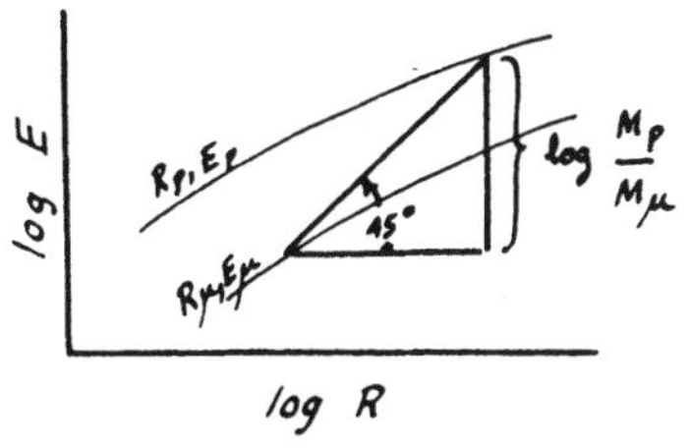


FIG. VII.2 Range-energy curves.

So $\frac{R_1}{E_{01}} = \frac{R_2}{E_{02}}$

But this is precisely the relation between all points lying on any line with unit slope on a log E vs. log R plot (see FIG. VII.2

The reader can easily check that $\frac{M_2}{M_1}$ is given by the line indicated in the figure.

The ratio $M_\pi/M_\mu$ can be found in a similar fashion. In fact, both of these methods will give the ratio $M_\pi/M_\mu$ (= 1.32 ± 0.01) with an accuracy considerably greater than that of the absolute masses.

Determination of the neutrino rest energy

We now have enough information to solve the equation for the conservation of (energy + mass).

$$W_\pi = W_\mu + W_\nu . \qquad \begin{cases} W_\mu = M_\mu c^2 + T_\mu , & \text{both known} \\ W_\nu & \text{, unknown} \end{cases}$$

$$= W_\mu + \sqrt{M_\nu^2 c^4 + p_\nu^2 c^2}$$

$$= W_\mu + \sqrt{M_\nu^2 c^4 + p_\mu^2 c^2} \quad \text{by conservation of momentum,}$$

$$= W_\mu + \sqrt{M_\nu^2 c^4 + W_\mu^2 - M_\mu^2 c^4}$$

$M_\nu c^2 < 15$ Mev, using the values and probable errors given in Table VII.I, p.133.

***************

PROBLEM 2 Assume that a neutron and a proton have an interaction

$$U = \pm g^2 \frac{e^{-\kappa r}}{r} ;$$

calculate the differential and total scattering cross-sections using the Born approximation.

Solution We shall use the notation of Schiff (26.18)

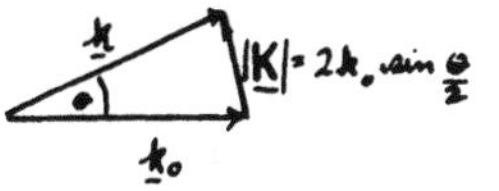


$$f(\theta,\phi) = -\frac{2\mu}{4\pi\hbar^2}\int U(r)\, e^{i\mathbf{K}\cdot\mathbf{r}}\, d\tau ; \qquad \mu = \textit{reduced mass.}$$

$$\sigma(\theta,\phi) = |f(\theta,\phi)|^2$$

Since $\sigma$ involves a square modulus, we can forget about the sign of the interaction. We have cylindrical symmetry, so $d\tau \equiv 2\pi r^2 dr \sin\theta' d\theta'$. ($\theta' \neq \theta$).

$$f(\theta) = 2\pi \frac{2\mu g^2}{4\pi\hbar^2}\int_0^\infty \frac{e^{-\kappa r}}{r} r^2\, dr \int_{-1}^{+1} e^{iKr\cos\theta'}\, d(\cos\theta')$$

$$= \frac{\mu g^2}{\hbar^2}\int_0^\infty \frac{e^{-\kappa r}}{r} r^2\, \frac{2\sin Kr}{Kr}\, dr$$

$$= \frac{2\mu g^2}{\hbar^2 K}\frac{1}{2i}\int_0^\infty \left[ e^{(-\kappa + iK)r} - e^{(-\kappa - iK)r}\right] dr$$

$$= \frac{2\mu g^2}{\hbar^2}\frac{1}{K^2 + \kappa^2}$$

*FOOTNOTE: since the non-minimum-ionization part of the path is traversed by the $\pi$ in $< 10^{-10}$ sec ($\ll \tau_\pi$), $\pi$'s at the end of such paths will almost surely be at rest.

$$f(\theta) = \frac{2\mu g^2}{\hbar^2} \frac{1}{4k^2 \sin^2 \frac{\theta}{2} + K^2}$$ Now set $k = \frac{\mu V}{\hbar}$, then

$$= \frac{1}{2} \frac{g^2}{\mu V^2} \frac{1}{\frac{K^2}{4k^2} + \sin^2 \frac{\theta}{2}}$$

$$\sigma(\theta) = |f(\theta)|^2 = \left(\frac{g^2}{2\mu V^2}\right)^2 \frac{1}{\left[\frac{K^2}{4k^2} + \sin^2 \frac{\theta}{2}\right]^2}$$

For $\frac{K}{k} \gg 1$ (low-energy collisions), we have isotropic scattering

$$\sigma_{low-E} = \left(\frac{2\mu g^2}{K^2 \hbar^2}\right)^2$$

while, for the other extreme,

$$\sigma_{hi-E} = \left(\frac{g^2}{2\mu V^2}\right)^2 \frac{1}{\sin^4 \theta/2}$$

is identical with the Rutherford formula for Coulomb scattering except that $g^2$ replaces $Z_1 Z_2$.

For the intermediate case, there is a straight-forward integration over $\Theta$, the result of which is

$$\sigma = 4\pi \left(\frac{2\mu g^2}{K^2 \hbar^2}\right)^2 \frac{1}{1 + 4\frac{k^2}{K^2}}$$

PROBLEM 3 Plan an experiment to observe the $\pi$'s produced by a 400 Mev proton beam in a cyclotron.

References: will be found on p. 131, footnote.

REFERENCES on Meson Theory are found on page 137

# CHAPTER VIII NUCLEAR REACTIONS

## A. Notation

The nuclear reaction $A + \alpha \longrightarrow B + p + Q$ is symbolized by

$$A(\alpha,p)B$$

Particles are symbolized by: $\alpha$ alpha, p proton, d deuteron, $\gamma$ gamma ray, and f for fission.

Q is (+) for an "exothermic" reaction, (-) for "endothermic".

The threshold is the minimum energy of the bombarding particle in order for the reaction to occur. Threshold is measured in the laboratory system, and therefore is not necessarily equal in magnitude to Q. If Q is positive, the threshold is, in principle, 0. If Q is negative, and if the bombarded particle A is approximately at rest, then (see Ch. I, page 5)

$$\text{Threshold energy} = (-Q) \times \frac{\text{Mass of incident particle}}{\text{Reduced mass of system}}$$

$$= (-Q) \times \frac{M_\alpha + M_A}{M_A} \qquad \text{VIII.1}$$

for the reaction symbolized above.

## B. General Features of Cross-sections for Nuclear Reactions.

The following considerations apply to cross-sections for nuclear reactions in the absence of resonances. Resonance phenomena are discussed in section D.

Consider the transition $A + a \longrightarrow B + b + Q$, where the nucleus "A" and the particle "a" become the nucleus "B" and particle "b" Both the initial and final states of the system consist of a pair of unbound particles; therefore the transitions is to one of a continuous distribution of states. The initial state also has a continuous range of possible energies, but the experiment itself specifies a particular initial energy.

There are similar situations in atomic physics. For example, in emission of a photon by an excited atom, the transition is from a single state to one of a continuum of states: Conservation of energy selects the final state.

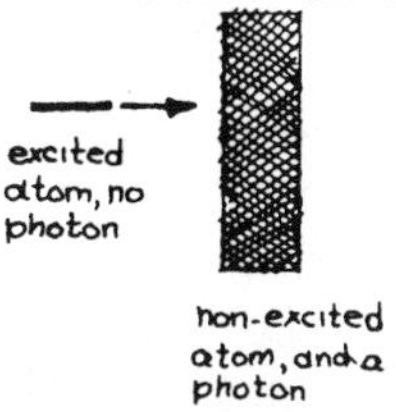


Another atomic example is the non-radiative or Auger transition. An excited atom may have two possible modes of decay. In addition to photon emission, the atom may decay by emission of an electron. Suppose, for example, the excitation corresponds to one missing electron in the K shell. The energy made available when an electron falls into this hole may be greater than the ionization energy, in which case an electron may be emitted from the atom. Again the final system consists of two unbound particles having a continuous range of possible energies.

Returning to the nuclear reaction $A + a \longrightarrow B + b$, we use a general principle of quantum mechanics to derive some essentially statistical results on the variation of the cross-section.

From quantum mechanics, the probability per unit time of

transition = number of transitions per unit time = w is given by "Golden Rule No. 2": *

$$w = \frac{2\pi}{\hbar} |\mathcal{H}|^2 \frac{dn}{dE} \qquad \text{VIII.2}$$

where $\mathcal{H}$ is the matrix element of the perturbation causing the transition, and dn/dE = energy density of final states, counting each degenerate state separately.

$|\mathcal{H}|^2$ may be the same for all energetically possible final states; more often it depends on the state. (For instance, $|\mathcal{H}|^2$ may depend on the direction of emission.) Then $|\mathcal{H}|^2$ in the formula is a suitable average over the possible final states.****

dn/dE = $\infty$ for a continuum of states. But in that case $|\mathcal{H}| \longrightarrow 0$, so that the expression $|\mathcal{H}|^2$ dn/dE has the indeterminate form 0 x $\infty$. This difficulty is removed by limiting space to a box of volume $\Omega$. $|\mathcal{H}|$ is then small but finite and dn/dE large but finite. $\Omega$ drops out of the result. The number of final states equals the number of states of the emitted particle. This is because a change in momentum of one particle compels a change in momentum of the other, by conservation of linear and angular momentum of the system.

It was shown in Chapter IV, p. 76 that the number of states available to a free particle, "b", with momentum between p and p + dp, confined to a box of volume $\Omega$, is

$$dn = \frac{4\pi p_b^2 \, dp_b \, \Omega}{(2\pi\hbar)^3} \qquad \text{VIII.3}$$

This must be multiplied by the multiplicity in the final state** caused by spin orientation, which is given by the factor $(2I_b+1)$x $(2I_B+1)$, where $I_b$ is the spin of the emitted particle and $I_B$ the spin of the nucleus. If b is a photon, $(2I_b+1)$ is put equal to two.***

$$dE = v_b \, dp_b \quad \text{(true relativistically)} \qquad \text{VIII.4}$$

where $p_b$ and $v_b$ are the momentum and velocity in the center of mass frame of reference of the final (B+b) state. Since "B" is usually massive compared with "b", $p_b$ and $v_b$ can usually be measured in the laboratory frame. Combining these two equations:

$$\frac{dn}{dE} = \frac{4\pi p_b^2 \, \Omega}{(2\pi\hbar)^3 v_b} (2I_b+1)(2I_B+1) \qquad \text{VIII.5}$$

From this and VIII.2 we get

$$\text{No. transitions per unit time} = \frac{1}{\pi\hbar^4} \frac{p_b^2}{v_b} \Omega |\mathcal{H}|^2 (2I_b+1)(2I_B+1) \qquad \text{VIII.6}$$

The following equation is essentially a definition of the cross-section $\sigma_{A \to B}$ per A nucleus:

---

* Derived in Schiff, Quantum Mechanics, p. 193. ("Golden Rule No. is on page 148 of this text).

** This is discussed in greater detail in section C, this chapter.

*** This point is discussed by Bethe and Placzek, Phys.Rev. 51 450, Appendix, p. 483. Multiplicity is caused by the two possible independent polarizations.

**** See page 214 for more complete discussion.

$$\text{No. transitions/sec per "A" nucleus} = n_a \times v_{\text{"a" rel. to "A"}} \times \sigma_{A\to B} \qquad \text{VIII.7}$$

where A and B refer to the (A+a) and (B+b) states respectively, and $n_a$ is the density of particles "a". Take $n_a$ to be $1/\Omega$ $\text{cm}^{-3}$ (one particle in the volume). Then

$$\frac{1}{\Omega} \times v_{\text{"a" rel. to "A"}} \times \sigma_{A\to B} = \frac{1}{\pi\hbar^4}\frac{p_b^2}{v_b}\Omega|\mathcal{H}|^2(2I_b+1)(2I_B+1) \qquad \text{VIII.8}$$

Since nucleus "A" is often massive compared to "a", $v_{\text{"a" rel. to "A"}}$ is often nearly equal to $v_a$ in the center of mass frame. In any case, these two velocity magnitudes are related by a constant factor. Writing $v_{\text{"a" rel. to "A"}} = v_a$,

$$\boxed{\sigma_{A\to B} = \frac{1}{\pi\hbar^4}|\Omega\mathcal{H}|^2\frac{p_b^2}{v_a v_b}(2I_b+1)(2I_B+1)} \qquad \text{VIII.9}$$

In general, $\mathcal{H}$ is unknown. It has the form $\int d\tau\, \psi^*_{final}\, U\, \psi_{initial}$ where U is the interaction energy. If the wave functions used to compute $\mathcal{H}$ are normalized in volume $\Omega$, $\Omega$ disappears from the expression $|\Omega\mathcal{H}|$ in VIII.9. This is seen as follows: Let $\Psi$ have the form, at large distances, N exp(ikz). Then $\int|\psi|^2 d\tau = N^2\Omega$ Setting $N^2\Omega = 1$, we get $N = 1/\sqrt{\Omega}$

If $\Psi_{initial}$ and $\Psi_{final}$ now mean the un-normalized plane wave functions, the matrix element factor in VIII.9 becomes

$$\Omega\mathcal{H} = \int d\tau\, \Psi^*_{final}\, U\, \Psi_{inital} \qquad \text{VIII.10}$$

(This may be looked upon as taking $\Omega = 1$.) Henceforth we use $\mathcal{H}$ for $\Omega\mathcal{H}$. In order to show the meaning of this expression, we write it as

$$|\mathcal{H}| = \bar{U} \times \text{Volume of nucleus} \times \overline{|\Psi_{initial}\Psi_{final}|} \qquad \text{VIII.11}$$

where $\overline{|\Psi_{in.}\Psi_{fin.}|}$ is a suitable average of the product of the wave functions over the volume of the nucleus. U, and hence the integrand, is zero outside the nucleus. $\bar{U}$ = average interaction energy $\approx$ depth of potential well. For our purposes here the important feature of VIII.11 is its dependence on the charge of the participating particles. If "a", say, is positively charged, its wave function will be reduced in <u>amplitude</u> at the nucleus by the barrier factor $\exp(-G_a/2)$, where, by III.3, p. 58.

$$\frac{G_a}{2} = \sqrt{\frac{2M_a}{\hbar^2}}\int\sqrt{U_a - E_a}\, dr \longrightarrow \approx \frac{\pi Z_A\, z_a e^2}{\hbar v_a} \text{ for high barriers} \qquad \text{VIII.12}$$

$U_a$ denotes the charge of "a" times the Coulomb potential of "A". Physically this factor represents Coulomb repulsion. The wave function of an outgoing particle at the nucleus is also reduced by such a barrier factor. The result for the <u>squared</u> matrix element is:

For neutral particles: $|\mathcal{H}|^2 \propto (\bar{U} \times \text{Vol. of nucleus})^2$ VIII.13

For + charged particles: $|\mathcal{H}|^2 \propto (\bar{U} \times \text{Vol.})^2 \times \exp(-G_a - G_b)$

(emission of negatively charged particles (electrons)is treated in Ch. IV)

For endothermic reactions there is a threshold energy for the bombarding particle. For exothermic reactions in which the energy liberated is much larger than the energy of the bombarding particl there are two simplifications in equation VIII.9: 1) the barrier factor $\exp(-G_b)$ for the outgoing particle is almost constant because it is a function of energy of the emitted particle "b", which is almost constant; 2) $p_b$ and $v_b$ are almost constant and therefore the statistical weight factor in VIII.9, $p_b^2/v_a v_b$, is proportional to $1/v_a$.

These results are now applied to specific cases to deduce the general features of the $\sigma$ vs. energy and $\sigma$ vs. velocity curves.

1) **ELASTIC** (n,n) (both particles uncharged)
$v_a = v_b$, therefore $p_b^2/v_a v_b = (M_{neut})^2$, a constant
At low energy $|\mathcal{H}|$ is approximately constant, therefore

$\sigma \approx$ constant at low energy.

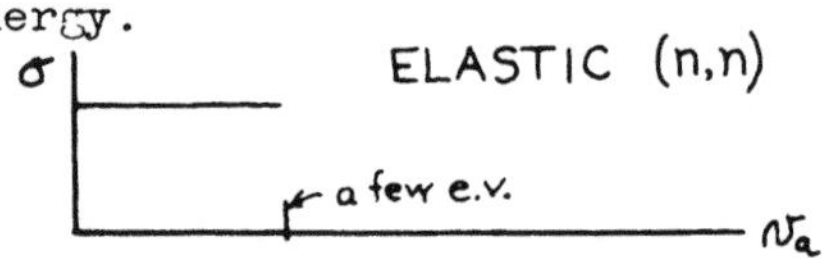


2) **EXOTHERMIC**, low energy **UNCHARGED** bombarding particle, as in (n,α), (n,p), (n,γ), (n,f). **Q** is usually ~ Mev. while neutron energy is ~ e.v., therefore $v_b \approx$ constant. Therefore $p_b^2/v_a v_b \approx 1/v_a$. $|\mathcal{H}|^2 \propto \exp(-G_n-G_b)$. $\exp(-G_b)$ is $\approx$ constant, since it depends on the almost constant energy of the outgoing particle, or, in the case of an uncharged "b", is 1 exactly. Also $\exp(-G_n) = 1$. Therefore
$\sigma \sim 1/v_n$ (the "1/v" law)

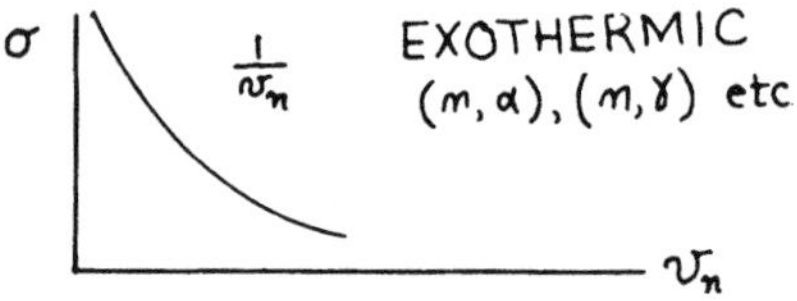


3) **INELASTIC** (n,n')
The nucleus is left in an excited state. The process is endothermic and **-Q** is the excitation energy of the nucleus. For incident neutron energies slightly above the threshold, $v_n \approx$ constant, since the fractional change in incident energy is small. But $v_{n'}$ changes relatively greatly in this region: $v_{n'}^2 \propto$ excess of energy above the threshold. Therefore $p_{n'}^2/v_n v_{n'} \propto v_{n'} \propto \sqrt{\text{energy excess}}$. Therefore near the threshold $\sigma \propto \sqrt{\text{energy excess}}$.

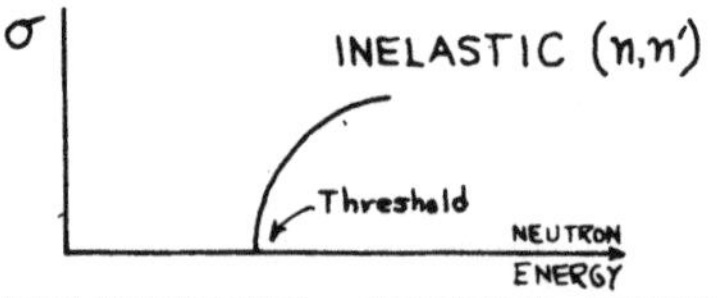


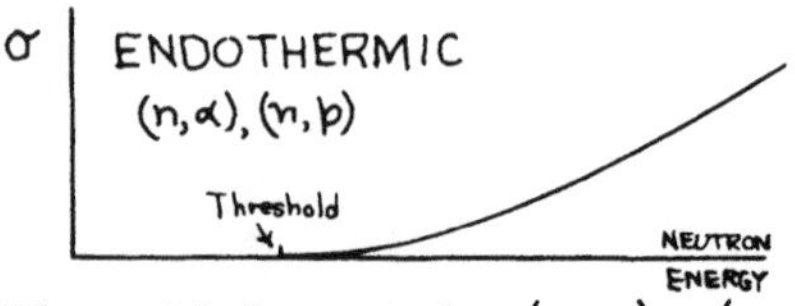


4) **ENDOTHERMIC, CHARGED OUTGOING** particle, as in (n,α), (n,p). Exactly as in case 3), except that the factor $\exp(-G_b)$ operates and is dominant. $\sigma \propto \sqrt{\text{energy excess}} \times \exp(-G_b)$

5) EXOTHERMIC, CHARGED INCOMING particle, as in (p,n), (α,n), (α,γ), (p,γ). For incident energies << Q, the factor $p_b^2/v_a v_b \propto 1/v_a$. The barrier factor $\exp(-G_a)$ operates on the incoming particle.

$$\sigma \propto 1/v_a \exp(-G_a)$$

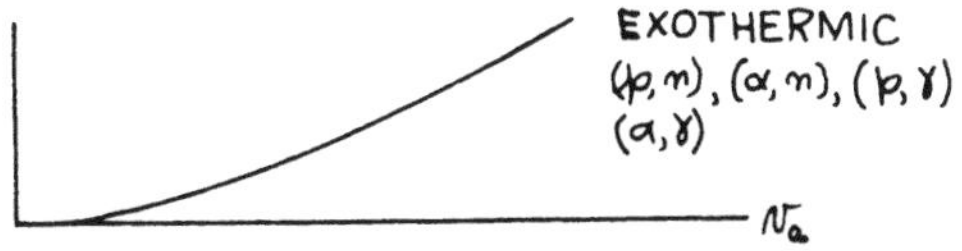


In all of the above, no account has been taken of resonance phenomena.

C. Inverse Processes

Consider the transition A + a → B + b, where "A" and "B" are nuclei and "a" and "b" are, in general, lighter particles. From equation VIII.9 the cross section for this transition is (neglecting spins):

$$\sigma_{A\to B} = \frac{1}{\pi\hbar^4} |\Omega\mathcal{H}|^2 \frac{p_b^2}{v_a v_b} \qquad \text{VIII.9'}$$

The inverse reaction is B + b → A + a. Its cross section is

$$\sigma_{B\to A} = \frac{1}{\pi\hbar^4} |\Omega\mathcal{H}|^2 \frac{p_a^2}{v_a v_b} \qquad \text{VIII.9''}$$

$|\Omega\mathcal{H}|^2$ is the same in both cases, because the operator of the perturbation is Hermitian, i.e., $\left|\int \psi_B^* \mathcal{H} \psi_A d\tau\right| = \left|\int \psi_A^* \mathcal{H} \psi_B d\tau\right|$

therefore,

$$\frac{\sigma_{A\to B}}{\sigma_{B\to A}} = \frac{p_b^2}{p_a^2} \qquad \text{(neglecting spin)} \qquad \text{VIII.14}$$

The same result may be looked at from a different aspect. Suppose we have a box filled with arbitrary numbers of particles "A", "a", "B", "b". The transitions A + a ⇄ B + b occur. Statistical mechanics asserts that at equilibrium all possible states of the system consistent with the specification of the energy of the system are occupied with equal probability. If a state consisting of a pair of particles A + a is called an "A" state, and similarly for "B" state, then the occupied states in the energy range ΔE may be divided into the two types, A and B. Since all states in ΔE are equally probably occupied, this division is such that

$$\frac{\text{No. occupied A states}}{\text{No. occupied B states}} = \frac{\text{No. possible A states in } \Delta E}{\text{No. possible B states in } \Delta E} \qquad .15$$

The number of possible A states = maximum number of (A + a) pairs times the number of states in ΔE for one pair $= \eta \frac{4\pi p_a^2 \Omega}{(2\pi\hbar)^3 v_a} \Delta E$

where $\eta$ = maximum number of (A + a) pairs formable with the particular numbers of particles put into the box initially. Similarly, the number of possible B states $= \eta \frac{4\pi p_b^2 \Omega}{(2\pi\hbar)^3 v_b} \Delta E$

where $\eta$ is the same. Therefore

$$\frac{\text{No. of occupied A states}}{\text{No. of occupied B states}} = \frac{p_a^2 v_b}{p_b^2 v_a} \qquad \text{VIII.16}$$

Now at equilibrium the number of transitions $A \rightarrow B$ equals the number of transitions in reverse, per unit time.

$$\text{No. transitions } A \rightarrow B/\text{sec} = (\text{No. A states occupied})\,\sigma_{A\rightarrow B}\, v_a$$
$$\text{No. transitions } B \rightarrow A/\text{sec} = (\text{No. B states occupied})\,\sigma_{B\rightarrow A}\, v_b \qquad \text{VIII.17}$$

Combining VIII.16 with VIII.17,

$$\frac{\sigma_{B\rightarrow A}\, v_b}{\sigma_{A\rightarrow B}\, v_a} = \frac{p_a^2 v_b}{p_b^2 v_a} \qquad \text{VIII.14'}$$

as before.

If the particles have spins, the density of states is increased. If the spins are $I_A$, $I_a$, $I_B$, $I_b$, the density of A states is increased by the factor $(2I_A+1)(2I_a+1)$, and similarly for B states. Then the rate of transition $A \longrightarrow B$ is proportional to

$$(2I_A+1)(2I_a+1)p_a^2\sigma_{A\rightarrow B}$$

and $B \longrightarrow A$ to

$$(2I_B+1)(2I_b+1)p_b^2\sigma_{B\rightarrow A}$$

therefore

$$\boxed{(2I_A+1)(2I_a+1)p_a^2\sigma_{A\rightarrow B} = (2I_B+1)(2I_b+1)p_b^2\sigma_{B\rightarrow A}} \qquad \text{VIII.18}$$

Note that in this formula, $\sigma$ is an <u>average</u> over the various kinds (spin orientations) of A states, and a <u>sum</u> of partial $\sigma$'s for various possible final states. *

---

* This may be elucidated by writing $\sigma_{A\rightarrow B}$ more explicitly. Divide $\sigma$ into contributions $\sigma(S)$ due to various relative orientations of $I_A$ and $I_a$. The number of states represented by each relative orientation is $2S+1$, where $S$ = resultant angular momentum of particles "A" and "a". In this discussion, orbital angular momentum is neglected. It is included in a discussion in the appendix of Bethe and Placzek's paper, <u>Phys.Rev.</u> <u>51</u> 450. The total number of A states is $(2I_A+1)(2I_a+1)$. The total cross section for transition to any B state is $\sigma_{A\rightarrow B} = \frac{1}{(2I_A+1)(2I_a+1)}\sum_S (2S+1)\sigma(S)$ which is an <u>average</u> over spin states. ($S$ takes on $2I_a+1$ values if $I_a < I_A$; $(2I_A+1)$ if $I_A < I_a$.) Now $\sigma(S)$, the partial cross section for various initial values of $S$, may be written as a <u>sum</u> of contributions to various possible final spin states, i.e.,

$\sigma(S) = \sum_i \sigma(S)_i$, where $i$ denotes a particular final spin state of the B + b system. $\sigma(S)_i$ contains in addition to the density of states in energy, the squared matrix element for the particular transition represented by $\sigma(S)_i$. For transitions not conserving total vector angular momentum, $\sigma(S)_i = 0$. For example,

Problem: Design an experiment to detect the inverse reaction to

$$Be^9 + H^1 \longrightarrow Li^6 + He^4.$$

(Design of the alpha particle source will depend on the threshold energy for the inverse reaction. From Allison, Skaggs and Smith, Phys.Rev. 57 550, or from Hornyak and Lauritsen, Rev. Mod.Phys. 20, 202, we find that Q for the forward reaction is 2.115 Mev. In the reverse reaction, in order to get 2.115 Mev into the center of mass coordinate system we must give the alpha an energy of about 3.5 Mev, and this is the threshold for the inverse reaction (See section A). Design of the $Li^6$ target and of the detector, and determining the required alpha beam strength require knowing the cross section. This is got by detail balancing arguments from $\sigma_{Be^9(p,\alpha)Li^6}$, taking into account a spin factor of 8/3. This cross section is found in Livingston and Bethe, C, Rev.Mod.Phys. 9 245, p. 310, or in the original source, Allen, Phys.Rev. 51 182 (1937), and is $5 \times 10^{-29}$ $cm^2$ at 0.1 Mev. The cross section for the inverse reaction increases rapidly as the volume of phase space available to the proton is increased, therefore it is advantageous to use alpha energies an Mev or more above the threshold of 3.5 Mev. Higher energy protons also penetrate the Coulomb barrier readily, and are easier to detect. A qualitative curve of cross-section for the forward reaction as a function of energy is given in Hornyak and Lauritsen, Rev. Mod.Phys. 20 191, p. 201.

## D. The Compound Nucleus

In the diagrams of section B it was assumed the $|H|^2$ was approximately constant, except for the Coulomb barrier factor. Often, perhaps in most cases, the matrix element has irregular variations. This phenomenon is called resonance. For example, in the $(n,\gamma)$ process in indium, there is an extremely pronounced peak in $\sigma$ at a neutron energy of 1.44 e.v. $\sigma$ reaches 27,000 barns at this energy. (one barn is $10^{-24}$ $cm^2$.) The half-width of this resonance peak is 0.042 e.v. $\equiv \Gamma_{1/2}$ Near the resonance, the curve of $\sigma$ vs. energy has the form $1/(E-E_R)^2$. Another example is the resonance at $E_R = 5.2$ e.v. for the $(n,\gamma)$ reaction in silver. In this case $\sigma$ reaches 24,000 barns, and the peak has a half-width $\Gamma_{1/2} = 0.063$ e.v.

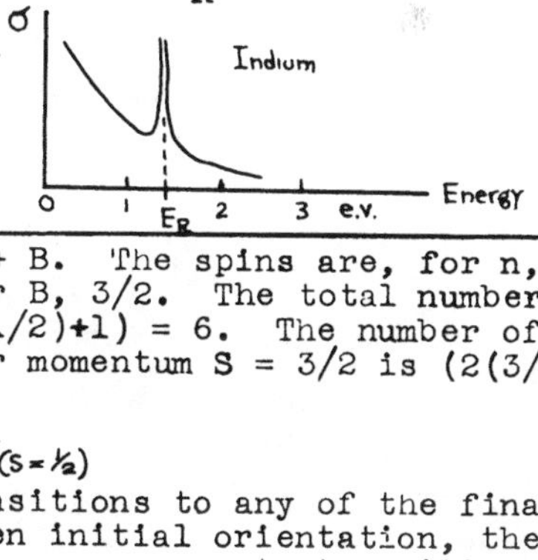


consider the reaction $n + A \longrightarrow \alpha + B$. The spins are, for n, 1/2; for $\alpha$, 0; assume for A, 1; and for B, 3/2. The total number of initial spin states = $(2(1)+1)(2(1/2)+1) = 6$. The number of initial spin states for total angular momentum $S = 3/2$ is $(2(3/2)+1) = 4$; for $S = 1/2$, $(2(1/2)+1) = 2$.

$$\sigma_{A(n,\alpha)B} = \frac{4}{6} \times \sigma_{(S=3/2)} + \frac{2}{6} \times \sigma_{(S=1/2)}$$

Now the first term represents transitions to any of the final spin states having $S = 3/2$. For a given initial orientation, there is only one. Similarly, the second term represents transitions to any final state having total angular momentum 1/2. But, since the spin of the $\alpha = 0$, there are none, so $\sigma(1/2) = 0$. When orbital angular momentum is involved, there may be more than one way in which the given initial state can form a final state, so that $\sigma(3/2)$, for example, is a sum over the various possibilities. See Bethe and Placzek, Phys. Rev. 51 450, appendix.

The explanation of this phenomenon is based on the assumption that the transition $A + a \rightarrow B + b$ occurs through an intermediate state C:

$$A + a \longrightarrow C \longrightarrow B + b$$

State C is the "Compound nucleus". The idea of the compound nucleus is due to Bohr.*

The idea of how resonances in cross section result from this assumption can be obtained from the quantum mechanics of second order transitions. The probability of transition, per unit time, is given by "Golden Rule No. 1":**

$$\text{trans. prob./sec} = \frac{2\pi}{\hbar}\left|\frac{\mathcal{H}_{CA}\mathcal{H}_{BC}}{E_A - E_C}\right|^2 \times \begin{pmatrix}\text{energy}\\ \text{density of}\\ \text{states}\end{pmatrix} \qquad \text{VIII.19}$$

provided there are no direct transitions from A to B. The cross section is, from VIII.9,

$$\sigma_{A\to B} = \frac{1}{\pi\hbar^4}\left|\Omega\mathcal{H}\right|^2 \frac{p_b^2}{v_a v_b} \qquad \text{VIII.9'}$$

which becomes, analogously,

$$\sigma_{A\to B} = \frac{1}{\pi\hbar^4}\left|\frac{\mathcal{H}_{CA}\mathcal{H}_{BC}}{E_A - E_C}\right|^2 \frac{p_b^2}{v_a v_b} \qquad \text{VIII.20}$$

Near $E_A = E_C$, (resonance), $\sigma$ is large. This formula gives infinite $\sigma$ at the resonance energy, but the formula does not take into account the short lifetime of the compound state. A correct formula is derived in section F.

The life-time of the compound state is long enough for the nucleus C to "forget" how it was formed,***and this results in a basic simplification in the interpretation.

From the Heisenberg relation $\Delta t \Delta E \gtrsim \hbar$, the lifetime of the compound nucleus and the uncertainty $\Gamma$ in its energy are related by

$$\Gamma \gtrsim \frac{\hbar}{\text{lifetime}} \qquad \text{VIII.21}$$

- - - - - - - - - - - - - - - - - -

The reasons why the compound nucleus has a lifetime greater than zero are the following:

1) For charged particle decay, the barrier factor (VIII.12) reduces the rate of decay.

2) Decay by $\gamma$ radiation is very slow compared to the times in which the nucleus changes its organization: the lifetime against $\gamma$ emission is $\sim 10^{-13} - 10^{-14}$ sec. The characteristic time of the nucleus, i.e., the time for a nucleon to cross the nucleus, is $\sim$(size)/(velocity) $\approx 10^{-13}/10^{9}$, or about $10^{-22}$ sec.

3) A particularly important reason is the tendency toward equipartition of energy in the nucleus. The excess energy due to the absorption of the bombarding particle is distributed among all the nucleons. It is rare that there is a fluctuation in which a large fraction of the excess energy is on one nucleon.

4) Selection rules forbid some modes of decay.

---

* Bohr, Nature 137 344 (1936)

** Schiff, p. 196, eq. (29.20)

*** Discussed in Peierl's review article in Reports on the Progres in Physics VIII (1940), Phys. Soc. of London, 1941.

E. Example of an Unstable Nucleus

An example of a nucleus which plays the role of an intermediate-state compound-nucleus for several well known nuclear reactions is $Be^8$.

The ground state $Be^8$ decays as follows:

$$Be^8 \longrightarrow 2\, He^4 + 110 \text{ Kev.*}$$

The reaction is barely exothermic. The Gamow exponent for decay into α's is low due to low nuclear charge, see equation VIII.57, p. 163. The theoretical estimate of the lifetime is $10^{-16}$ sec.**, corresponding to a width of between 1 and 100 e.v. This time is long compared with the nuclear characteristic time of $10^{-22}$ sec.; hence the width of the level is small.

Information on the excited levels of $Be^8$ can be obtained from study of those nuclear reactions for which $Be^8$ is the intermediate compound nucleus state, such as $Li^7(p,\gamma)Be^8 \longrightarrow 2\alpha$, $Li^7(p,n)Be^7$. These reactions are discussed here. The energy levels are plotted in FIG. VIII.1.

1) α-α scattering. For two Coulomb centers, the total scattering cross section is ∞. We may study the scattering at some angle not near 0 (90° in center of mass system is best). We expect peaks in the value of σ when the incident relative energy equals the energy of excitation of an excited state. For α's scattered on α's, the first such resonance should come at 0.110 Mev (in center of mass system), corresponding to the $Be^8$ ground state. This resonance is presumably very sharp, a few e.v. wide, as mentioned above. It has never been observed experimentally.

Problem. Discuss the possibility of experimentally observing the resonance expected in alpha-helium scattering at an energy corresponding to the $Be^8$ ground state, i.e., 0.110 Mev in the center of mass frame.
(The Coulomb barrier keeps alphas of this energy at least $5 \times 10^{-12}$ cm apart classically, so the effect of nuclear forces is probably undetectable. Also the experiment is difficult because the range of 200 Kev alphas is so short that it is hard to shoot them through an appreciable number of scattering centers and detect them. Any attempt to detect a resonance might be guided by the experimental procedure of Devons (Proc.Roy.Soc. A 172 127 and 559 (1939)), who investigated alpha-helium scattering at higher energies. The theory of α-α scattering and its relation to the $Be^8$ nucleus is given in Wheeler, Phys.Rev. 59 16 and 27, (1941).)

A second resonance, this one experimentally observed, is at ~3 Mev. The barrier factor is lower at 3 Mev, hence the state has shorter lifetime and greater width. The half-width is estimated to be 0.8 Mev.

Further resonances in α-α scattering are so broad as to be scarcely recognizable as resonances. All the resonances mentioned so far correspond to states of even parity. This is because α's obey Bose-Einstein statistics and have symmetric wave

* Hemmendinger; quoted in Seaborg and Perlman table of isotopes, Rev. Mod. Phys. 20 585.

** Wheeler, Phys.Rev. 59 27.

functions.* The incident α's will have angular momentum 0,2,4,.. with respect to a target α particle. Therefore states of $Be^8$ detectable by α scattering in helium are even states.

Not all states of $Be^8$ are even. Odd states of $Be^8$ cannot decay directly into two α's or into the even $Be^8$ states mentioned above. Emission of electromagnetic radiation must occur first, because an $\substack{\text{odd}\\ \text{even}}$ state cannot change to an $\substack{\text{even}\\ \text{odd}}$ state by "mechanical", i.e., non-radiative, interactions. Change of parity occurs in emission of photons.

2) $Li^7(p,\gamma)Be^8$. There is a prominent and narrow resonance at a proton energy of 440 Kev. This indicates that the lifetime of the excited $Be^8$ is long and thus that it is an odd state. It decays through the relatively slow process of γ emission to the much lower even $Be^8$ states. The energy of a $Li^7$ and a proton separated and at rest is 17.2 Mev higher than that of the ground state of $Be^8$. The γ's given off in decay from the excited $Be^8$ state produced in the $Li^7(p,\gamma)$ reaction have energies of 17.5 Mev and ≈ 14.5 Mev, indicating γ decay to the two even states mentioned in the paragraph on α-α scattering.

That the excited $Be^8$ state produced in $Li^7(p,\gamma)$ is odd accords with the following considerations. The most probable case is for the bombarding proton to be in a high S state, which is even. The $Li^7$ is odd, as is suggested by the arguments in the following paragraph. Then $Li^7$(odd) + proton (S state) is an odd $Be^8$ state.

The picture of a nucleus as built up of "shells" of protons and neutrons, somewhat like atomic electron shells, suggests that $Li^7$ in the ground state is odd.

Suppose the average potential for the nucleons is a square well. The single particle approximate quantum mechanical solution to the problem leads to orbits which may be designated: 1s, 1p, 1d, etc. The 1s orbit accomodates 2 neutrons (spins opposed) and two protons (spins opposed); 1p accomodates 6 neutrons and 6 protons, etc. ${}_3Li^7_4$ would have the configuration:

protons: $1s^2\ 1p^1$
neutrons: $1s^2\ 1p^2$

or a total configuration: $(1s^2 1p; 1s^2 1p^2)$, which has $\Sigma|l| = 3$, and is hence an odd state. This model of the nucleus is discussed in

---

* For two identical particles, parity of the state and symmetry of the wave functions are simply related: If the wave function $\Psi\left(\substack{\text{changes}\\ \text{does not change}}\right.$ sign when space is inverted by the operation $x\rightarrow -x, y\rightarrow -y, z\rightarrow -z$, then the Ψ has $\left(\substack{\text{odd}\\ \text{even}}\right.$ parity. If $\Psi \substack{\text{changes}\\ \text{does not change}}$ sign when the two particles are interchanged in position, then Ψ is is $\left(\substack{\text{antisymmetrical}\\ \text{symmetrical}}\right.$.

Operation of inversion of space: $\Psi(\underline{r}_1,\underline{r}_2) \rightarrow \pm\Psi(-\underline{r}_1,-\underline{r}_2)$
Operation of particle interchange: $\Psi(\underline{r}_1,\underline{r}_2) \rightarrow \pm\Psi(\underline{r}_2,\underline{r}_1)$

But for identical particles, $\underline{r}_1 = -\underline{r}_2$, $\underline{r}_2 = -\underline{r}_1$, so that $\Psi(-\underline{r}_1,-\underline{r}_2) = \Psi(\underline{r}_2,\underline{r}_1)$, and inversion is equivalent to particle interchange.

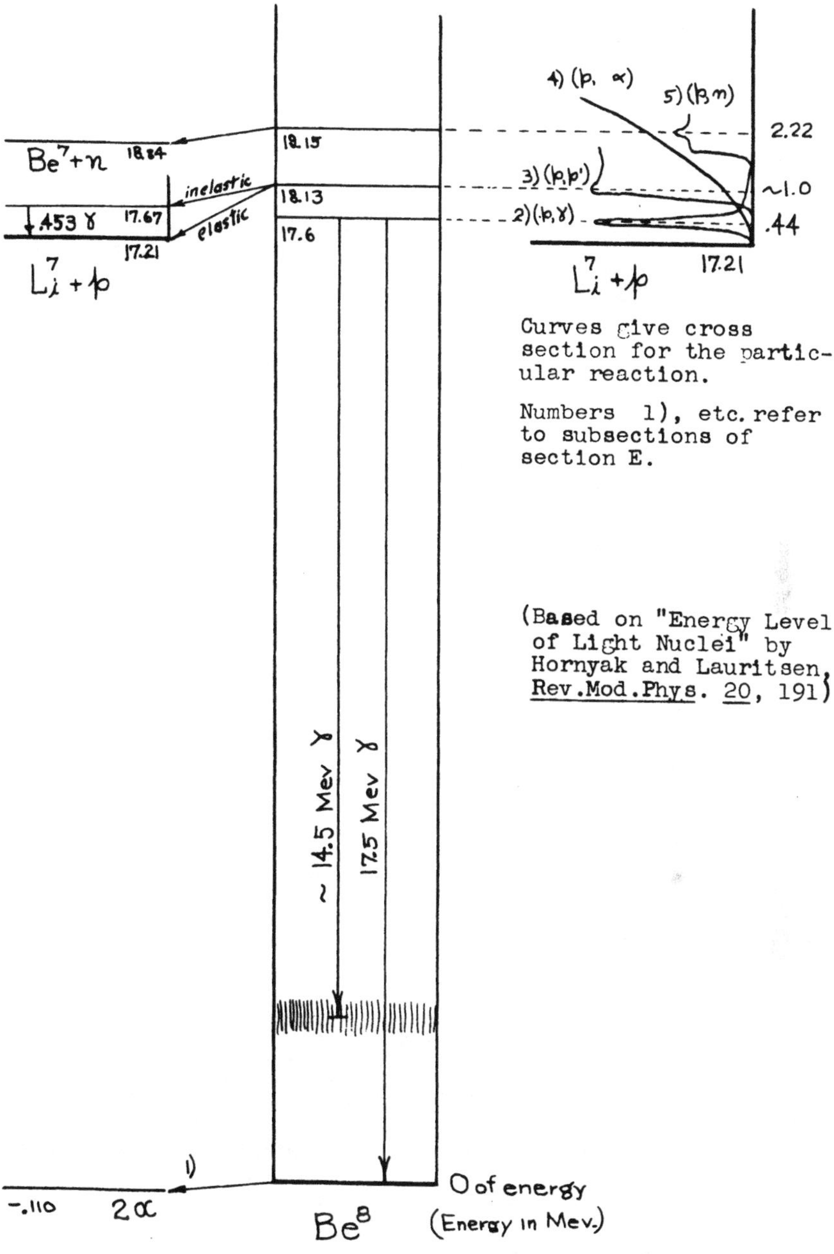


Curves give cross section for the particular reaction.

Numbers 1), etc. refer to subsections of section E.

(Based on "Energy Levels of Light Nuclei" by Hornyak and Lauritsen, Rev. Mod. Phys. 20, 191)

FIG. VIII.1

greater detail in section K.

3) $Li^7(p,p')Li^{7*}$. This is similar to 2) except that a proton is emitted having less energy than the incident proton, leaving $Li^7$ in an excited state. The resonance in $\sigma$ is observed at a proton energy of $\sim 1.05$ Mev. $Li^{7*}$ decays by emitting a $\gamma$ of about 0.45 Mev. The 0.45 Mev splitting between this excited state and the ground state $Li^7$ may be due to energy difference between $p_{1/2}$ and $p_{3/2}$ states of $Li^7$, on the nuclear shell model:

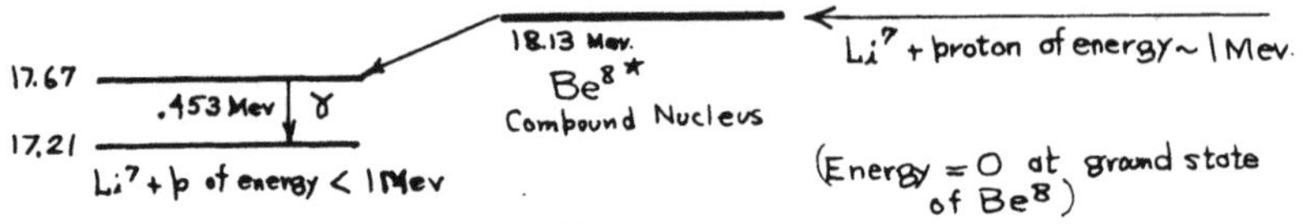


$$Li^7 + p(1\text{ Mev}) \rightarrow Be^{8*} \rightarrow Li^{7*} + p' \rightarrow Li^7 + \gamma + p'$$

4) $Li^7(p,\alpha)He^4$. Since two $\alpha$'s are in an even state, and since $Li^7$ is odd, the incident proton must be in an odd state, probably a p state with respect to the $Li^7$ nucleus. No resonances are observed. None is to be expected, since all even resonances are extremely broad. Note that all the observed resonances are odd levels.

5) $Li^7(p,n)Be^7$. A resonance is observed at a proton energy of $\sim 2.22$ Mev, corresponding to an odd $Be^8$ state 19.15 Mev above the ground state.

6) $Li^6(d,\alpha)He^4$ No resonances. Evidently $Li^6 + H^2$ form an even state, and quickly decay to two $\alpha$'s.

> Problem: Design an experiment to observe the famous 440 Kev resonance in the reaction $Li^7 + p \rightarrow Be^{8*} \rightarrow \gamma + Be^8 \rightarrow 2\alpha$ (This experiment has been performed by Walker and McDaniel, using a gamma ray spectrometer which measures the energy of pairs produced by the gamma ray (Phys.Rev. 74 315) and by Delsasso, Fowler and Lauritsen (Phys.Rev. 51 (1937)) using a cloud chamber to detect the gammas by means of electrons produced in the chamber by pair production and Compton collisions. Recent electrostatic accelerators have been equipped with electrostatic velocity selectors which provide an energy spread in the proton beam of less than 300 e.v. at 1 Mev. In order to take advantage of this narrow energy range, very thin targets must be used. These problems are discussed in "Gamma-Radiation from Excited States of Light Nuclei" by Fowler, Lauritsen and Lauritsen, Rev.Mod.Phys. 20 236 (1948)).

## F. Quantitative Development of Resonance Theory; Breit-Wigner Formula.

In this discussion we use as an example the $(n,\gamma)$ reaction, i.e., radiative capture of neutrons, which is an important reaction.

As in the preceding sections, the energy levels of the initial and final states form a continuous distribution. The experment picks out the particular initial state.

The resonance phenomenon that we wish to describe is very energy-sensitive. We shall attribute it to the existence of a compound nucleus state C, connecting with the initial and final states by matrix elements $\mathcal{H}_{ac}$ and $\mathcal{H}_{bc}$, where the notation is given in FIG.VIII.2 We shall neglect matrix elements of the form $\mathcal{H}_{ao'}$, $\mathcal{H}_{bb'}$ and $\mathcal{H}_{ab}$, which connect the initial state with itself, the final state with itself, and the initial state with the final state without the intermediate C state; i.e., no direct transitions from A to B.

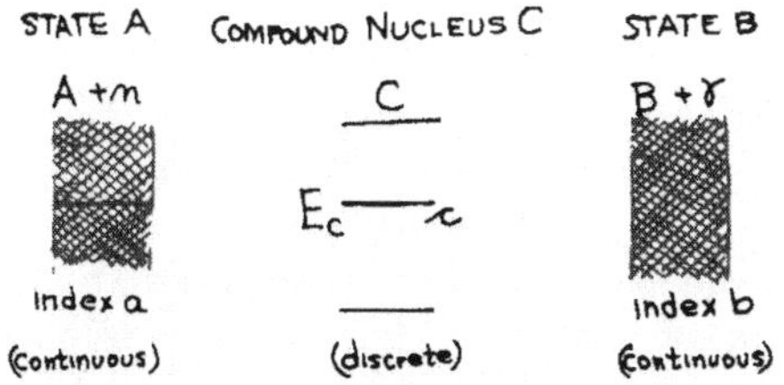


FIG. VIII.2

The problem of computing transition probabilities can be done by using the complete machinery of time dependent perturbation theory. In summary, if $u_i$ are the time independent eigenfunctions of the unperturbed states having energies $E_i$, the true eigenfunction for the perturbed state is

$$\Psi = \sum a_i u_i e^{-\frac{i}{\hbar} E_i t} \qquad \text{VIII.22}$$

where $a_i$ are the amplitudes of the unperturbed states in this expansion. For the method to be useful, the true eigenfunction must not differ greatly from one of the unperturbed eigenfunctions. If there is no perturbation, $\dot{a}_n = 0$ for all n. With the perturbation, $\mathcal{H}$, the a's change according to the equation

$$\dot{a}_n = -\frac{i}{\hbar} \sum_m \mathcal{H}_{nm} a_m e^{\frac{i}{\hbar}(E_n - E_m)t}, \quad \left(\mathcal{H}_{nm} \equiv \int u_n^* \mathcal{H} u_m \, d\tau\right) \qquad \text{VIII.23}$$

If, just before the perturbation is applied, the system is in a state represented by $u_k$, i.e., $a_i = \delta_{ik}$, then some time afterward there is some probability of finding the system in states other than the $k^{th}$. For states differing in energy greatly from state number k, the exponential in the above equation oscillates rapidly and the change in $a_n$ tends to average to 0.

Applied to the present problem, we denote amplitudes of A states by $a_a$, of B states by $a_b$, and of the compound nucleus at $E_c$ by $a_c$ (see FIG.VIII.2). Assume that only one intermediate state is near enough to be important. (The result is then called the one-level Breit-Wigner formula.) The differential equations for the amplitudes are

$$\dot{a}_a = -\frac{i}{\hbar} \mathcal{H}_{ac} e^{\frac{i}{\hbar}(E_a - E_c)t} a_c$$

$$\dot{a}_b = -\frac{i}{\hbar} \mathcal{H}_{bc} e^{\frac{i}{\hbar}(E_b - E_c)t} a_c \qquad \text{VIII.24}$$

$$\dot{a}_c = -\frac{i}{\hbar} \sum_a \mathcal{H}_{ca} e^{\frac{i}{\hbar}(E_c - E_a)t} a_a - \frac{i}{\hbar} \sum_b \mathcal{H}_{cb} e^{\frac{i}{\hbar}(E_c - E_b)t} a_b$$

The initial A state, say $a_{a_o}$, is chosen by experiment. At t = 0, $a_{a_o} = 1$, $a_a = 0$ for $a \neq a_o$, $a_c = a_b = 0$.

At $t = 0$, $\dot{a}_a = \dot{a}_b = 0$, but $\dot{a}_c \neq 0$. $a_c$ increases, representing the build up of probability that the system is in intermediate state C. When $a_c$ becomes larger than 0, the amplitudes $a_a$ and $a_b$ begin to increase. The amplitude that arises in $a_b$ denotes the forward reaction, i.e., $A + n \rightarrow B + \gamma$. The amplitude that arises in $a_a$ denotes scattering (perhaps inelastic), i.e., $A + n \rightarrow A + n'$.

The rigorous way to solve for the transition probabilities * is to carry through the solution of the system of differential equations.

The following is a practical way to get the answer quickly.

Assume for the sake of argument that somehow the intermediate compound nucleus state is built up so that $a_c = 1$. It can disintegrate either to A or to B. The transition probabilities are given by "Golden Rule No. 2" (VIII.2). Ignore spin, and let $\tau$ be the lifetime of the state against a certain mode of decay. Then

$$\text{Probability of transition to B (per sec.)} = \frac{1}{\tau_\gamma} = \frac{2\pi}{\hbar}\left|\mathcal{H}_{bc}\right|^2 \frac{4\pi p_\gamma^2}{8\pi^3\hbar^3 v_\gamma} \text{ sec}^{-1}$$
$$= \frac{1}{\pi\hbar^4}\left|\mathcal{H}_{bc}\right|^2 \frac{\hbar^2\omega^2}{c^3} \qquad \text{VIII.25}$$

where we have put $\hbar\omega/c = p_\gamma$, $v_\gamma = c$.

$$\text{Probability of transition to A (per sec.)} = \frac{1}{\tau_n} = \frac{M^2}{\pi\hbar^4}\left|\mathcal{H}_{ac}\right|^2 v_n \qquad \text{VIII.26}$$

(The wave functions are normalized in a volume $\Omega = 1$.) The decay from the state C would follow the equation:

$$\text{Prob. of occupation of state C} = e^{-t\left(\frac{1}{\tau_n} + \frac{1}{\tau_\gamma}\right)} \qquad \text{VIII.27}$$

Actually the probability of occupation of a state = $|\text{amplitude}|^2$. If we ignore the phase factor, we write

$$\text{amplitude} = \sqrt{\text{probability}}$$

or

$$a_c = e^{-\frac{t}{2}\left(\frac{1}{\tau_n} + \frac{1}{\tau_\gamma}\right)} \qquad \text{VIII.28}$$

More thorough treatment confirms the result of this step.

$$\dot{a}_c = -\left(\frac{1}{2\tau_n} + \frac{1}{2\tau_\gamma}\right)a_c \; ** \qquad \text{VIII.29}$$

Defining

$$\boxed{\frac{\Gamma_n}{\hbar} \equiv \frac{1}{2\tau_n}, \quad \frac{\Gamma_\gamma}{\hbar} \equiv \frac{1}{2\tau_\gamma}} \qquad \text{VIII.30}$$

$\Gamma \equiv \Gamma_n + \Gamma_\gamma = (\hbar/2)$ times the total probability of destruction of the intermediate state, per unit time.

($\Gamma$ is sometimes defined as $\hbar$, rather than $\hbar/2$, times the total probability of decay per unit time.) Then VIII.29 can be written

---

* Breit and Wigner, Phys.Rev. 49 519 (1936)

** From now on, $a_c$ is not necessarily equal to one.

$\dot{a}_c = -\Gamma/\hbar\, a_c$. We must add a term to account for the accretion of the state C. State C is filled from state A according to the appropriate term from perturbation theory quoted in equation VIII.24, namely $-\frac{i}{\hbar}\mathcal{H}_{ca_0}\exp(\frac{i}{\hbar}(E_c-E_{a_0}))t$ ($a_{a_0}=1$, state kept filled up)

Therefore we try

$$\dot{a}_c = -\frac{\Gamma}{\hbar}a_c - \frac{i}{\hbar}\mathcal{H}_{ca_0}e^{\frac{i}{\hbar}(E_c-E_{a_0})t} \qquad \text{VIII.31}$$

The solution of this differential equation satisfying the initial condition that $a_c = 0$ at $t = 0$ is

$$a_c = \frac{-\frac{i}{\hbar}\mathcal{H}_{ca_0}\left\{e^{\frac{i}{\hbar}(E_c-E_{a_0})t} - e^{-\frac{\Gamma}{\hbar}t}\right\}}{\frac{i}{\hbar}(E_c-E_{a_0}) + \Gamma/\hbar} \qquad \text{VIII.32}$$

Using the uncertainty principle and our/experimental knowledge of the energy widths, the dispersal time for the compound state C, $\hbar/\Gamma$, turns out usually to be less than $10^{-14}$ sec. After about $10^{-13}$ seconds the term $\exp(-\Gamma t/\hbar)$ is nearly 0. Therefore

$$|a_c|^2 = \frac{|\mathcal{H}_{ca_0}|^2}{\Gamma^2 + (E_c - E_{a_0})^2} \qquad \text{VIII.33}$$

The C state builds up to the amount represented by this value of $|a_c|^2$. Then

$$\text{Number of reactions to the right/sec. } (A + n \rightarrow C \rightarrow B + \gamma) = |a_c|^2\frac{1}{\tau_\gamma}$$

$$\text{Number of reactions to the left/sec. } (A + n \rightarrow C \rightarrow A + n') = |a_c|^2\frac{1}{\tau_n} \qquad \text{VIII.34}$$

Before writing the above equations in the final form, we examine $|a_c|^2$. It clearly has the form of a resonance curve: By definition, $\tau$ and $\Gamma$ are connected by $2\Gamma\tau = \hbar$. Since $2\Gamma$ turns out to be approximately the uncertainty in energy of the state, this is an aspect of the Heisenberg uncertainty relation $\Delta E\tau \gtrsim \hbar$.

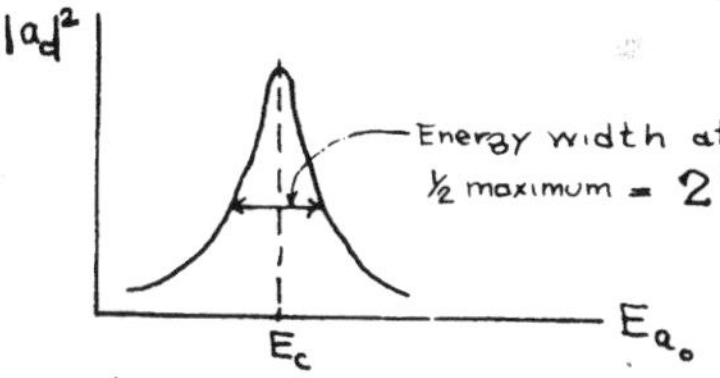


Returning to equations VIII.34, writing them explicitly:

$$\text{No. reactions/sec } A \rightarrow C \rightarrow B \text{ (per unit beam density)} = \frac{|\mathcal{H}_{ca_0}|^2}{\Gamma^2 + (E_c - E_{a_0})^2}\left(\frac{2\Gamma_\gamma}{\hbar}\right) = \sigma_{(n,\gamma)} v_n$$

$$\text{No. reactions/sec } A \rightarrow C \rightarrow A \text{ (per unit beam density)} = \frac{|\mathcal{H}_{ca_0}|^2}{\Gamma^2 + (E_c - E_{a_0})^2}\left(\frac{2\Gamma_n}{\hbar}\right) = \sigma_{(n,n)} v_n \qquad \text{VIII.35}$$

where the definition of $\sigma$ in equation VIII.7 is used, and we assume the problem is normalized to volume $\Omega = 1$ and density of neutrons = 1.

(In deriving this result more carefully, one should substitute $a_c$ into expressions for $a_a$ and $a_b$, then put these $a_a$'s and $a_b$'s into the correction expression for $a_c$, and check that the result is consistent.)

$$\sigma_{(n,\gamma)} = \frac{2\Gamma_\gamma}{\hbar v}\,\frac{|\mathcal{H}_{ca_0}|^2}{\Gamma^2 + (\delta E)^2} \qquad \text{VIII.36}$$

$$\sigma_{(n,n)} = \frac{2\Gamma_n}{\hbar v}\,\frac{|\mathcal{H}_{ca_0}|^2}{\Gamma^2 + (\delta E)^2}$$

The next step is to replace $|\mathcal{H}|^2$ with more useful parameters. We use equation VIII.26 which may be written

$$\Gamma_n = \frac{M^2}{2\pi\hbar^3}\,\overline{|\mathcal{H}_{ca_0}|^2}\,v \qquad \text{VIII.26'}$$

If $\Gamma_n^R$ denotes $\Gamma_n$ for $v_n$ = velocity of neutron if it had the right energy to hit the resonance exactly $\equiv v_R$, then

$$\frac{\Gamma_n}{v} = \frac{\Gamma_n^R}{v_R} = \frac{1}{2\pi\hbar^3}\,|\mathcal{H}_{ca_0}|^2 M^2 \qquad \text{VIII.37}$$

provided $|\mathcal{H}_{ca_0}|$ varies little with change in velocity. Rewriting VIII.36 in terms of the parameters $v_R$ and $\Gamma_n^R$,

$$\sigma_{(n,\gamma)} = \frac{2\Gamma_\gamma}{\hbar v}\,\frac{1}{\Gamma^2 + (\delta E)^2}\,\frac{2\pi\hbar^3\Gamma_n^R}{M^2 v_R} \qquad \text{VIII.38}$$

$$\sigma_{(n,n)} = \frac{2\Gamma_n}{\hbar v}\,\frac{1}{\Gamma^2 + (\delta E)^2}\,\frac{2\pi\hbar^3\Gamma_n^R}{M^2 v_R} \qquad \text{VIII.39}$$

Now use the relation $\Gamma_n = \Gamma_n^R(v/v_R)$, where v is, as always, the actual velocity of the neutron. Also, $\lambda\!\!\!^{-} \equiv \hbar/Mv$, $\lambda\!\!\!^{-}_R = \hbar/Mv_R$. M = mass of neutron.

$$\sigma_{(n,\gamma)} = 4\pi\,\lambda\!\!\!^{-}\,\lambda\!\!\!^{-}_R\,\frac{\Gamma_\gamma\Gamma_n^R}{\Gamma^2 + (\delta E)^2} \qquad \text{VIII.40}$$

$$\sigma_{(n,n)} = 4\pi\,\lambda\!\!\!^{-2}_R\,\frac{(\Gamma_n^R)^2}{\Gamma^2 + (\delta E)^2} \qquad \text{VIII.41}$$

v = actual velocity, $v_R$ = velocity of neutron to hit resonance exactly. The general nature of VIII.40 and VIII.41 is shown in FIG. VIII.3. Note that for energies below the resonance, the equations above behave as described in section B, p. 144.

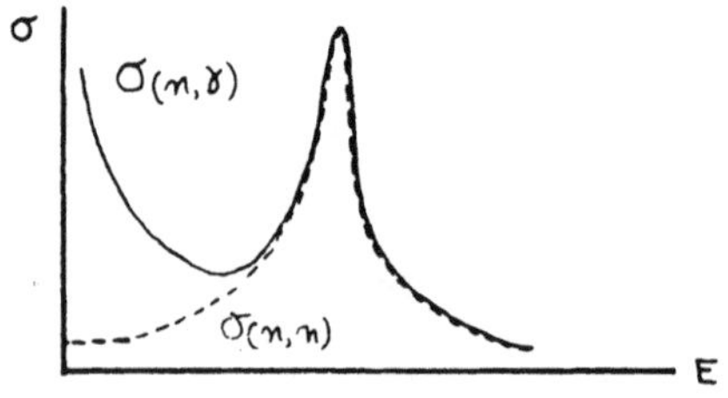


FIG. VIII. 3

So far, angular momentum multiplicity has been neglected. For the (n, γ) reaction the spins are: Neutron, $I_n = 1/2$

Initial Nucleus, $I_A$

Compound State Nucleus, $I_C$

Final Nucleus, $I_B$
Photon of fixed momentum has two degrees of freedom and is like a particle of spin 1/2, (see ref., footnote ***, p.142)

In slow neutron reactions the neutron usually has no orbital angular momentum with respect to the nucleus. Then only compound nucleus states such that $I_C = I_A \pm 1/2$ resonate. The cross section must be multiplied by the probability that, for given $I_A$ and $I_C$ differing by 1/2, the neutron has the right orientation. There are $2(I_A + 1/2) + 1$ initial states having total angular momentum $= I_A + 1/2$, (spins parallel), and $2(I_A - 1/2) + 1$ states having total angular momentum $= I_A - 1/2$, (spins antiparallel). The total number of states is $2(I_A+1/2)+1 + 2(I_A-\frac{1}{2})+1 = 4I_A + 2$. The probability that an incoming neutron will form a spin-parallel state with the nucleus is

$$\frac{2(I_A+\frac{1}{2})+1}{4I_A+2} = \frac{I_A+1}{2I_A+1} \qquad \text{VIII.42}$$

and that the neutron forms a state with spins anti-parallel is

$$\frac{2(I_A-\frac{1}{2})+1}{4I_A+2} = \frac{I_A}{2I_A+1} \qquad \text{VIII.43}$$

These two factors can be combined, giving for the final equations:*

$$\sigma_{(n,\gamma)} = 4\pi \lambda\!\!\!^{-} \lambda\!\!\!^{-}_R \frac{\Gamma_\gamma \Gamma_n^R}{\Gamma^2+(\delta E)^2}\, g \qquad \text{VIII.44}$$

$$\sigma_{(n,n)} = 4\pi \lambda\!\!\!^{-\,2}_R \frac{(\Gamma_n^R)^2}{\Gamma^2+(\delta E)^2}\, g \qquad \text{VIII.45}$$

$$g \equiv \frac{1}{2}\left(1 \pm \frac{1}{2I_A+1}\right)$$

where the + is used if $I_C = I_A + 1/2$, and - if $I_C = I_A - 1/2$. There is little or no resonance if $I_C \neq I_A \pm 1/2$, for slow neutrons.

## G. Discussion of Observed Resonance Phenomena

Cross section curves for neutron absorption and scattering in most elements are given in Goldsmith, Ibser and Feld, Rev. Mod.Phys. 19 259 (1947) or in Goodman, et.al., "Science and Engineering of Nuclear Power", Vol. I.

At the middle of the periodic table, A = 100 to 150, $(n,\gamma)$ processes are prominent. Often several resonances occur in a range of a few e.v. (Equations VIII.44 and 45 do not take into account more than one resonance.) The states of the compound nucleus are evidently close together.

Since in this region of the periodic table, the binding energy for a neutron is ~ 8 Mev, the compound nucleus is excited by about this amount. At this degree of excitation, the levels must be closely spaced to account for the closely spaced resonances, FIG. VIII.4. But at low excitations we find that the energy levels

* These cross sections do not include non-resonance processes.

are of the order of 0.1 Mev apart.* Evidently there is a great increase in energy level density with increase in excitation energy.

The increase in level density has been lucidly explained by Bohr. Suppose that the nucleus resembles somewhat a collection of harmonic oscillators of different frequencies $\omega_i$. Each has energy $\hbar\omega_1 n_1$ (ignoring zero point energy). The total energy is $E = \hbar\omega_1 n_1 + \hbar\omega_2 n_2 + \cdots + \hbar\omega_N n_N$. For low excitation, few of the $n_i$ differ from 0, and the energy changes in jumps of $\sim\hbar\omega$, which we may take as $\sim$ 0.1 Mev. For large excitation, say $10^7$ e.v., many $n_i$'s are large; there are in general many sets of $n_i$'s that give a total energy near to a large value of excitation. Each such set of $n_i$'s will in general give a slightly different total energy, therefore the density of levels will be large. This resul clearly depends on having many nucleons. We expect few energy levels and few low energy resonances in the light elements, the number of resonances increasing with A. This expectation is fulfilled to some extent, because there are few low energy ($<$1000 e.v.) resonances in elements lighter than manganese.

For very high atomic number, the density of levels accessible to $(n,\gamma)$ processes does not increase; altho higher A means more degrees of freedom and denser levels, the binding energy of the neutron decreases from $\sim$ 8 Mev to 5 or 6 Mev, and thus the states for formation of compound nuclei by absorption of a neutron are at lower excitation energy and the density of levels is correspondingly less. These opposing trends tend to cancel.

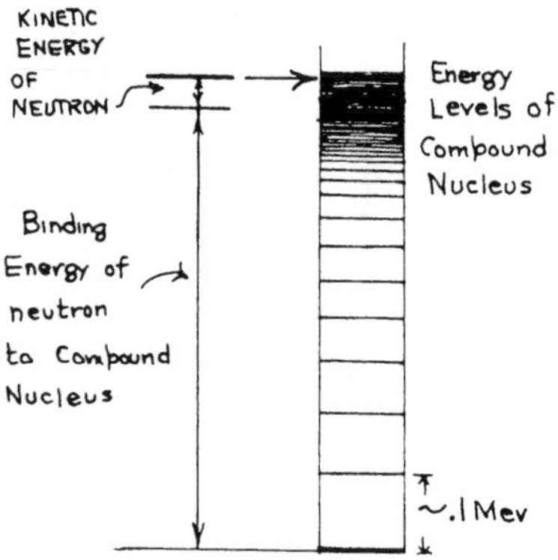


FIG. VIII.4

Not all levels are detectable by $(n,\gamma)$ processes. Those for which the spin of the compound nucleus state is not equal to $I_A \pm 1/2$, and/or parity is not conserved, do not give rise to resonances in $(n,\gamma)$ cross section.

Throughout the periodic table $\Gamma_\gamma$ is roughly constant at a value 0.1 e.v. with variation by a factor of 10 either way.

For medium and heavy elements, $\Gamma_n \approx 10^{-3}$ e.v. This corresponds to a life time against neutron decay of $10^{-12}$ seconds. This is large compared to the nuclear periods of about $10^{-21}$ sec; the captured neutron moves around $\sim 10^9$ times before a neutron escapes This very low probability of escape can be understood by considering partition of energy among the many nucleons. (See p.148.) This process to make neutron emission slow operates better as A becomes larger.

In light nuclei, $\Gamma_n$ is larger than in heavy nuclei, as one expects from the partition of energy argument. $\Gamma_n$ large means large probability of scattering compared to capture. In manganese, for example, the resonance at 280 e.v. is almost pure scat-

* Known from gamma emission spectra and from complex alpha spectra

tering.* The peak in absorption is only about 1% as large as the scattering peak. The Breit-Wigner formulas VIII.44 and 45 are fitted to this resonance by putting $\Gamma_\gamma \approx 0.1$ e.v. and $\Gamma_n \approx 10$ e.v.

In very light elements, there are no $(n,\gamma)$ resonances at all. There are, however, $(n,p)$ and $(n,\alpha)$ resonances.

The Breit-Wigner theory may be extended to reactions involving charged particles by including the barrier factor due to Coulomb forces.** For energies above the height of the Coulomb barrier resonances are observed. For example, the reaction $Li^6 + n \rightarrow He^4 + H^3$ has a broad resonance at 0.27 Mev.

Problem. Assuming that $\Gamma_n >> \Gamma_\gamma$, find the possible values for $\sigma$ for the 280 e.v. resonance in manganese. ($\lambda\!\!\!^{-}$ at 280 e.v. is $2.73 \times 10^{-11}$ cm. The spin of $Mn^{55}$ is 5/2.)
From equation VIII.45,

$$\sigma_{(n,n)} = 2\pi\left(1 \pm \frac{1}{6}\right)\lambda\!\!\!^{-}{}_R^2 \frac{\Gamma_n^2}{\Gamma^2 + (\delta E)^2}$$

At resonance $\delta E = 0$. Then ignoring $\Gamma_\gamma$ compared to $\Gamma_n$, we get

$$\sigma_{(n,n)} = 2\pi\left(1 \pm \frac{1}{6}\right)\lambda\!\!\!^{-}{}_R^2 = 4670\left(1 \pm \frac{1}{6}\right) \text{ barns}$$

If we know that $\Gamma_n \approx 100\, \Gamma_\gamma$, we can also find $\sigma_{(n,\gamma)}$.

## H. Statistical Gas Model of the Nucleus

Various models of the nucleus emphasize various different features of the nucleus. No single simple model explains all nuclear properties. We shall consider the statistical or gas model, then the liquid drop model applied to fission, and finally the nuclear shell model.***

The gas model pictures the nucleus as a gas of protons and neutrons. This model ignores surface effects like capillarity, a serious omission. The volume of the gas is $4\pi/3(1.5 \times 10^{-13})^3 A$. Due to this restriction to a small volume, the energy level spectrum for a particle has widely spaced levels.**** Unless very high excitation energies are postulated, particles will occupy the lowest available states. For the usual nuclear excitation energies of 10 Mev or so, the nucleon gas is almost completely degenerate.

We shall use the gas model to compute an approximate nuclear potential well depth, to explain semi-quantitatively the increase in nuclear level density with energy, and to consider emission of particles as an evaporation process.

The excitation of the nucleus, the extent to which higher states are occupied, may be expressed by attributing a temperature to the gas. In the literature this temperature, T, is often measured on the "dynamic" scale, where Boltzmann's k is 1 and kT is written simply as T. We will write kT in the conventional way. T = 0 corresponds to complete degeneracy. At T = 0 the number of states up to the highest one occupied just equals the number of particles, either Z or A-Z, depending on whether the proton or neutron part of the gas is in question.

---

* Seidl, Harris and Langsdorf, Phys.Rev. 71 65.
** See Bethe B, Ch. XIII, page 186.
*** Various nuclear models are discussed by L. Rosenfeld, Nuclear Forces II, p. 185, and by Bethe B, p. 79
**** Equation VIII.5, for energy density of states, contains the volume.

The number, n, of states of momentum less than $p_{max}$ of a proton confined to a volume $\Omega$ is

$$n = 2\,\frac{4\pi p^3 \Omega}{3(2\pi\hbar)^3} \quad \text{(factor 2 is for spin, see p.142)}$$

At complete degeneracy n = Z, therefore

$$p_{max}^{proton} = (3\pi^2)^{1/3}\hbar\left(\frac{Z}{\Omega}\right)^{1/3} \qquad \text{VIII.46}$$

Similarly

$$p_{max}^{neutron} = (3\pi^2)^{1/3}\hbar\left(\frac{A-Z}{\Omega}\right)^{1/3} \qquad \text{VIII.47}$$

In the crude approximation, number of neutrons = number of protons = N = A/2,

$$p_{max} = (3\pi^2)^{1/3}\hbar\left(\frac{A}{2\,\frac{4\pi}{3}(1.5\times10^{-13}A^{1/3})^3}\right)^{1/3} = 1.05\times10^{-14} \text{ cgs units} \qquad \text{VIII.48}$$

$\sim$ independent of A. The corresponding kinetic energy is $\sim$ 21 Mev $\equiv \mu_0$. $\mu_0$ is the kinetic energy of the highest occupied neutron state. This energy is measured from the bottom of the potential well. Further more, $\mu_0$ is about 8 Mev below 0, as shown in FIG. VIII.5 This fixes the depth of the well for neutrons. Actually there are fewer protons than neutrons in the nucleus, but the topmost proton level must have energy $\mu_0$, otherwise there would be β decay. Therefore the depth of the well for protons is somewhat less than for neutrons, by this model. Also, the Coulomb potential acts on protons.

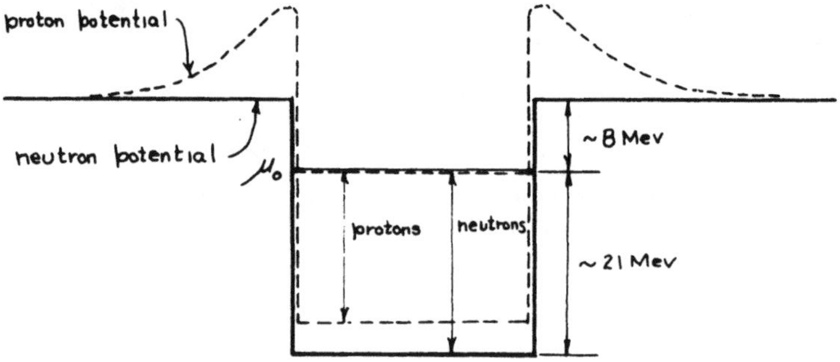


FIG. VIII.5

The following is a calculation of T corresponding to a usual value of nuclear excitation. From the statistical mechanics of a degenerate Fermi-Dirac neutron gas,* the total energy of excitation (measured above the T = 0 level) is, for neutrons

$$E = \frac{\pi^2}{4}(A-Z)\frac{(kT)^2}{\mu_0} = a(kT)^2 \text{ defining } a \quad \text{(For T small)} \qquad \text{VIII.49}$$

(FIG. VIII.6)

* Mayer and Mayer, Statistical Mechanics, p. 374. The total ener measured from the 0 of kinetic energy is

$$E' = \frac{3}{5}N\mu_0\left[1+\frac{5\pi^2}{12}\left(\frac{kT}{\mu_0}\right)^2+\cdots\right] \qquad \text{VIII.50}$$

Now put in for $\mu_0 = p_{max}^2/2M$, and then for $p_{max}$ put VIII.47. The result is

$$E_{neutrons} = \left(\frac{\pi^2}{72}\right)^{1/3} \frac{\Omega M}{\hbar^2} \left(\frac{(A-Z)}{\Omega}\right)^{1/3} (kT)^2 \qquad \text{VIII.51}$$

Similarly

$$E_{protons} = \left(\frac{\pi^2}{72}\right)^{1/3} \frac{\Omega M}{\hbar^2} \left(\frac{Z}{\Omega}\right)^{1/3} (kT)^2 \qquad \text{VIII.52}$$

where $\Omega$ = nuclear volume, T = nuclear temperature, and M = mass of a nucleon. The total (neutron plus proton) excitation energy, $E_t = E_n + E_p$, is

$$\underset{(Mev)}{E_t} = .07\, A^{2/3}\left(Z^{1/3} + (A-Z)^{1/3}\right) \underset{(Mev)}{(kT)^2} \qquad \text{VIII.53}$$

where $E_t$ and (kT) are in Mev. If A = 100 and Z = 44, $E_t = 11\ (kT)^2$. Then (kT) = 1 Mev corresponds to $E_t$=11 Mev. If we attribute to each kT a "degree of freedom" in analogy to classical statistics, then in this case there are only 11 degrees of freedom, whereas there would be 3A degrees of freedom in the classical gas.

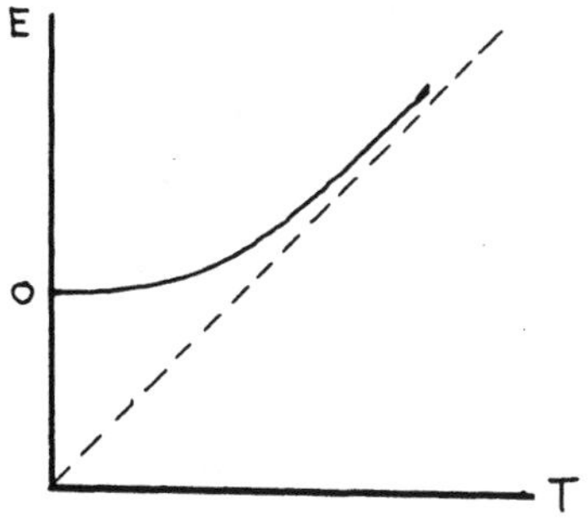

FIG.VIII.6

The increase of energy level density with energy of excitation can be computed using the statistical-mechanical definition of entropy:

$$\text{Entropy} = S(T) = k(\ln w(T) - \ln w(0)) \qquad \text{VIII.54}$$

where w is the total number of quantum states available to the system at the specified temperature. From thermodynamics,

$$S(T) = \int_0^T \frac{dE}{T} = \int_0^T \frac{2a k^2 T\, dT}{T} = 2a k^2 T \quad \text{since } E = a(kT)^2 \qquad \text{VIII.55}$$

So $S = 2\sqrt{E a k^2}$

From the definition of entropy,

$$e^{S/k} = \frac{\omega(T)}{\omega(0)} = \frac{\text{density of states at temp. T}}{\text{density of states at temp. 0}} \qquad \text{VIII.56}$$

$$e^{S/k} = e^{2\sqrt{Ea}}$$

For A = 100 and Z = 44 (same example as above), E = $11(kT)^2$, and at E = 8 Mev, the excitation of a compound nucleus after capturing a thermal neutron, $e^{S/k} \approx e^{19} \approx 10^8$. Therefore, if near excitation 0 the level spacing is ~100 Kev, at 8 Mev excitation it is ~ 1 millivolt, by this very approximate calculation. This spacing is probably much too small, although one expects this calculation to give a greater level density than observed by resonances because resonance experiments detect only levels with spin compatible with those of the initial particles.

Nuclear evaporation. The emission of a neutron from a nucleus may be considered an evaporation of a particle from a statistical group of particles held in a potential well.* In such an evaporation, the particle carries away an energy of order of magnitude kT, which is, in general, much less than the total excitation energy of the group of particles.**. For example, suppose a nucleus of A = 100, Z = 44 has excitation energy, $E = 20$ Mev. The binding energy for one neutron is about 8 Mev. The temperature will be about 1.3 Mev. A neutron will, on the average, have kinetic energy of about 2.6 Mev after escaping. Therefore the nucleus is left with an excitation of about 10 Mev., which is sufficient to emit another neutron. After the excitation is reduced below the binding energy of a neutron, the nucleus may decay by gamma emission.

Protons encounter the Coulomb barrier, which has a height of ~ 5-10 Mev for medium weight elements. The probability that 8 + ~ 5 Mev is concentrated on one proton is small if the total excitation energy is of the order of 20 Mev, therefore protons are less likely to escape. At small Z, the Coulomb barrier is relatively small, and p emission processes compete with n emission processes. For large Z, n emission dominates.

When the energy delivered to a nucleus is very much larger than the binding energy of a particle, as in nuclei excited by cosmic ray particles, α's and even larger nuclear fragments are "evaporated". In photographic emulsion or in a cloud chamber. a "star", having 3 to ~20 or so prongs, is observed, (See p. 177)

Problem. Plot the probable number of neutrons that an excited nucleus will emit, as a function of excitation energy. The plot should take the form:

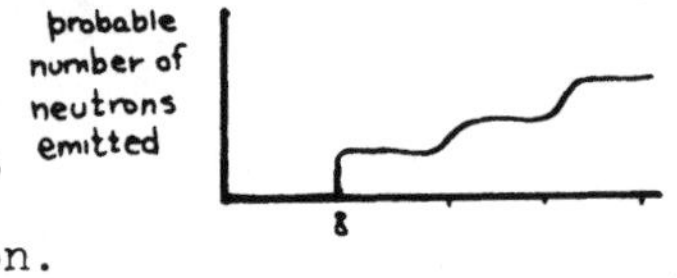


Plot up to energies such that four neutrons have probably evaporated. Discuss the probability for emission of a proton at this excitation. Assume A = 130, Z = 54, and that the binding energy of a neutron or of a proton is 8 Mev.
(The temperature of the nucleus changes only a few percent when a neutron leaves, so the error is not great in assuming that T is constant. Gamma decay may be assumed negligible for energies of excitation of over a few Kev above 8 Mev (see Bethe B, p. 160). Considering emission of neutrons first: For the energy region 0-8 Mev, no neutron can be emitted. For the region 8+a few Kev) to 16 Mev, one neutron is emitted. For 16-24 Mev there are competing modes of decay, namely: 1) one neutron may take enough energy to prohibit further evaporation, 2) two neutrons may leave. The state population function for the nuclear gas has the form:

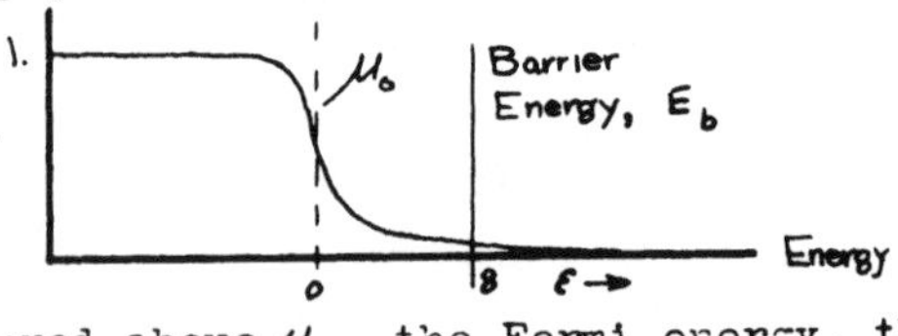


We may approximate the tail of this function by an exponential, and say that the differential probability for escape $= C v\sqrt{\mu_0+\varepsilon}\, e^{-\varepsilon/kT}$; $\varepsilon > E_b$ where $C$ is a normalization factor and $\varepsilon$ is energy measured above $\mu_0$, the Fermi energy, the

* Frenkel, Phys.Zeits.Sowjetunion 9 533 (1936) (In English)
** According to Weisskopf, Phys.Rev. 52 295, the average energy of evaporated neutron is about 2kT.

energy of the highest occupied state at $T = 0$. Let $\delta = E - 16$; $E$ is the excitation energy. Then the probability that one neutron takes out so much energy that less than 8 Mev is left is

$$P_1 = C\int_{8+\delta}^{16+\delta} v\sqrt{\mu_0+\epsilon}\, e^{-\epsilon/kT}\, d\epsilon \approx C'\int_{8+\delta}^{16+\delta} e^{-\epsilon/kT}\, d\epsilon \quad \left(\begin{matrix}\text{because } e^{-\epsilon/kT} \text{ varies} \\ \text{more rapidly than } v\sqrt{\mu_0+\epsilon}\end{matrix}\right)$$

$$= C'\int_{8+\delta}^{\infty} e^{-\epsilon/kT}\, d\epsilon, \quad \because kT << 8 \text{ Mev.}$$

$$P_1 = C'kT e^{-\frac{8+\delta}{kT}} = C'' e^{-\frac{\delta}{kT}} \quad \text{where}$$

$C''$ determined by setting $P_1 = 1$ at $\delta = 0$; $C'' = 1$. $P_1 = e^{-\frac{\delta}{kT}}$

Probable number evaporated, excitation 8-16 Mev = $P_1 + 2P_2 = P_1 + 2(1-P_1) = 2 - e^{-\delta/kT}$

For total excitation 24-32 Mev, there are three competing process: 1) one neutron leaves not enough energy for further evaporation. $P_1 = e^{-(8+\delta)/kT}$ by calculation similar to above. This is negligible. 2) the first neutron evaporated leaves enough energy for just one further emitted neutron. The probability for this is

$$P_2' = C\int_{8+\delta}^{16+\delta} e^{-\frac{\epsilon}{kT}}\, d\epsilon = e^{-\frac{\delta}{kT}}$$

3) the first neutron leaves enough energy for two more neutrons, and either one or two more are emitted. The probability that case 3) occurs with just one neutron being evaporated subsequently can be shown to be $P_2'' = \frac{\delta}{kT}\left(e^{-\frac{\delta}{kT}}\right)$. Then the probability for case 3) with two subsequent neutrons, i.e., three altogether, is

$$P_3 = 1 - P_2 = 1 - P_2' - P_2'' = 1 - e^{-\frac{\delta}{kT}}\left(1 - \frac{\delta}{kT}\right)$$

The average number for the 24-32 Mev range is $2P_2 + 3P_3 =$

$$= 3 - e^{-\frac{\delta}{kT}}\left(1 + \frac{\delta}{kT}\right)$$

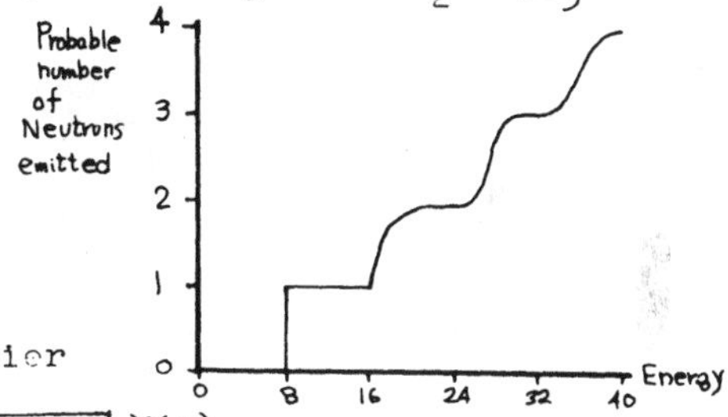


The plot turns out as given: For protons, the barrier is effectively higher, by an amount that is approximately equal to 0.9 times the peak height of the Coulomb potential. The probability for penetration of a Coulomb barrier by a charged particle is

$$P = e^{-\frac{2}{\hbar}\sqrt{A_1 Z_1}\sqrt{2Z_2 e^2 R}\,\gamma(x)} \qquad \text{VIII. 57}$$

where $x$ = Energy/Barrier height and $\gamma(x) = x^{-1/2}\cos^{-1} x^{1/2} - (1-x)^{1/2}$ (after Bethe, B, p. 167)

$A_1$ and $Z_1$ are mass and charge of particle, $Z_2$ is charge giving rise to the barrier. $A_2$, the atomic weight, enters through the nuclear radius, $R_2$. For a proton, $A_1 Z_1 = 1$. Curves of probability of penetration for three values of $A_2$, for protons, are given:

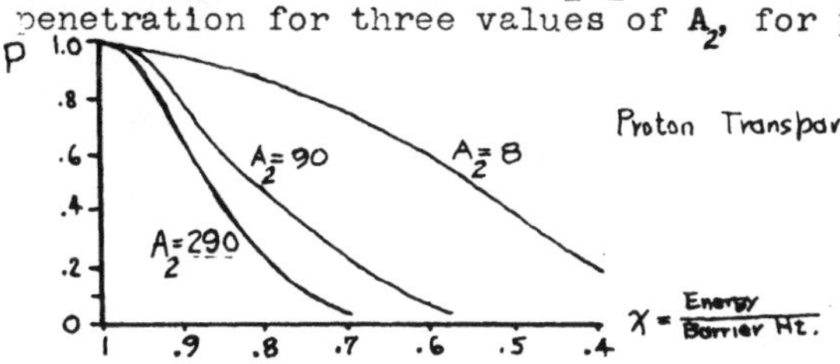

The ratio of probabilities for emission of a neutron and of a proton is

$$\frac{\omega(P)}{\omega(N)} = \frac{Z}{N} e^{-\frac{9\,\mathrm{Mev}}{1.65\,\mathrm{Mev}}} = .0037$$

where 9 Mev = 0.9 of the 10 Mev Coulomb barrier. 1.65 is kT corresponding to an excitation of 38 Mev, average number of neutrons emitted = 4.

## J. Fission **

The most useful model for explaining fission phenomena is the liquid drop model (see chapter I,C, p. 6 ). This model permits calculation of the change in potential energy when the nuclear drop suffers an ellipsoidal deformation from spherical shape*. If the potential energy increases, spherical shape is a stable configuration.

The two contributions to the potential energy are 1) capillary energy, 2) electrostatic energy. We will calculate the change in these contributions to the potential energy for a constant volume prolate ellipsoidal deformation given by the equations

major semi-axis = $a = R(1+\varepsilon)$

minor semi-axis = $b = R/\sqrt{1+\varepsilon}$ VIII.58

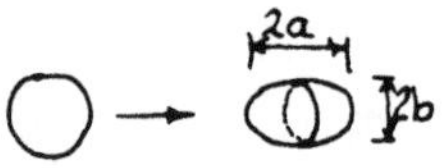


where R = initial radius, $\varepsilon$ = parameter giving extent of deformation. ($3\varepsilon$ approaches the square of the eccentricity of the elliptical section as both approach 0.) Volume is invariant: $V = (4\pi/3)ab^2 = (4\pi/3)R^3$.

1) The capillary energy is proportional to the surface area.

$$\text{Ellipsoidal surface} = 4\pi R^2 \left(1 + 2/5\,\varepsilon^2 + \ldots\right) \qquad \text{VIII.59}$$

The capillary energy was computed in Ch. I, p. 7 , and found to be $0.014\,A^{2/3}$ for an unexcited (spherical) nucleus, therefore

$$\underline{\text{Capillary energy}} = 0.014\,A^{2/3}\left(1 + 2/5\,\varepsilon^2 + \ldots\right) \quad \text{(mass units)} \qquad \text{VIII.60}$$

2) The electrostatic energy $= (3/5)(e^2Z^2/R)(1 - 1/5\,\varepsilon^2)$. At sphericity the energy is, from Ch. I, p. 6 , $0.000627\,Z^2/A^{1/3}$, therefore

$$\underline{\text{Electrostatic energy}} = 0.000627(Z^2/A^{1/3})(1 - 1/5\,\varepsilon^2) \qquad \text{VIII.61}$$

This is evidently maximum at sphericity. The total change is

$$\varepsilon^2\left(2/5 \times 0.014\,A^{2/3} - 1/5\;0.000627\,Z^2/A^{1/3}\right) \qquad \text{VIII.62}$$

Spherical shape is stable if this is +; unstable if -. Roughly, electrical energy is proportional to $A^{5/3}$ , capillary energy to $A^{2/3}$, Therefore the electrostatic energy term dominates at high A. The expression VIII.62 gives a criterion for stability for given Z and A, namely,

$$\text{Spherical nuclear drop is stable if } \frac{Z^2}{A} < 45 \qquad \text{VIII.63}$$

---

* Bohr and Wheeler, Phys.Rev. 56 426 (1939); Frenkel, J.of Physics, Akad.Sci.U.S.S.R. Vol. 1 No.2 (1939) (In English)

**General reference: W.E.Stephen,"Nuclear Fission and Atomic Power," Lancaster, 1949.

The plot of $Z^2/A$ shows that elements up to $U^{235}$ at least, are stable by this criterion; they are in reality.

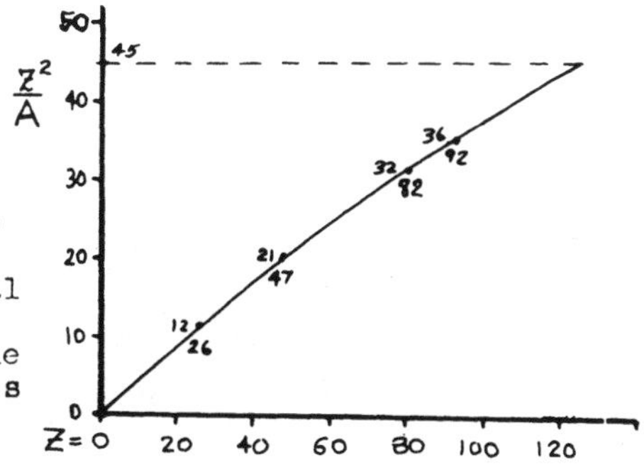


Although spherical shape of the nuclear drop may be a relative minimum of potential energy (i.e., metastable), the potential energy may be even lower for completely separated halves of the drop. To investigate this, we can compute the energy of two separated spherical drops of equal volume having a total volume equal to that of the combined sphere, and compare this with the energy of the original drop. The mass formula I.8, p. 7, may be used for this purpose. If M(Z,A) denotes the mass of combined nucleus, then we are interested in the difference

$$M(A,Z) - 2\,M\left(\frac{A}{2},\frac{Z}{2}\right) = 0.014\,A^{2/3}\,(1-2^{1/3}) + 0.000627\,\frac{Z^2}{A^{1/3}}\,(1-2^{-2/3})$$

This gives the difference in energy for infinitely separated half-volume fragments. At closer distances, still not touching each other, the potential increases due to Coulomb repulsion. These facts permit drawing an approximate potential energy curve. For $U^{236}$ the separated halves have energy -169 Mev relative to the combined non-excited nucleus, FIG. VIII.7. In FIG. VIII.7, the early part of the curve represents deformation of the sphere into a prolate ellipsoid of small eccentricity, and is known from equation VIII.62; it is quadratic in the parameter $\varepsilon$. For fully separated fragments, the curve is $\left(\frac{Z}{2}\right)^2\frac{e^2}{r}$.

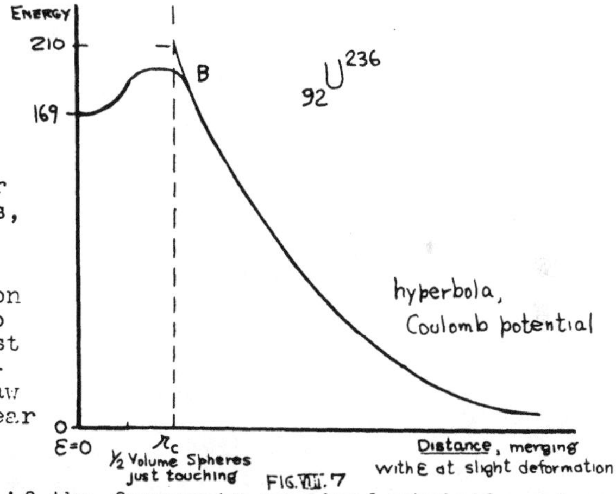

FIG. VIII.7

Near to $r_c$, the separation distance at which the two half-volume fragments just touch, the potential deviates from the Coulomb law due to the onset of nuclear forces. (In the case of $U^{236}$, if the Coulomb law held right up to $r_c$, and if the fragments remained strictly spherical up to this point, the potential energy would be 210 Mev.) The curve in the region near $r_c$ is complicated. Calculations of potential energy for large deformations (up to about point B on FIG.VIII.7) of the nucleus have been performed by Frankel and Metropolis (Phys.Rev. 72 914 (1947)) using the "Eniac" computer.

Due to zero point energy characteristic of quantum-mechanical systems, the system has energy slightly above the bottom of the potential well.

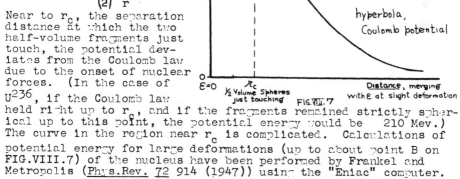

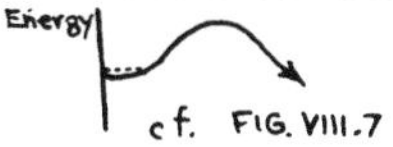

cf. FIG. VIII.7

There is some chance that the nucleus will undergo "spontaneous" fission by tunneling through the barrier. The probability is low, because the masses are large and the system approaches being classical so far as tunneling through barriers is concerned. Therefore, effectively, there is a threshold excitation for fission.

In 1939 Bohr and Wheeler* deduced a rule that has been verified experimentally. It is that the reaction (n,f) is produced in elements having an odd number of neutrons by thermal neutrons, but with elements having an even number of neutrons, fission is induced only by high energy neutrons. This is because changing from an odd number of neutrons to an even number of neutrons releases one or two Mev. This rule agrees with the table given in Ch. I, p. 7 , part of which is reproduced here. The rule is exemplified by $U^{235}$ and $U^{238}$. The binding energy of a neutron to the former may be something like 7 Mev; to the latter, which already has an even number of neutrons, only about 5 Mev. Fission induced by thermal neutrons occurs in $U^{235}$ but not in $U^{238}$. In this case, the barrier to fission is presumably around 6 Mev. in height.

Increasing Stability ↑ (upward)

| Z | N | Possible Binding Energy: Neutron Removed | Possible Binding Energy: Neutron Added |
|---|---|---|---|
| even | e | ~ 7 | ~ 5 |
| even | o | ~ 5 | ~ 7 |
| odd | e | — | — |
| odd | o | — | — |

Asymmetry of Fission Fragment Distribution. Nuclei tend not to split into equal fragments. Fragments tend to cluster in two zones, as shown in FIG.VIII.8 and 9. Referring to FIG.VIII.8, the fragments, after formation, move toward the line of maximum stability by β decay. There is no adequate theory for this asymmetry.

Extensive data on nuclear species produced in fission appeared in Rev.Mod.Phys. 18 513 (1946).

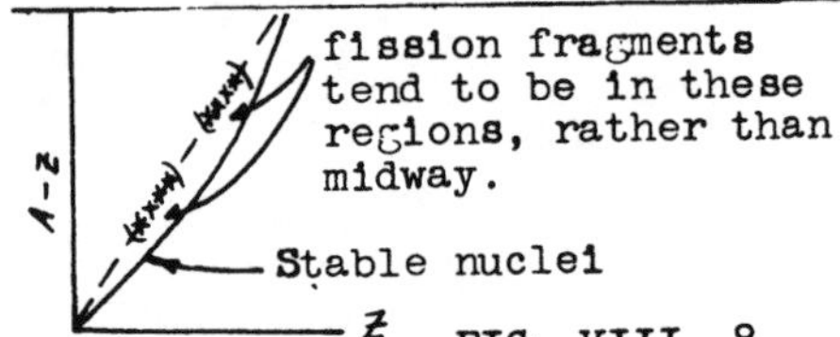


FIG. VIII, 8

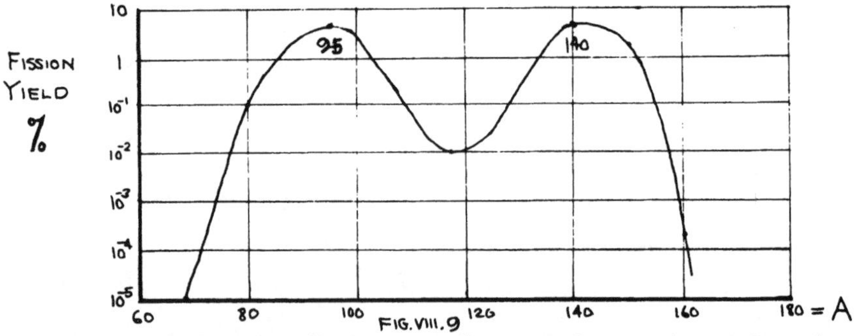


FIG.VIII.9

Remarks on Fission Fragments. Fission fragments produce heavy cloud chamber tracks about 2 cm. long in air at NTP. In contrast to ionization produced by α particles, the density of ionization here decreases with distance travelled, as shown in FIG.VIII.10. The explanation is that the fragment, initially to a large extent stripped of electrons, gains electrons as it slows down. Its

* Bohr and Wheeler, Phys.Rev. 56 426 (1939)

effective Z decreases, and ionization depends on $Z^2$ (equation II.10, p. 30 ). (The $\alpha$ picks up charge also, of course, but the effect is smaller.)

Emission of Neutrons in Fission. Immediately after fission, when the fragment "drops" are somewhat like:  , each fragment possesses considerable excitation energy. This energy is used to a large extent to evaporate neutrons. There is approximately one neutron produced per fragment.

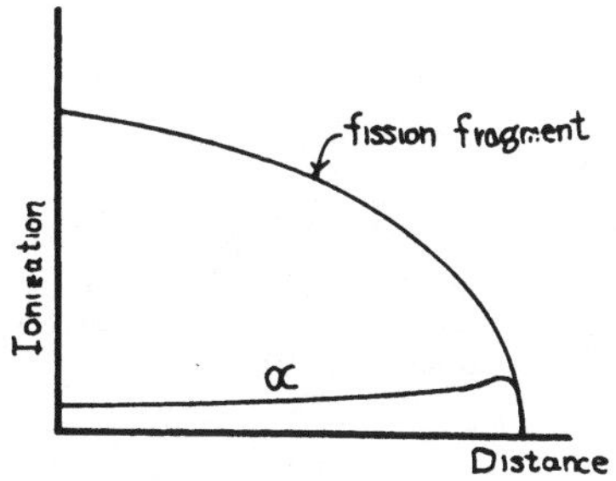


FIG.VIII.10

Delayed Neutrons. About 1% of the neutrons emitted by fission fragments are emitted at relatively long times after fission, i.e., from 55 seconds to a fraction of a second. The explanation is as follows. In the $\beta$ decay by which fission fragments become stable nuclei, it may happen that $\beta$ decay from a neutron-rich nucleus Z to the ground state of Z + 1 is forbidden. Then the nucleus Z+1 is produced in an excited state, and may, if it has sufficient energy, transform to (A-1, Z+1) by evaporating a neutron. The neutron binding energy is relatively small for these neutron-rich nuclei. It can, of course, decay by gamma emission. The time delay comes in the $\beta$ processes leading to the nucleus $(Z+1)^*$. *denotes excited state.

Triple Fission. Fission of a large nucleus into three fragments---two major fragments plus one alpha particle---is known.

Problem. Draw quantitatively as well as possible the curve FIG.VIII.7, the energy of a nucleus as a function of some parameter giving the extent of deformation. Invent a suitable parameter to measure distortion.
( It is probably simplest to describe the deformation by surface harmonics, i.e.,

$$r = a + \sum_n b_n S_n$$

where a is the initial radius and $S_n$ are surface harmonics; for axial symmetry $S_n = P_n(\cos\theta)$. Volume invariance is thus automatically provided for* Good approximation to experiment is obtained by using $r = a + b_2 P_2(\cos\theta)$. The capillary energy is obtained by computing the surface of the distorted drop. An approximation to the Coulomb energy of the slightly deformed sphere can be obtained by thinking of the spheroid as a sphere plus a thin surface layer of additional charge. When the fragments are not joined, but are still close together, they are deformed something like:  The energy can be crudely approximated by assuming each to be an ellipsoid. The papers of Bohr and Wheeler, l,c., Frankel and Metropolis, l.c., and of Frenkel, J.of Phys.Acad.Sci.USSR 1, No. 2 (1939) (in English) are pertinent to this problem.

## K. Orbit Model of the Nucleus

This model describes the nucleus in terms of nucleon orbits somewhat like the description of the atom in terms of electron orbits. The orbit picture is valid if collisions are rare enough so that a nucleon may travel at least across the nucleus without collision. This requirement seems at first not to be fulfilled

*As far as terms linear in the $b_n$

in a nucleus, for at ≈ 20 Mev the n-p scattering cross-section is of the order of 0.3 barn, and, for the known density of nucleons, the mean free path is only about 1/3 or so the radius of the nucleus. However, there are two factors which this calculation ignores, and these make the orbit picture appear not so untenable.

1) When one nucleon passes another, it passes through a potential well. If the nucleon is constantly passing other closely spaced nucleons , the wells may be so closely spaced so as to blend together to form a roughly uniform potential.

2) The nucleus is a degenerate system in which the lowest energy states are, for the most part, filled. A collision can occur between nucleons only if the collision results in transferring both the nucleons to empty states. The Pauli exclusion principle prevents two nucleons of the same kind in the same state. Diagrammatically,

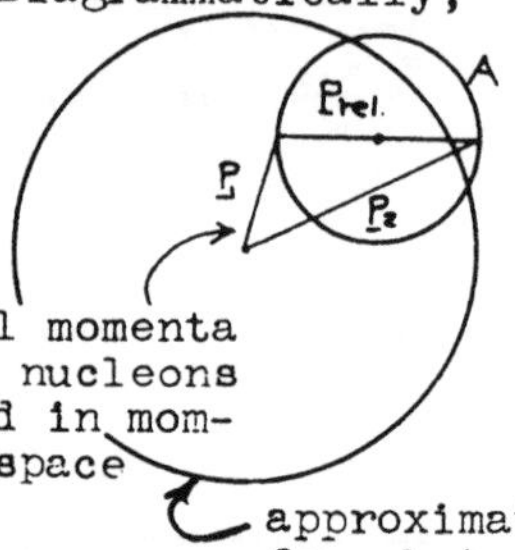


collision causes $\underline{P}_{rel}$ to change direction. Both new end points must represent previously empty states. If there are no empty states on the locus A, the collision does not occur. Thus, if occupied states described a completely filled perfect sphere there could be no collisions at all.

Neither of these two ideas has been fully investigated. To the extent that they represent the true situation, the orbit model is justified.

The orbit model has been explored with some success. In the absence of accurate information, a square well, or a square well with rounded corners, is adopted as the form of the potential. The depth of the potential is assumed not to change much with A. This is justified by the computation at equation VIII.48 showing that the kinetic energy of the highest occupied state, $\mu_0$ , is, to fair approximation, independent of A, according to the gas model. The requirement that the binding energy for a neutron be about 8 Mev then fixes the total depth of the well.

Quantum mechanical calculations for the square well give levels which may be denoted as follows:

| 1s | 1p | 1d | 2s | 1f | 2p | 1g | 2d | 3s | 1h | 2f | 3p | 1i | 2g | 3d | 4s | state |
|---|---|---|---|---|---|---|---|---|---|---|---|---|---|---|---|---|
| 2 | 6 | 10 | 2 | 14 | 6 | 18 | 10 | 2 | 22 | 14 | 6 | 26 | 18 | 10 | 2 | capacity |
| 2 | 8 | 20 | | 40 | | 70 | | | 112 | | | 168 | | | | summed no. of occupants |

The letter gives $\ell$ in the usual way, that is, s means $\ell = 0$, f means $\ell = 3$, etc. The number gives the number of radial nodes. "1" means no node, "2" means one node, etc. (This differs from atomic spectra notation, where the number of radial nodes is $n - \ell - 1$.)

For a sharp cornered well, the states listed are about equally spaced in energy. When the corners are rounded off, the states shift so as to clump into the groups of states given in the list above;* for example, 1d shifts closer to 2s.

As mentioned above, the depth of the well is approximately constant throughout the periodic table. The radius R increases with A according to $R = 1.5 \times 10^{-13} A^{1/3}$ . For small radius, only the 1s states are bound. As the radius increases, bound states of higher $\ell$ become possible, in the order listed above.

The Pauli exclusion principle prescribes the maximum number of one kind of particle that can occupy a particular space state, just as in electron configuration of atoms. There may be $2(2\ell+1)$ identical particles in the state having angular momentum $\ell$ . The factor 2 represents the two possible spin orientations.

We may attempt to build up the species of nucleus out of nuclear shells, just as the periodic table is obtained in atomic theory:

<u>Presumed</u> <u>configuration</u> (P;N)

1) $H^2$,D (1s;1s) meaning a proton in 1s, a neutron in 1s. Could have spin 0 or 1 and still accord with Pauli principle. Actually observed spin 1.

2) $He^3$ ($1s^2$;1s) Spin 1/2 necessarily, since the proton spins must be opposed.

3) $H^3$ (1s;$1s^2$) Spin 1/2 necessarily, similar to above.

4) $He^4$ ($1s^2$;$1s^2$) Spin 0, a "closed shell" nucleus.

The 1s orbit for both neutrons and protons is now filled. The next orbit, 1p, accomodates six neutrons and six protons, giving $O^{16}$ as the next closed shell nucleus.

5) $He^5$ does not exist; evidently in $He^5$, ($1s^2$;$1s^2$1p), the 1p state is not bound. $He^5$ is the compound nucleus

---

* This may be understood by the following argument. Consider a sharp cornered square well having eigenstate functions $\Psi_1$. The perturbation in energy of state k when the corners of the well are rounded can be found from first order perturbation theory, using $\Psi_1$ as the unperturbed states. The shift in energy is

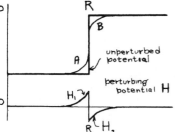

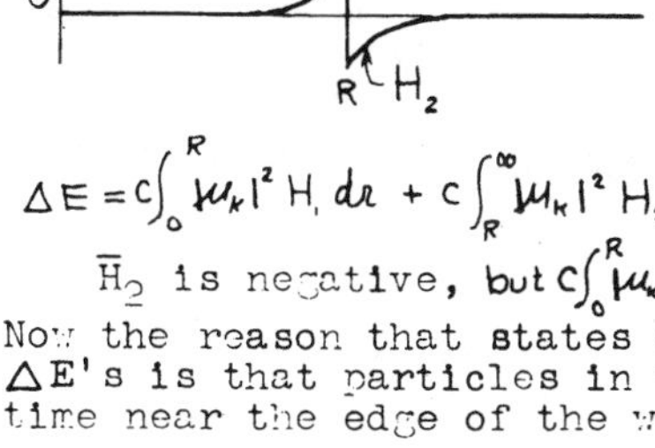


$$\Delta E = \int \psi_k^* H \, \psi_k \, d\tau$$
$$= c\int u_k^* H u_k \, dr, \qquad c = 4\pi,$$

where $u_1$ is the radial part of $\Psi_1$, H is the perturbation of the potential that occurs when the corners are rounded. Let $H = H_1 + H_2$. $H_1$ gives the rounding on the inside, at A, $H_2$ gives the rounding outside, at B.

$$\Delta E = c\int_0^R |u_k|^2 H_1 \, dr + c\int_R^\infty |u_k|^2 H_2 \, dr = \bar{H}_1 c\int_0^R |u_k|^2 dr + \bar{H}_2 c\int_R^\infty |u_k|^2 dr$$

$\bar{H}_2$ is negative, but $c\int_0^R |u_k|^2 dr > c\int_R^\infty |u_k|^2 dr$; therefore $\Delta E$ is +. Now the reason that states of higher angular momentum have larger $\Delta E$'s is that particles in these states spend relatively more time near the edge of the well, in region A.

for the reaction $n + He^4 \rightarrow n + He^4$. A scattering resonance is observed at about 1 Mev, presumably representing the 1p level of $He^5$. In fact, it is claimed that two peaks are observed. These presumably would be the $p_{1/2}$ and $p_{3/2}$ states:

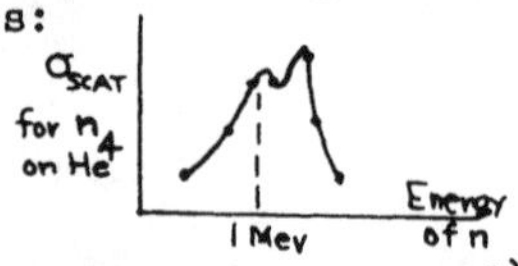


(Hall and Koontz, Phys.Rev. 72 196)

6) $Li^6$ $(1s^2, 1p; 1s^2, 1p)$ There are two independent systems, each consisting of a spin 1 (orbital) and of a spin 1/2 (intrinsic spin). There are many possible total angular momenta, including the observed, namely 1.

The end of this "period" is $O^{16}$, the next closed shell nucleus. It has spin 0 and is unusually stable.

On the orbit model as given so far, the closed shells for either the proton or neutron configurations are at 2, 8, 20, 40, 70, 112, 168,... neutrons and/or protons. For large nuclei the Coulomb energy makes the number of protons less than the number of neutrons, and the nucleus cannot have closed shells of both at the same time.

Empirically, the closing numbers appear to be*

2, 8, 20, 50, 82, 126 (Magic numbers) VIII.64

Nuclei having either such a number of protons, or of neutrons, are unusually stable.

These are not the same as the closed shell numbers given by the preceding development. This discrepancy has been interpreted by M.G.Mayer as follows. Suppose that spin-orbit coupling splits the energy levels corresponding to different J values, that is, 1g splits into $1g_{9/2}$ and $1g_{7/2}$. **Assume** that the level with larger J is more stable, i.e. lies lower. This assumption is not contrary to any known facts about the nucleus. Then the former closed shell number 40, for example, must be altered as follows:

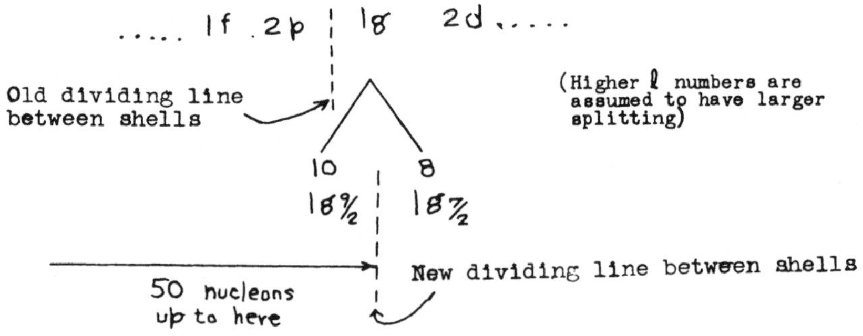


---

* M.G.Mayer, Phys.Rev. 74 235 (1948)

The next shells are

$$\ldots\ldots 1g_{7/2}\ 2d\ 3s\ 1h_{11/2} \big| \ldots\ldots$$

82 nucleons to here

$$\ldots\ldots 1h_{9/2}\ 2f\ 3p\ 1i_{13/2} \big| \ldots\ldots$$

126 nucleons to here

Thus the altered nuclear shell theory gives closed shell numbers that agree with the numbers deduced from experiment.

| Problem. Look up the nuclei having closed shells of either neutrons or protons, and note to what extent they tend to be relatively stable. |
|---|

## L. Capture of Slow Neutrons by Hydrogen

This nuclear reaction is one of the few that can be calculated with some precision.

Only the S wave component of the resolution of the neutron plane wave into radial functions is important for the low neutron energies to be considered here, i.e., thermal energies. $\ell > 0$ waves have angular momentum with respect to the proton so great as to keep a slow neutron beyond the reach of nuclear forces. (See chapter VI, section c. , p.118-19) The only source of angular momentum is intrinsic spin. The final state is the deuteron $^3S$ state (the $^1S$ state is not bound).

$$n(\text{slow}) + H^1 \longrightarrow D + \gamma \ , \quad \hbar\omega = 2.19 \text{ Mev.}$$

The conceivable transitions are

$$^1S \text{ (continuum)} \xrightarrow{\text{(magnetic dipole)}} {}^3S$$

$$^3S \text{ (continuum)} \longrightarrow {}^3S$$ (is not possible, as shown just after eq.VIII.78)

The parity selection rule excludes electric dipole for $S \rightarrow S$ transitions, Ch. V, p. 100 .

The physical mechanism, in classical terms, is the following. Since the magnetic moments of the proton and neutron are not equal, $\mu_P \neq \mu_N$ , the total $\mu$ for the deuteron does not have the same direction as the angular momentum vector, FIG. VIII.11. Therefore the system is a rotating magnetic dipole. The formula for magnetic dipole radiation is almost identical to that for electric dipole radiation, differing only in that the magnetic moment matrix element is used (see Ch. V, p. 95 )

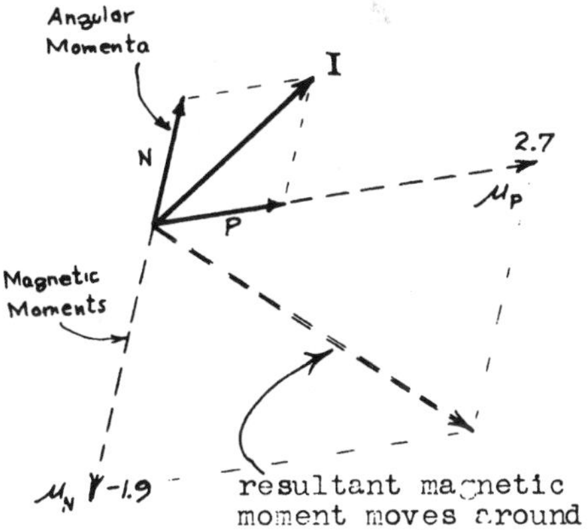


FIG.VIII.11

$$\text{Probability of transition per unit time} = \frac{4\omega^3}{3\hbar c^3}\left|\mu_{\text{final, Initial}}\right|^2 \qquad \text{VIII.65}$$

where $\omega$ = (energy of transition)/$\hbar$. The energy of the transition is 2.23 Mev, the binding energy of the deuteron, provided the energy of the incident neutron is small.

The wave functions are

$$(\text{final}) \quad \psi_{(^3S)} \equiv \frac{u(r)}{r}\begin{cases}(++) & S_z = +1\\ (\ (+-) + (-+)\ )1/\sqrt{2} & S_z = 0\\ (--) & S_z = -1\end{cases}$$

$$(\text{initial}) \quad \psi_{(^1S)} \equiv \frac{j(r)}{r}\left[(+-)-(-+)\right]1/\sqrt{2} \qquad \text{VIII.66}$$

Notation for the spin functions: (+-) means proton spin up,(+); neutron spin down, (-). The space parts of the wave functions are independent of angle, being S functions.

u(r)/r is the solution of the Schrödinger equation for the known potential, $U_3$, the triplet S state neutron-proton potential. The radial equation is

$$u''(r) + \frac{2}{\hbar^2}\left(\frac{M}{2}\right)\left(W_3 - U_3(r)\right)u = 0 \qquad \text{VIII.67}$$

$W_3$ = -2.23 Mev; $U_3$ = -21 Mev for r smaller than the range of nuclear forces, and 0 beyond. (See theory of the deuteron in Ch.VI, p. 115 .) u(r) has the form:

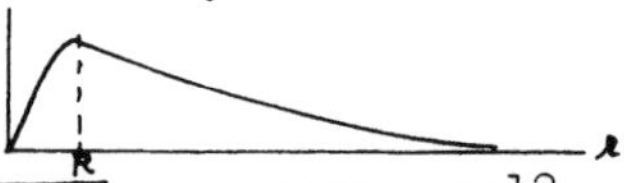


$u(r) \propto e^{-k_3 r}$ for r > R. $k_3 = \sqrt{\frac{M|W_3|}{\hbar^2}} = 2.26 \times 10^{12}\ \text{cm}^{-1}$

The error committed in using $u(r) = C\,e^{-k_3 r}$ for all r is tolerable since R < $1/k_3$ and we shall correct partly as follows. When u(r) is set = $C_3\,e^{-k_3 r}$, we get somewhat too large values in integrations over r. If we compute the normalization C using the same approximation, an error is made which tends to compensate. Normalizing so that

$$\int|\psi_{\text{final}}|^2 d\tau = 4\pi C^2\int_0^\infty |e^{-k_3 r}|^2 dr = 1 \qquad \text{VIII.68}$$

we get $\qquad C = \sqrt{\frac{k_3}{2\pi}}$ $\qquad$ VIII.69

So $\qquad u(r) = \sqrt{\frac{k_3}{2\pi}}\ e^{-k_3 r}$ $\qquad$ VIII.70

j(r) is the S wave component of the resolution of a plane wave $e^{ikz}$ into polar eigenfunctions, but perturbed by the potential well of the proton, The potential well is $U_1$, the potential existing between a neutron and a proton in a singlet S state. $e^{ikz}$ is already normalized to one particle per unit volume in the absence of the perturbing potential well. The unperturbed plane wave can be expanded as follows*

$$e^{ikz} = \sum(2l+1)i^l\left(\frac{\pi}{2kr}\right)^{1/2} J_{l+1/2}(kr)\,P_l(\cos\theta) \qquad \text{VIII.71}$$

---

* Stratton, "Electromagnetic Theory," p. 408.

The S wave is

$$\left(\frac{\pi}{2kr}\right)^{1/2} J_{1/2}(kr) = \frac{\sin kr}{kr} \qquad \text{VIII.72}$$

We must now find how $U_1$ perturbs this. The curvature of the wave function is greatly increased in the region of the well, and is nearly independent of neutron energy, since the neutron energy is small compared to the well depth. From the theory of scattering of neutrons by protons, Ch. VI, we know that for slow neutrons the tangent to the radial function j(r) at the potential well edge R intersects the axis at a distance $a_1$ from r = 0. Therefore j(r) must have the form:

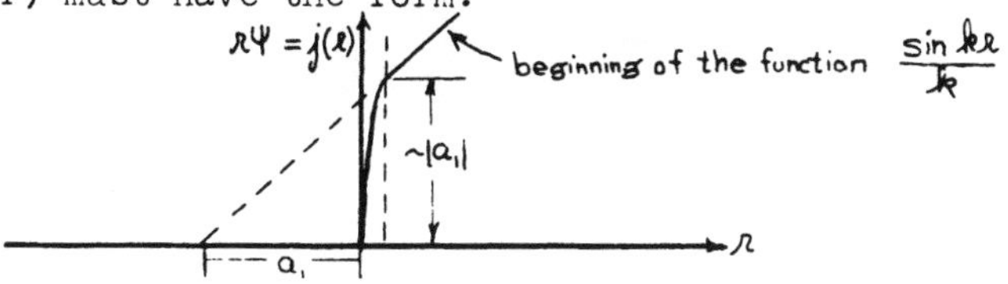


VIII.73

It is evident from the figure that $j(r) \approx (1/k)\sin(kr + |a_1|)$ a fairly good approximation to the true j(r). $j(r) \approx |a_1| + r$ for small r.

Computation of the matrix element. The matrix element is

$$\mu_{fin.\,in.} = \int \psi^*_{final}\, \mu\, \psi_{initial}\, d\tau \qquad \text{VIII.74}$$

The magnetic moment vector is

$$\mu = \mu_P(\sigma_x^p \mathbf{i} + \sigma_y^p \mathbf{j} + \sigma_z^p \mathbf{k}) + \mu_N(\sigma_x^n \mathbf{i} + \sigma_y^n \mathbf{j} + \sigma_z^n \mathbf{k}) \qquad \text{VIII.75}$$

where $\sigma_x^p$ etc. are the Pauli spin operators for proton and neutron. These operators act as follows:

$$\sigma_x(+) \rightarrow (-) \qquad \sigma_y(+) \rightarrow i(-) \qquad \sigma_z(+) \rightarrow (+) \qquad \text{VIII.76}$$

$$\sigma_x(-) \rightarrow (+) \qquad \sigma_y(-) \rightarrow -i(+) \qquad \sigma_z(-) \rightarrow -(-)$$

For example, $\sigma_x^p(+-) = (--)$, leaving the neutron spin symbol unchanged. We calculate $(\mu_{f,i})_z$ as an example.

$$(\mu_{f.i})_z = \int \psi_{final}\,(\mu_P \sigma_z^p + \mu_N \sigma_z^n)\, \psi_{initial}\, d\tau \qquad \text{VIII.77}$$

For this we need $(\mu_P \sigma_z^p + \mu_N \sigma_z^n)$, the spin parts of which give

$$\sigma_z^p \left(\frac{(+-) - (-+)}{\sqrt{2}}\right) = \frac{(+-) + (-+)}{\sqrt{2}}$$

$$\sigma_z^n \left(\frac{(+-) - (-+)}{\sqrt{2}}\right) = \frac{-(+-) - (-+)}{\sqrt{2}}$$

Therefore

$$(\mu_P \sigma_z^p + \mu_N \sigma_z^n)(\text{Spin part of } \psi_{initial}) = (\mu_P - \mu_N)\,\frac{(+-) + (-+)}{\sqrt{2}}$$

(Note that if $\mu_P = \mu_N$ the matrix element would be 0. Physically this corresponds to the case that the resultant magnetic moment has the same direction as the angular momentum, which is invariable and the average magnetic moment would be constant.)

The term above has the same form as the spin function for the final $S_z = 0$ state, and is multiplied by

$$\psi_{final} = \frac{j(r)}{r} \begin{cases} (++) & S_g = +1 \\ [(+-)+(-+)]\,1/\sqrt{2} & S_g = 0 \\ (--) & S_g = -1 \end{cases}$$

Due to orthogonality of the spin functions, only the $\Psi_{final}$ for $S_z = 0$ gives a contribution. Therefore

$$(\mu_{f.i})_g = \int \psi^*_{final} \left(\mu_p \sigma_g^p + \mu_N \sigma_g^n\right) \psi_{initial}\, d\tau$$

reduces to $(\mu_{f.i})_g = (\mu_p - \mu_N) \int_0^\infty \frac{j(r)u(r)}{r^2}\, 4\pi r^2 dr$

Therefore

$$|(\mu_{f.i.})_g|^2 = (\mu_p - \mu_N)^2 \left(\int_0^\infty j(r)u(r)\, 4\pi\, dr\right)^2 \qquad \text{VIII.78}$$

It can be seen now that $^3S$(continuum) ⟶ $^3S$ (bound) does not occur, for in such a transition the matrix element depends on the integral over two S wave functions for the same potential, and this is 0, by orthogonality of eigenfunctions of the same Hamiltonian. In the transition considered above, $^1S \rightarrow {}^3S$, the S functions are for different potential functions.

Since there is no preferred axis, $|(\mu_{f.i})_x|^2 = |(\mu_{f.i})_y|^2 = |(\mu_{f.i})_g|^2$

$$|\mu_{f.i}|^2 = 3\,|(\mu_{f.i})_g|^2 \qquad \text{VIII.79}$$

From the definition of the cross section $\sigma$, and from VIII.65, we get

$$\sigma_{(n,\gamma)} = \frac{1}{v}\, \frac{16\pi^2 \omega^3}{\hbar c^3} (\mu_p - \mu_N)^2 \left(\int_0^\infty j(r)u(r)\, dr\right)^2 \qquad \text{VIII.80}$$

where a factor 1/4 comes from the fact that only 1/4 of the initial continuum states of the neutron and proton are singlet S.

$$\int_0^\infty j(r)\, u(r)\, dr = \sqrt{\frac{k_3}{2\pi}} \int_0^\infty (|a_1| + r)\, e^{-k_3 r}\, dr = \sqrt{\frac{k_3}{2\pi}} \left(\frac{|a_1|}{k_3} + \frac{1}{k_3^2}\right)$$

$$\sigma_{(n,\gamma)} = \frac{1}{v}\, \frac{16\pi^2 \omega^3}{\hbar c^3} (\mu_p - \mu_N)^2\, \frac{k_3}{2\pi} \left(\frac{|a_1|}{k_3} + \frac{1}{k_3^2}\right)^2 \qquad \text{VIII.81}$$

Using the values: $k_3 = 2.26 \times 10^{12}\ \text{cm}^{-1}$
$a_1 = -2.32 \times 10^{-12}$ cm

$$\sigma_{(n,\gamma)} = \frac{6.4 \times 10^4}{\text{velocity (cm/sec)}}\ \text{barns} \qquad \text{VIII.82}$$

For thermal neutrons, $V = 2.2 \times 10^5$ cm/sec, $\sigma_{(n,\gamma)} = 0.29$ barn. The experimental value is 0.30 barn. An elaborate theory including tensor forces gives 0.31 barn.

Remarks on neutron capture in light nuclei. $\sigma_{(capt)}$ is in the range .001-.01 barn, with some exceptions. H, with $\sigma_{(capt)} = .3$, is one of them. This relatively high $\sigma$ is due to the fact that the $^1S$ deuteron state is almost bound, i.e., because $|a_1|$ is exceptionally large. There are other cases of large $\sigma_{abs.}$ in the light elements, but for the reaction (n,particle):

$$\begin{aligned} B^{10},\ \sigma(n,\alpha) &= 3800 \text{ barns} \\ He^3,\ \sigma(n,p) &= 6000 \\ Li^6,\ \sigma(n,\alpha) &= 8000 \\ N^{14},\ \sigma(n,p) &= 1.7 \end{aligned}$$

$\sigma$ for $N^{14}$ is small, compared to the others, because, in contrast to the very light elements, Z is large enough to make the barrier factor exp(-G) * important for emission of a charged particle. The probability for emission of a proton is reduced by a factor of the order of 100.

## M. Photonuclear Reactions

These are mainly $(\gamma,n)$, although $(\gamma,2n)$, $(\gamma,f)$, etc. are known also. The threshold for $(\gamma,n)$ is the binding energy of the neutron, usually about 8 Mev. Exceptions to this figure are D and $Be^9$, having thresholds of 2.2 and 1.71 Mev, respectively.

Strong $\gamma$ absorption resonances have been observed in $_6C$ at 30 Mev (threshold 18.7 Mev), in $_{29}Cu$ at 22 Mev, and in $_{73}Ta$ at 16 Mev. Goldhaber and Teller ** have interpreted these resonances as due to an electric dipole interaction in which all the protons in the nucleus move as a unit with respect to all the neutrons.

Photodisintegration of the deuteron. This is the only photonuclear reaction having a reasonably detailed theory. The transition $\gamma + D \longrightarrow n + p$ (unbound) occurs in two ways:

1) $^3S$ (D ground state) $\xrightarrow{\text{mag dipole}}$ $^1S$ (continuum)

2) $^3S$ (D ground state) $\xrightarrow{\text{elect. dipole}}$ $^3P$ (continuum)

Except at low energies (up to a few tenths of an Mev above threshold), the latter is the dominant reaction. The angular momentum of the P state requires too high a velocity to be attained by low energy neutrons.

The first mode, $^3S \longrightarrow {}^1S$(continuum), is the inverse to the reaction discussed in the preceding section. Therefore its cross section is obtainable from detailed balance arguments, (see section C, p.145), provided equation VIII.81 is first generalized ***to be valid for all neutron energies. The result is

$$\sigma_{(\gamma,n)}(\text{mag. dipole}) = \frac{2\pi}{3}\frac{e^2\hbar}{M^2c^3}(\mu_P-\mu_N)^2 \frac{\sqrt{W_0}\sqrt{\omega-\omega_0}\left(\sqrt{\hbar\omega_0}+\sqrt{W_0}\right)^2}{\omega\left(\hbar\omega-\hbar\omega_0+W_0\right)} \qquad \text{VIII.83}$$

where $\omega_0$ is the angular frequency at threshold, namely (binding energy)/$\hbar$, and Wo is the magnitude of the fictitious binding energy of the deuteron singlet state.****

---

* For definition of G see VIII.12 or III.3, p. 58.
** Goldhaber and Teller, Phys.Rev. 74 1046 (1948).
*** Bethe, D, p. 58.
**** $W_0$ is really a parameter such that if the singlet deuteron state were bound, its energy would have to be $W_0$ in order to give the observed singlet scattering cross section, as shown on the FIGURE in this footnote, which is continued on the next page.

The cross section for the second type of transition, $^3S \rightarrow {}^3P$, can be computed using the formula for gamma ray absorption.*

$$\sigma = 2\,\frac{\omega \mu^2 v}{\hbar^3 c}\,|M|^2 \qquad \text{VIII.84}$$

$\mu$ is the reduced mass, and $v$ the velocity of the emitted particle. M is the matrix element for the electric dipole moment

$$M = \frac{e}{2}\int \psi_{(3S)}\, r\, \psi_{(3P)}\, d\tau \qquad \text{VIII.85}$$

r is the coordinate of the proton with respect to the center of mass. The factor 1/2 results because the neutron is uncharged, and the system is like half a dipole.

We will not carry out the calculation, but merely describe the approximate wave functions.

Except when highest accuracy is wanted, $\Psi_{(^3S)}$ may be taken as $\frac{u(r)}{r} = \sqrt{\frac{k_3}{2\pi}}\, e^{-k_3 r}/r$

(from equation VIII.70). In the absence of the nuclear potential, $\Psi_{(^3P)}$ is the P wave component of a plane wave (see equation VIII.71), and has the form $\frac{1}{kr}\left(\frac{\sin kr}{kr} - \cos kr\right)$ from the properties of half-odd Bessel functions.** At low energies, (A) in FIG. VIII.12, $\Psi_{(^3P)}$ is nearly 0 in the region of the short range nuclear potential well, and is therefore little disturbed by it, and is like the P wave of a free particle. At higher energies $\Psi_{(^3P)}$ has shorter period, and is perturbed more by the well, (B) in FIG. VIII.12.

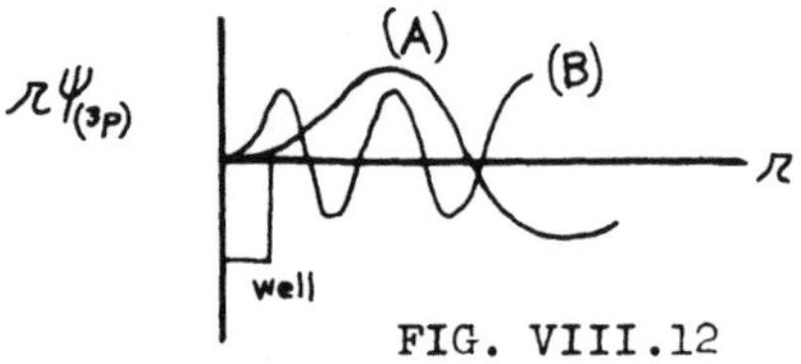


FIG. VIII.12

$r\Psi$ is plotted in FIG. VIII.12 because the behavior of $\Psi$ is seen best when matrices are written in the form, $M \propto \int (r\psi_f^*)\, r\, (r\psi_i)\, dr$

The result is:

$$\sigma_{(\gamma,n)}\,(\text{electric dipole}) = \frac{8\pi}{3}\,\frac{e^2}{Mc\omega_0}\left\{\frac{\omega_0(\omega-\omega_0)}{\omega^2}\right\}^{3/2} \qquad \text{VIII.86}$$

where $\omega_0$ is the frequency at threshold, as before. The two cross sections $\sigma_{(\gamma,n)}$ (mag.dip.) and $\sigma_{(\gamma,n)}$ (elect.dip.) are plotted in FIG.VIII.13

(cont.).

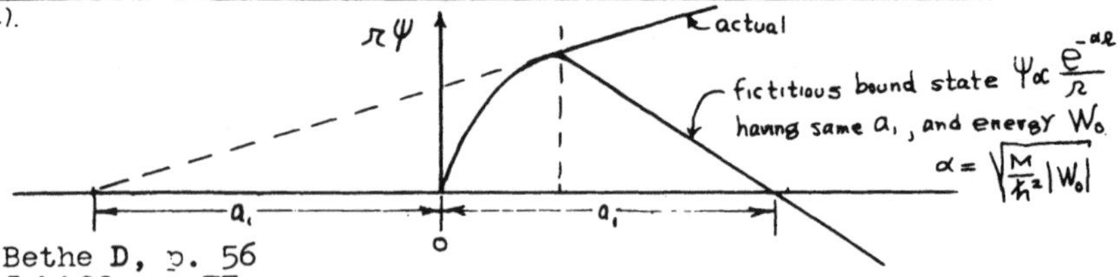


* Bethe D, p. 56
** Schiff, p. 77

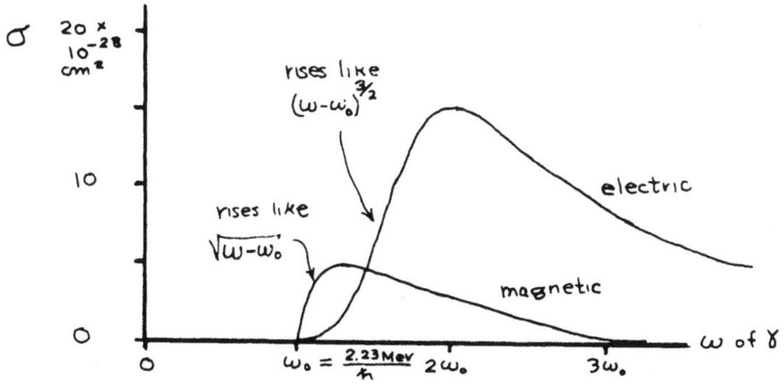


The reason that $\sigma_{(\gamma,n)(mag)}$ rises so sharply is that 0 energy above the threshold (0 excitation) is almost a resonance energy. As can be seen from equation VIII.83, if $W_0$ were zero, there would be a resonance at 0 excitation energy.

N. Remarks on Very High Energy Phenomena

a) Stars. A prominent feature of very high energy phenomena is the production of stars. These are seen in photographic emulsions and in cloud chambers. The prongs are due to ionizing particles, protons, alphas, or larger fragments. Neutrons leave no trace. The quantitative interpretation is rudimentary. Star production can be interpreted roughly as evaporation of particles from a very high temperature nucleus. M. Goldberger* has added to the evaporation model consideration of the situation when a very high energy nucleon, of energy around 100 Mev, has first entered the nucleus. At this time, when its wavelength is very small and its energy is still undistributed among the other nucleons, it has collisions which knock other nucleons out of the nucleus immediately. Soon, however, the energy not carried off by these quickly escaping nucleons is distributed in the form of statistical excitation of the nucleus. Then the escape of nucleons is described by the slower process of evaporation. Goldberger calculates that about 1/2 of the initial 100 Mev leaves immediately, and about 1/2 is trapped and produces evaporation.

STAR

b) Deuteron stripping.** Deuterons, given an energy of 200 Mev in the Berkeley 184" cyclotron, impinge on a target inside the cyclotron, and a beam of very high energy neutrons emerges. (FIG. VIII.14) The interpretation is that the proton of the loosely bound deuteron hits and is caught by a nucleus, and the neutron flies on. Ideally, the neutron would have about 1/2 the energy of the deuteron. However the neutron has a velocity with respect to the center of mass of the deuteron. After the proton is removed, the neutron's total velocity vector is the vector sum of the velocity of the center of mass of the deuteron and the velocity of the neutron relative to the center of mass at the instant the proton is removed. Therefore the neutrons emerge with a spread in energy and in angle.

---

* M.L.Goldberger, Phys.Rev. 74 1269 (1948)

** Helmolz, McMillan and Sewell, Phys.Rev. 72 1003 (1947)

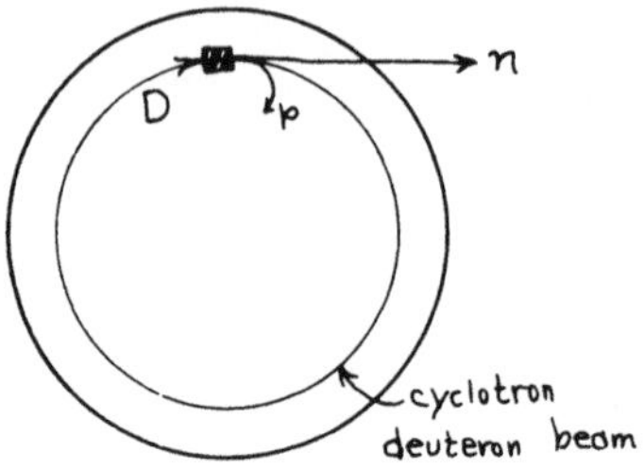


FIG.VIII.14

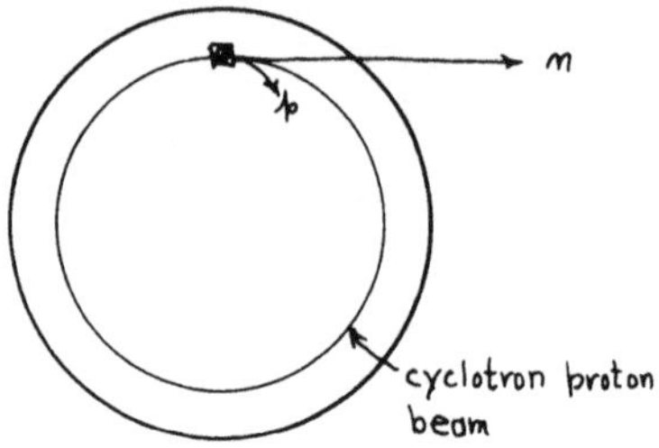


FIG.VIII.15

c) Observed exchange in p-n scattering. When a beam of 350 Mev protons hits a scatterer placed inside the Berkeley cyclotron, FIG.VIII.15, high energy neutrons (as well as protons) emerge. This shows that in many collisions the neutron and proton exchange roles:

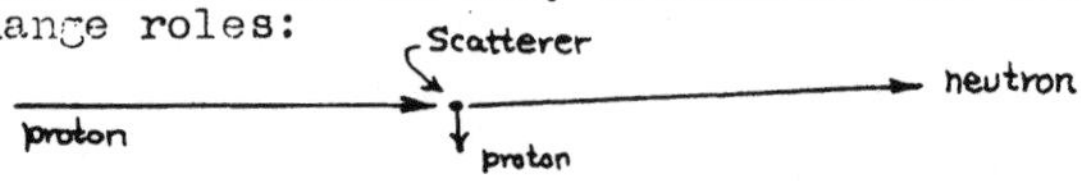


A similar effect has been observed in n-p scattering.*

> Problem: Compute the distribution in energy and angle of scattering of neutrons resulting from the stripping of fast deuterons, using the simple model described above.
> (This calculation is done in the first part of Serber's paper on the theory of stripping of deuterons, Phys.Rev. 72 1007 (1947). Although the most probable neutron energy is 1/2 that of the deuteron, the energy spread is larger than the binding energy of the deuteron by a factor of $(1/2\ E_d/BE(d))^{1/2} \approx (100/2)^{1/2} \approx 7$. This may seem like violation of conservation of energy. However, the velocity of the neutron relative to the laboratory system $= v_{deut.} + v_{n\ rel.\ deut.}$, and the energy of the neutron in the lab. system therefore is
>
> $$\frac{1}{2} M \left( v_d^2 + (v_{n\ rel.\ to\ deut.})^2 + 2 v_d v_{n\ rel.\ to\ deut.} \right)$$
>
> where the last term is the main contribution to the spread.

For a summary of very high energy phenomena, see Chew and Moyer, Am. J. Physics 18 125-135, 19 17, 19 203.

*This is discussed further under "Exchange Forces" in Ch. VI, p. 122

## Chapter IX
## NEUTRON PHYSICS

### A. NEUTRON SOURCES

1. Radioactive Sources

The most common laboratory neutron source is the radium-beryllium source. This consists of about five parts of Be to one of Ra and makes use of the reaction:

$$Be^9 + \alpha \rightarrow C^{12} + n + 5.5 \text{ Mev}$$

Due to the long range $\alpha$'s (7.68 Mev) from Ra decomposition products, the energy spectrum of neutrons obtained extends all the way up to 13 Mev. Slow neutrons are usually obtained by surrounding the source with paraffin (see section B.). The disadvantage of this source is the strong gamma radiation from the radium. In cases where such a strong background cannot be tolerated, a Po - Be source can be used. The relative strengths are

| | | | |
|---|---|---|---|
| Ra-Be: one curie*(1 gm.) of Ra | yields | 10 to 15 x $10^6$ | neutrons/sec |
| Po-Be: one curie of Po | yields | 2.8 x $10^6$ | " |
| B($\alpha$,n)N: one curie of Ra | " | 2 x $10^6$ | " |
| F($\alpha$,n)Na: one curie of Ra | " | 2 x $10^5$ | " |

2. Photo Sources

This type of source makes use of ($\gamma$, n) reactions. Two of these reactions have low enough $\gamma$ thresholds for use with natural $\gamma$ sources. They are

$Be^9 + \gamma \rightarrow Be^8 + n$ threshold = 1.6 Mev**

$H^2 + \gamma \rightarrow H^1 + n$ threshold = 2.23 Mev***

An advantage to this type of source is that monochromatic neutrons may be obtained. A practical rule for calculating the strength of a particular source is that 1 gm. of Ra at 1 cm. from 1 gm. of Be gives 30,000 neutrons/sec. Disadvantages are the strong gamma background, smaller yield than the ($\alpha$,n) reactions, and that the source is bulkier. The following table gives the monochromatic energies obtained using various sources:

| $\gamma$ Source | Lifetime | Target | E in Mev |
|---|---|---|---|
| Ra | | Be | .12, .50 |
| Mesothorium | 6.7y | Be | .16, .88 |
| " | | $D_2O$ | .22 |
| $Sb^{124}$ | 60d | Be | .03 |
| $Ga^{72}$ | 14h | $D_2O$ | .12 |
| $Na^{24}$ | 15h | $D_2O$ | .24 |
| $La^{140}$ | 40h | Be | .60 |
| $Na^{24}$ | 15h | Be | .3 |

* Ch. I, page 18
**Rasetti, page 253
***Ch.VIII, page 175.

3. Artificial Sources *

One of the simplest of these is the bombardment of heavy ice or heavy paraffin with a deuteron beam of energy in the hundreds of kev. The reaction is:

$$D + D \rightarrow He^3 + n + 3.2\ \text{Mev}$$

or $$D + D \rightarrow H^3 + H^1$$

Both reactions are about equally probable. Another reaction used in artificial production of neutrons is a proton beam (usually from a Van de Graaf machine) on a lithium target:

$$Li^7 + H^1 \rightarrow Be^7 + n - 1.647\ \text{Mev}$$
(endothermic)

Yields for a Thick Target of Heavy Ice

| Deuteron Beam Energy | Yield in N's per Deuteron |
|---|---|
| 0 kev | 0 x $10^{-7}$ |
| 50 | 0.2 |
| 100 | 0.68 |
| 200 | 3.0 |
| 300 | 6.9 |
| 500 | 19 |
| 1000 | 81 |
| 2000 | 400 |

The threshold is 8/7 x 1.647 = 1.882 Mev in the lab system. For a proton at the threshold energy, the collision looks as follows:

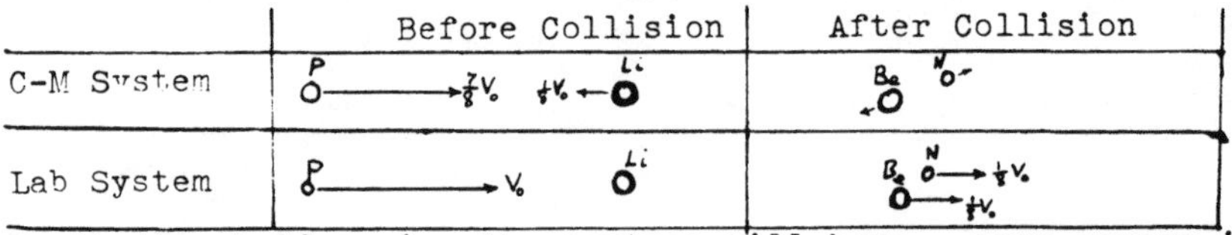


In the threshold region the neutron will have ~zero energy in the c-m system, but 30 kev in the forward direction in the lab system. If the proton beam energy is such that in the c-m system the neutron has 1/8 $V_0$, then in the lab system the neutrons coming off in the back direction will have zero velocity.

The following reaction is useful for obtaining high energy neutrons:

$$H^3 + H^2 \rightarrow He^4 + n + 17.6\ \text{Mev}$$

Either a $H^3$ or a D beam may be used. In the case of the $H^3$ beam, heavy ice is used as the target, while for the D beam the target is a chamber of $H^3$ gas. This latter method conserves $H^3$ which can be obtained from Oak Ridge only in small amounts.

$D(t,n)He^4$

| Beam Energy | $\sigma$ in barns |
|---|---|
| 24 kev | .01 |
| 50 | .3 |
| 100 | 2.6 |
| 200 | 4.5 |
| 500 | 1.5 |
| 1000 | .4 |
| 2000 | .17 |
| 4000 | .15 |

Neutrons can be produced by using a deuteron beam in cyclotrons. The data for 10 Mev beams against various targets is

| Target | D | Au | Be | $B_4C$ | C | $H_2O$ | $D_2O$ | Al | Ni | Cu | Pt |
|---|---|---|---|---|---|---|---|---|---|---|---|
| Yield in Neutrons per $\mu$ coulomb of D | $2.1 \times 10^{10}$ | 0.5 | 3.7 | 2.0 | 1.3 | 0.5 | 1.3 | .99 | .36 | .62 | .07 |

Cyclotrons can give beams of 200 $\mu$ amps. or $\sim 7 \times 10^{12}$ N/sec.

High energy neutron beams (~90 Mev) are produced in the largest cyclotrons by deuteron stripping (see Ch. VIII p.177).

The source which gives the greatest intensity of neutrons is the chain-reacting pile. More will be said about these piles at the end of this chapter. A pile which gave 2 neutrons per

* Rev. Mod. Phys. 21, 635 (49)

fission would then yield about one neutron for each 1.5 x $10^{-4}$ ergs, or 6 x $10^{13}$ neutrons per second for each kilowatt.

More detailed data on these and other reactions is given in "Preliminary Report #4", Nuclear Science Series, published by the Division of Mathematical and Physical Sciences of the National Research Council. The report is by A. Hanson and R. Taschek.

## B. SLOWING DOWN OF NEUTRONS

### 1. Inelastic Scattering

This is discussed in Ch. VIII, p. 144. The general reaction is

$$A + n \rightarrow A^* + n'$$

where $A^*$ is an excited state of A.

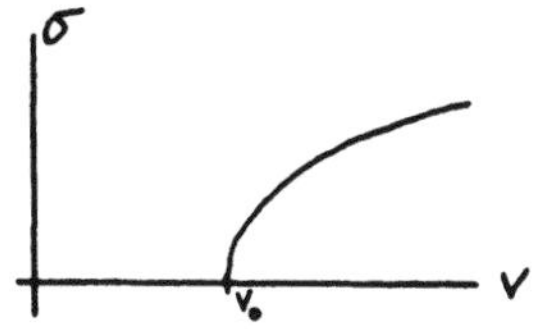

Endothermic σ

Then $\frac{1}{2}\mu V_{final}^2 = \frac{1}{2}\mu V_{initial}^2 -$ excitation E of $A^*$ - kinetic E of $A^*$

Thus inelastic scattering can't take place for incident neutrons whose $\frac{1}{2}\mu V_{initial}^2 <$ excitation energy.

This leaves only elastic scattering to reduce the energies of neutrons which are in the energy range of thousands of ev's down to thermal energies (1/40 ev).

### 2. Elastic Scattering

Consider a collision between a neutron of mass one and a nucleus of mass A. The collision will look as follows:

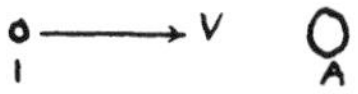

Lab System

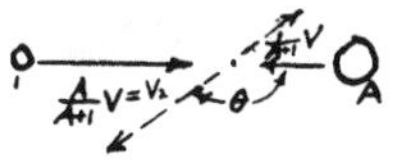

C-M System

$V_2$ is the velocity of the neutron in the c-m system $= \frac{A}{A+1} V$

Let $V_2'$ be the velocity of the neutron after collision in the lab system. It is seen that $V_2'$ is the resultant of the two vectors in FIG. 1.

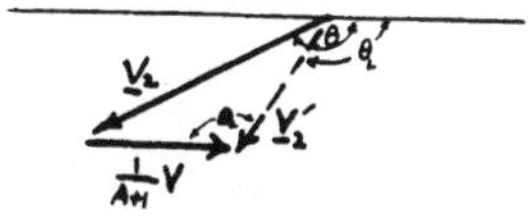
FIG. IX.1

From the law of cosines,

$$V_2'^2 = V_2^2 + \left(\frac{1}{A+1}V\right)^2 - 2V_2\left(\frac{1}{A+1}V\right)\cos(\pi-\theta)$$

$$= \frac{V^2}{(A+1)^2}\left(A^2+1+2A\cos\theta\right)$$

$$\left(\frac{V_2'}{V}\right)^2 = \frac{E}{E_0} = \frac{1}{(A+1)^2}\left(A^2+1+2A\cos\theta\right) \qquad \text{IX.1}$$

where E is the neutron energy after one collision and $E_0$ is the energy before the collision.

Thus $\left(\frac{A-1}{A+1}\right)^2 E_0 \le E \le E_0$ since $0 \le \theta \le \pi$

Also $V_2^2 = \left(\frac{1}{A+1}V\right)^2 + V_2'^2 - 2\left(\frac{1}{A+1}V\right)V_2' \cos\theta_L$

which gives $\cos\theta_L = \frac{A\cos\theta + 1}{\sqrt{A^2+1+2A\cos\theta}}$ IX.2

Next the energy distribution $\frac{dW}{dE}$ will be determined where dw is the probability that the energy lies between E and E + dE.

Assumption: The scattering is isotropic in the c-m system. This is true for S-wave scattering which is applicable here.

Then $dw = \frac{d\Omega}{4\pi} = \frac{2\pi \sin\theta\, d\theta}{4\pi} = \frac{1}{2}\sin\theta\, d\theta$

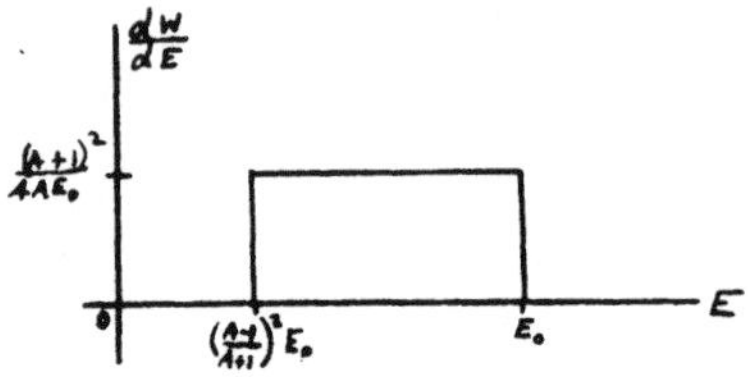

FIG. IX.2

But, from IX.1,

$$\frac{dE}{E_0} = \frac{2A\sin\theta\, d\theta}{(A+1)^2}$$

thus

$$\frac{dE}{E_0} = \frac{4A\, dW}{(A+1)^2}$$

$$\frac{dW}{dE} = \frac{(A+1)^2}{4A E_0} \qquad \text{IX.3}$$

$$\bar{E} = \frac{1}{2}E_0\left[1 + \left(\frac{A-1}{A+1}\right)^2\right]$$

$$\bar{E} = E_0\left[1 - \frac{2A}{(A+1)^2}\right] \qquad \text{IX.4}$$

For A = 1 (collisions with hydrogen) $\bar{E}_1 = \frac{1}{2}E_0$, where $E_n$ stands for the energy after n collisions.

One of the homework problems was to determine the energy distribution ($\frac{dW}{dE_n}$) after n collisions in hydrogen. The result is

$$\frac{dW}{dE_n} = \frac{1}{(n-1)!\, E_0}\left(\ln\frac{E_0}{E_n}\right)^{n-1} \qquad (\text{for } A=1) \qquad \text{IX.5}$$

Rather than take up space here with the derivation, we refer to an article by Breit and Condon, Phys. Rev. 49, 229 (1936). Using IX.5 one can show that the average energy after n collisions is

$$\bar{E}_n = \left(\tfrac{1}{2}\right)^n E_0 \qquad \text{IX.6}$$

This average however, is not very useful, since most of the neutrons will have a much smaller energy. See FIG. 3 where $E_m$ is the median value. The physical problem confronting us is: given neutrons initially at $E_0$, how many collisions (n) must be made for most of the neutrons to reach thermal energies? Clearly IX.6 will not do.

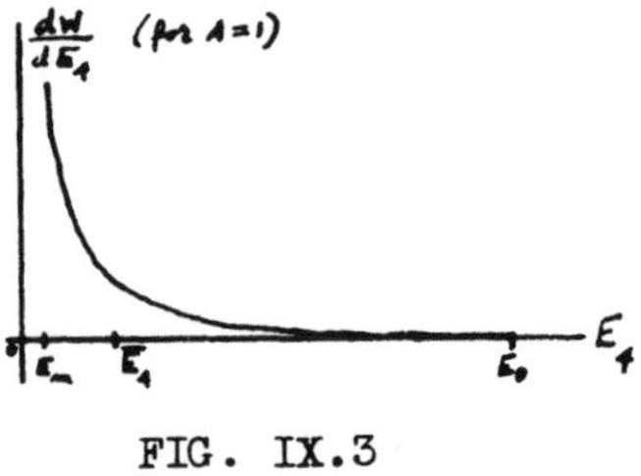

FIG. IX.3

The trick is to calculate the average value of ln(E) after the first collision and to use the relation

$$\ln E_n = \ln E_0 - \sum_{j=0}^{n-1} \ln\frac{E_j}{E_{j+1}}$$

(for any one neutron after n collisions)

$$\overline{\ln E_n} = \ln E_0 - \sum_{j=0}^{n-1} \overline{\ln \frac{E_j}{E_{j+1}}}$$

Let $$\xi \equiv \overline{\ln \frac{E_0}{E_1}} = \frac{\int_{\left(\frac{A-1}{A+1}\right)^2 E_0}^{E_0} \left(\ln \frac{E_0}{E_1}\right) \frac{dW}{dE_1}\, dE_1}{\int_{\left(\frac{A-1}{A+1}\right)^2 E_0}^{E_0} \frac{dW}{dE_1}\, dE_1} = \int_{\left(\frac{A-1}{A+1}\right)^2 E_0}^{E_0} \ln \frac{E_0}{E_1}\, dE_1$$

$$\xi = 1 - \frac{(A-1)^2}{2A} \ln\left(\frac{A+1}{A-1}\right) \qquad \text{IX.7}$$

To evaluate $\overline{\ln \frac{E_j}{E_{j+1}}}$ consider a group of neutrons which have had j collisions and which are in a small energy range $dE$. By IX.7, the average $\ln \frac{E_j}{E_{j+1}}$ for them is $\xi$. The same holds for all other groups of neutrons which have had j collisions. Thus $\overline{\ln \frac{E_j}{E_{j+1}}} = \xi$.

$$\therefore \quad \overline{\ln E_n} = \ln E_0 - n\xi \qquad \text{IX.8}$$

This formula can be used directly to get an idea of the number of collisions (n) required to reduce the neutrons to a median energy $E_n'$. We shall assume $\overline{\ln(E_n)}$ is close to $\ln(E_n')$. To this approximation

$$E_n' \approx E_0 e^{-n\xi} \qquad \text{IX.8}$$

For large A, $\xi$ is almost = 2/A since

| A | $\xi$ |
|---|---|
| 1 | 1 |
| 2 | .725 |
| 12 | .158 |
| 14 | .135 |
| large A | $\frac{2}{A}$ |

$$\ln\left(\frac{A+1}{A-1}\right) = \ln\left[(A+1)A^{-1}\left(1-\tfrac{1}{A}\right)^{-1}\right] \approx \ln\left(1+\tfrac{2}{A}\right) \approx \frac{2}{A}$$

$$\xi \approx \frac{2A - 2A + 4 - \frac{2}{A}}{2A}$$

$$\xi \approx \frac{2}{A}$$

For an $E_0$ of 1 Mev, IX.8 shows that 17.6 collisions in hydrogen are needed to reduce $E_n$ to thermal energy (1/40 ev). In carbon, on the other hand, 110 collisions are needed.

Actually this reasoning breaks down for low values of $E_n$, since the scattering nuclei already have thermal energies rather than being at rest. For substances other than monatomic gases the formula breaks down even before this due to the binding energy between atoms. In this energy region (~1/3 ev) the collisions cannot be considered as elastic. See section D.1 page 194.

3. Energy Distribution of Neutrons from a Mono-energetic Source

Let $\sigma_s$ be the scattering cross section per unit volume

$\sigma_s$ = atomic cross section x no. atoms per unit volume

(A plot of the atomic scattering cross section for hydrogen appears in FIG. IX.8 p 194 )

Let $\sigma_a$ be the absorption cross section per unit volume.

Let $\varepsilon \equiv \ln(E)$. $\varepsilon_0 = \ln(E_0)$.

Let $n(\varepsilon)\, d\varepsilon$ be the total number of neutrons in the energy range $d\varepsilon = \frac{dE}{E}$.

Then the mean reduction in $\varepsilon$ per collision $= \xi$

Let $\lambda(\epsilon) \equiv$ scattering mean free path.

$$\lambda = \frac{1}{\sigma_s}$$

Let $\Lambda(\epsilon) \equiv$ absorption mean free path.

$$\Lambda = \frac{1}{\sigma_a}$$

Let $q(\epsilon)$, the "current" along the $\epsilon$ axis, = the total number of neutrons per unit $\epsilon$ times their time rate of change of $\epsilon$.

$$\text{Then } q(\epsilon) = n(\epsilon)\frac{V}{\lambda}\xi = n(\epsilon)V\sigma_s\xi. \qquad \text{IX.9}$$

Let Q be the rate of production of neutrons at the source.

Case 1: No absorption ($\sigma_a = 0$)
Then $q(\epsilon) = Q$

or $\dfrac{nV\xi}{\lambda} = Q$

$$n(\epsilon)\,d\epsilon = \frac{Q\lambda}{V\xi}\frac{dE}{E} \propto \frac{dE}{E^{3/2}} \quad \text{in regions where } \sigma_s(\epsilon) \text{ is constant}$$

In most practical cases one is interested in the neutron flux nV rather than the neutron density.

$(nV)d\epsilon = \dfrac{Q\lambda}{\xi}\dfrac{dE}{E}$ = total neutron flux along $\epsilon$ axis in energy range $dE$

Case 2: With absorption

At equilibrium the (excess of q at $\epsilon + d\epsilon$ over q at $\epsilon$) = the number of neutrons per second absorbed in the energy interval $d\epsilon$ or

$$dq = \frac{nV}{\Lambda}d\epsilon$$

$$\xi\frac{d(nV\sigma_s)}{nV\sigma_s} = \frac{\sigma_a}{\sigma_s}d\epsilon$$

integrating,

$$\xi \ln(nV\sigma_s)\Big]_{\epsilon}^{\epsilon_0} = \int_{\epsilon}^{\epsilon_0}\frac{\sigma_a}{\sigma_s}d\epsilon$$

$$\xi\ln\left(\frac{Q}{\xi}\right) - \xi\ln(nV\sigma_s) = \int_{\epsilon}^{\epsilon_0}\frac{\sigma_a}{\sigma_s}d\epsilon$$

$$\xi\ln(nV\sigma_s) \doteq -\int_{\epsilon}^{\epsilon_0}\frac{\sigma_a}{\sigma_s}d\epsilon + \xi\ln\left(\frac{Q}{\xi}\right)$$

$$\therefore \qquad n(\epsilon)V = \frac{Q}{\xi\sigma_s}e^{-\frac{1}{\xi}\int_{\epsilon}^{\epsilon_0}\frac{\sigma_a}{\sigma_s}d\epsilon} \qquad \text{IX.10}$$

Problem: High energy neutrons are produced in the upper atmosphere. In the high energy ranges they are slowed down by both ineleastic and elastic scattering. Let Q be the number of these neutrons crossing E = $10^5$ ev per second. Now IX.10 may be applied to determine how many neutrons reach thermal energies before being absorbed by nitrogen.

We shall make use of the following atomic cross sections:

| | Nitrogen | Oxygen |
|---|---|---|
| $\sigma_s$ atomic | $11 \times 10^{-24}$ $cm^2$ | $5 \times 10^{-24}$ $cm^2$ |
| $\sigma_a$ atomic | $\frac{3.85\times10^{-19}}{V}$ $cm^2$ | 0 |

Thus the total effective cross sections for air are

$\sigma_a = \frac{3.85\times10^{-19}}{V} \times n_N = \frac{16.6}{V}\ cm^{-1}$ where $n_N$ is the no. of nitrogen atoms/$cm^3$

$\sigma_s = 5\times10^{-24} n_{oxy.} + 11\times10^{-24} n_N = 52.7\times10^{-5}\ cm^{-1}$

The effective $\xi \approx \frac{2}{A_{eff.}}$ where $A_{eff.} = .9\times14 + .1\times16 = 14.2$

$$\xi \approx .141$$

$$\int_E^{E_0} \frac{\sigma_a}{\sigma_s} dE = \int_E^{E_0} \frac{16.6}{52.7\times10^{-5}} \sqrt{\frac{M}{2E}} \frac{dE}{E} = 2.87\times10^{-8} \int_E^{E_0} E^{-\frac{3}{2}} dE \approx 5.74\times10^{-8} \frac{1}{\sqrt{E}}$$

$$n(E)V = \frac{Q}{.141\times52.7\times10^{-5}} e^{-\frac{5.74\times10^{-8}}{.141}\frac{1}{\sqrt{E}}} \quad cm\ sec^{-1} \text{ per unit } E \qquad \text{IX.10a}$$

$$= 1.35\times10^4 Q\, e^{-2.04} \qquad \text{for} \qquad E = \tfrac{1}{40}\ ev$$

$n(E)V = 1.75\times10^3 Q$ Thus at thermal energy the neutron flux is 1/7.7 of what it is at $10^5$ ev.

4. Mean Distance from a Point Source vs. Energy

In this section we shall calculate $\overline{R^2(E)}$, the mean square distance to all the neutrons of energy E generated by a mono-energetic point source.

The total number of collisions $\frac{1}{\xi}\ln(\frac{E_0}{E})$ by IX.8. If the neutrons travelled isotropically the exact distance $\lambda$ between collisions, this would be the ordinary random walk problem where the mean square distance travelled is given by

$$\overline{R^2} = n\lambda^2 = \frac{1}{\xi}\ln\left(\frac{E_0}{E}\right)\lambda^2$$

However $\lambda$ is so defined that the probability of having a collision in a distance dr is $\frac{1}{\lambda}dr$.

Let P(r) be the probability of not having had a collision after travelling a distance r.

Then $dP(r) = -P(r)\frac{dr}{\lambda}$

$$P(r) = e^{-\frac{r}{\lambda}} \qquad \text{IX.11}$$

$$\bar{r} = \frac{\int_0^\infty r P(r)\,dr}{\int_0^\infty P(r)\,dr} = \lambda$$ which is the mean free path by definition.

$$\overline{r^2} = \frac{\int r^2 e^{-r/\lambda}\,dr}{\int e^{-r/\lambda}\,dr} = 2\lambda^2$$

For isotropic scattering after n collisions,

$$\overline{R^2} = n\,\overline{r^2}$$

$$\overline{R^2} = 2\left(\frac{1}{\xi}\ln\frac{E_0}{E}\right)\lambda^2$$

However, in the lab system, the forward direction is preferred. This persistence of velocity modifies $\overline{R^2}$ as follows:*

$$\overline{R^2} = \frac{1}{1-\overline{\cos\theta}}\,2\left(\frac{1}{\xi}\ln\frac{E_0}{E}\right)\lambda^2$$

Using IX.2 it is easily seen that $\overline{\cos\theta} = \frac{2}{3A}$ ($\theta$ is now used as lab system angle rather than $\theta_L$ )

For $\lambda$ as a slowly varying function of E,

$$\overline{R^2} \approx \frac{2}{\left(1-\frac{2}{3A}\right)\xi}\int_E^{E_0} \lambda^2(E)\,\frac{dE}{E} \qquad \text{IX.12}$$

---

* Let $\underline{r}_j$ be the path after the jth collision.

$$R = \sum_{j=0}^{n} \underline{r}_j$$

$$R^2 = \sum_j r_j^2 + \sum_{j\neq k} \underline{r}_j \cdot \underline{r}_k$$

$$\overline{R^2} = n\,\overline{r_j^2} + \sum_{j=0}^{n-1}\sum_{k=j+1}^{n} 2\,\overline{r_j\, r_k \cos\theta_{jk}}$$

Since the average of the product of independent quantities equals the product of the averages,

$$\overline{r_j\, r_k \cos\theta_{jk}} = \bar{r}_j\,\bar{r}_k\,\overline{\cos\theta_{jk}}$$

It has been shown (II.67, page 51) that the average cosine of the resultant angle ($\theta_{jk}$ ) which is the sum of (k - j) deflections is

$$\overline{\cos\theta_{jk}} = \overline{\cos\theta\,\cos\theta_{j\,k-1}}$$

Since the distribution of $\theta$ is the same after each collision, the above equation is seen to be a recursion formula. Since each particular $\theta$ is independent of the others,

$$\overline{\cos\theta_{jk}} = \left(\overline{\cos\theta}\right)^{k-j}$$

$$\overline{R^2} = n(\overline{r_j^2}) + (n-1)\,2(\bar{r}_j)^2\left[\,\overline{\cos\theta} + (\overline{\cos\theta})^2 + (\overline{\cos\theta})^3 + \dots\,\right]$$

$$\approx n\,2\lambda^2\left[1-\overline{\cos\theta}\right]^{-1}$$

$$\overline{R^2} \approx \frac{2n}{1-\overline{\cos\theta}}\,\lambda^2$$

IX.12 is not quite exact. However, for particles of equal mass (neutrons in hydrogen) there exists an exact but lengthy expression. It is given in section 3 of Ch.VI of "Neutron Physics" (L.A. 255).

We shall now discuss experimental methods for determining spatial distributions of neutrons. A foil of indium sandwiched between two foils of cadmium makes a sensitive detector for 1.44 ev neutrons. This is easily seen from the absorption cross section curves given in FIG 4. A measurement of the activity of the indium (13s and 54m half lives) gives the value of the neutron flux. A compilation of cross section curves of the elements is given in Goodman and also in the Rev. Mod. Phys., Oct. 1947.

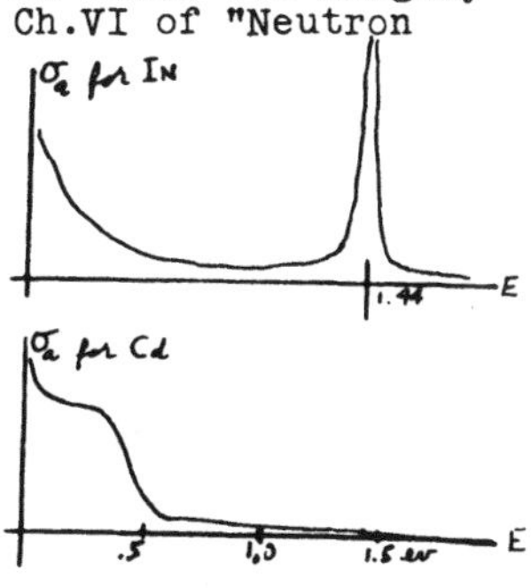

FIG. IX.4

The experimental distribution of neutrons around a mono-energetic point source is a gaussian-like distribution as shown in FIG. 5. A description of experimental techniques is given in a paper by Amaldi and Fermi, Phys. Rev. 50, 899 (1936). The theoretical verification of these results will now be demonstrated.

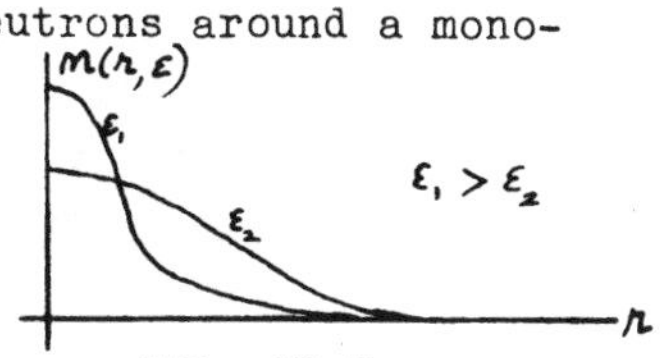

FIG. IX.5

## C. DIFFUSION THEORY *

1. The Age Equation

The two following assumptions must be made for this theory:

1. $\lambda(E)$ is slowly varying.
2. A large number of collisions take place.

This theory will be found more applicable to collisions with heavier nulcei rather than with protons for two reasons. First, $\lambda$ in hydrogen is much larger for the first few collisions (see FIG. IX.8, page 194 ) than for the others. Secondly, the law of large numbers is more applicable in the case of heavier nuclei ($\sim$100 collisions for carbon), than for hydrogen where 1 Mev neutrons become thermal in about 17 collisions.

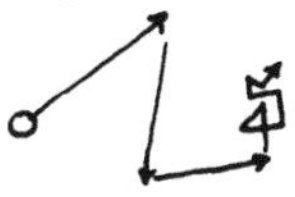
High Energy Neutron Path in Hydrogen

We shall now derive the neutron diffusion equation, which is also known as the Fermi age equation.

Let $n(\underline{r},\varepsilon)$ be the number of neutrons per unit volume per unit $\varepsilon$. This is the previous $n(\varepsilon)$, per unit volume.

Let $q(\underline{r},\varepsilon)$ be the current density along the $\varepsilon$ axis. This is the previous $q(\varepsilon)$, per unit volume. It is also called the slowing down density.

$$q(\underline{r},\varepsilon) = n(\underline{r},\varepsilon)\frac{v}{\lambda}\xi$$

$$\varepsilon = \ln E$$

*A review article based on lectures of R.E. Marshak is in the May - Aug. 1949 issues of Nucleonics.

Consider all neutrons between $\varepsilon$ and $\varepsilon + d\varepsilon$. If $\nabla n(\underline{r},\varepsilon)$ isn't zero, there will be a flow from the more concentrated regions to the less concentrated. Kinetic theory tells us that the number of neutrons in the energy interval $d\varepsilon$ flowing thru a surface element $d\underline{S}$ per second $= -D\,\nabla(n\,d\varepsilon)\cdot d\underline{S}$. This relation defines the diffusion coefficient, D, which can be shown to be $\frac{\lambda V}{3}$. In our case the persistence of velocity increases the mean square distance by a factor $\frac{1}{1-\overline{\cos\theta_L}}$. It will soon be shown that D and $\overline{R^2}$ are related (see IX.17). Since IX.12 contains the factor $\frac{1}{1-\overline{\cos\theta_L}}$, then D must also contain it when there is persistence of velocity.

$$D = \frac{\lambda V}{3(1-\frac{2}{3A})}$$

The number of neutrons in the energy interval $d\varepsilon$ which accumulate per second in a volume element $\Delta V$ is

$$\int_{\Delta V} D\nabla(n\,d\varepsilon)\cdot d\underline{S}$$

Applying the divergence theorem, one obtains

$$D\int_{\Delta V}\nabla\cdot\nabla(n\,d\varepsilon)\,dV \approx D\nabla^2(n\,d\varepsilon)\,\Delta V$$

Thus the increase in neutrons in the interval $d\varepsilon$ per unit time per unit volume by diffusion is $D\,\nabla^2 n\,d\varepsilon$. However there is another mechanism to contribute to this increase. It is the neutrons from a higher energy region dropping down to the energy region under consideration. The number of neutrons crossing the value $\varepsilon$ per second is the $\varepsilon$ current density. Thus the number accumulating per second per unit volume in $d\varepsilon = q(\varepsilon + d\varepsilon) - q(\varepsilon)$

$$\approx \frac{\partial q}{\partial \varepsilon}\,d\varepsilon$$

Since in a steady state there can be no total accumulation,

$$D\,\nabla^2 n(\underline{r},\varepsilon) + \frac{\partial q}{\partial \varepsilon} = 0$$

Let

$$\tau \equiv \int_{\varepsilon}^{\varepsilon_0} \frac{\lambda^2(\varepsilon)}{3\xi(1-\frac{2}{3A})}\,d\varepsilon = \int D\,dt \qquad \text{IX.13}$$

$$= \frac{\lambda^2}{3\xi(1-\overline{\cos\theta_L})}(\varepsilon_0-\varepsilon) = Dt \quad \text{for constant } \lambda$$

$\tau$ is called the "age" of the neutron. At $\varepsilon = \varepsilon_0$, $\tau = 0$, and as $(\varepsilon_0 - \varepsilon)$ increases, $\tau$ increases. Note that $\tau$ has the dimensions of $L^2$.

Using IX.13, $\frac{\partial q}{\partial \varepsilon} = \frac{\partial q}{\partial \tau}\frac{\partial \tau}{\partial \varepsilon} = \frac{\partial q}{\partial \tau}\left[-\frac{\lambda^2}{3\xi(1-\frac{2}{3A})}\right]$

Thus $\frac{\lambda V}{3(1-\frac{2}{3A})}\nabla^2 n - \frac{\lambda^2}{3\xi(1-\frac{2}{3A})}\frac{\partial q}{\partial \tau} = 0$

$$\boxed{\nabla^2 q(\underline{r},\tau) = \frac{\partial q(\underline{r},\tau)}{\partial \tau}} \qquad \text{The Age Equation} \qquad \text{IX.14}$$

Mathematically this is the same as the heat equation:

$$\frac{K}{\rho C}\nabla^2 T = \frac{\partial T}{\partial t}$$

Discussions of its solutions for various initial and boundary conditions are given in books on heat and diffusion.

We shall consider two such solutions here. The first will be the general case of an infinite medium given the initial condition that $q(\underline{r},0) = F(\underline{r})$. The result can immediately be applied to the point source and should give a gaussian distribution to check with the experimental results which have been previously discussed.

Consider a particular solution of the age equation:

$$q_m = Q_m(\underline{r})\, T_m(\tau)$$

Substituting into IX.14 gives

$$\frac{\nabla^2 Q_m}{Q_m} = \frac{\dot{T}_m}{T_m} = -k_m^2$$ where $k_m$ is the separation constant

$$Q_m = e^{i\underline{k}_m\cdot\underline{r}} \qquad T_m = e^{-k_m^2\tau}$$

$$q_m = e^{i\underline{k}_m\cdot\underline{r} - k_m^2\tau}$$

Any solution can be expressed as a sum of these $q_n$'s using Fourier integrals.

$$q(\underline{r},\tau) = \iiint_{-\infty}^{\infty} a(\underline{k})\, e^{i\underline{k}\cdot\underline{r} - k^2\tau}\, dk_x\, dk_y\, dk_z$$

$$q(\underline{r},0) \equiv F(\underline{r}) = \iiint a(\underline{k})\, e^{i\underline{k}\cdot\underline{r}}\, d^3k$$

$$a(\underline{k}) = \frac{1}{8\pi^3}\iiint_{-\infty}^{\infty} F(\underline{r}')\, e^{-i\underline{k}\cdot\underline{r}'}\, d^3r'$$

( $a(\underline{k})$ is the Fourier transform of $F(\underline{r})$ )

$$q(\underline{r},\tau) = \frac{1}{8\pi^3}\iiint d^3r' \left\{ F(\underline{r}') \iiint d^3k\; e^{-[\tau k^2 + i\underline{k}\cdot(\underline{r}'-\underline{r})]} \right\}$$

$$= \frac{1}{8\pi^3}\iiint d^3r' \left\{ F(\underline{r}')\, e^{-\frac{(\underline{r}'-\underline{r})^2}{4\tau}} \iiint d^3k\; e^{-[\sqrt{\tau}\,\underline{k} + \frac{i}{2\sqrt{\tau}}(\underline{r}'-\underline{r})]^2} \right\}$$

$$\iiint d^3k\; e^{-[\sqrt{\tau}\,\underline{k} + \frac{i}{2\sqrt{\tau}}(\underline{r}'-\underline{r})]^2} = \left\{ \int_{-\infty}^{\infty} dk_x\; e^{-[\sqrt{\tau}k_x + \frac{i}{2\sqrt{\tau}}(x'-x)]^2} \right\}^3$$

$$= \left(\sqrt{\frac{\pi}{\tau}}\right)^3$$

$$\therefore\; q(\underline{r},\tau) = \frac{1}{8(\pi\tau)^{3/2}} \iiint F(\underline{r}')\, e^{-\frac{(\underline{r}'-\underline{r})^2}{4\tau}}\, dx'\,dy'\,dz' \qquad \text{IX.15}$$

For a mono-energetic point source of strength Q, $F(\underline{r}')$ is the delta function $Q\,\delta(x')\,\delta(y')\,\delta(z')$

$$\therefore\; q(\underline{r},\tau) = \frac{Q}{(4\pi\tau)^{3/2}}\, e^{-\frac{r^2}{4\tau}} \qquad \text{for a point source at } r=0 \qquad \text{IX.16}$$

From this we obtain as the mean square distance

$$\overline{R^2} = \frac{\int_0^\infty r^2 e^{-\frac{r^2}{4\tau}} 4\pi r^2 dr}{\int_0^\infty e^{-\frac{r^2}{4\tau}} 4\pi r^2 dr} = 6\tau \qquad \text{IX.17}$$

This agrees with the previous result, IX.12, if the factor $\frac{1}{1-\overline{\cos\theta}}$ is included in D, the diffusion coefficient.

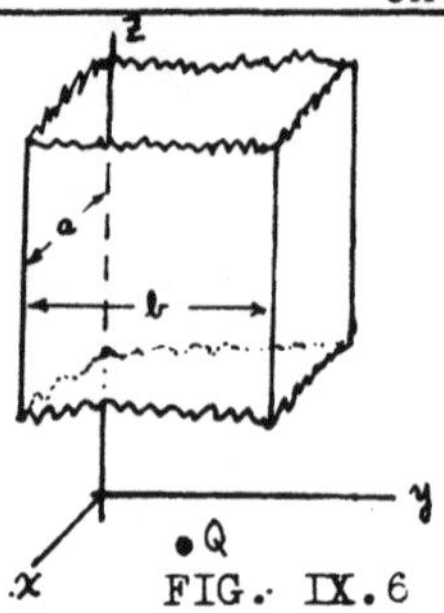

FIG. IX.6

Problem: The second solution to be considered here was assigned as a problem. The boundary conditions are that the medium is a rectangular prism of sides a, b, and $\infty$. It will be shown later (p. 191) that $q(\underline{r},\tau)$ approaches zero near the boundary. In this problem q will be assumed zero at the boundaries. The initial condition is that a monoenergetic point source of neutrons of strength Q is located at the center of the prism.

Solution: Choose the z axis as one of the edges (see FIG. 6).

The initial condition is $q(\underline{r},0) = Q\delta(x-\frac{a}{2})\delta(y-\frac{b}{2})\delta(z)$

At first the prism will be considered as of finite length. $-L \le z \le L$
Then L will be made to approach infinity.

Let $q(\underline{r},\tau) = \sum_{\substack{m,n,p \\ \text{odd}}} C_{m,n,p} \sin\frac{m\pi}{a}x \sin\frac{n\pi}{b}y \cos\frac{p\pi}{2L}z\, T_{m,n,p}(\tau) \equiv \sum C_{mnp} q_{mnp} T_{mnp}$ IX.18

where m, n, and p are odd integers only. Thus this expansion satifies the boundary condition and makes use of the fact that q must be symmetric with respect to the center of the prism. Putting $q_{mnp}T_{mnp}$ into the age equation yields

$$\left[-\frac{\pi^2}{a^2}m^2 - \frac{\pi^2 n^2}{b^2} - \frac{\pi^2 p^2}{4L^2}\right] T_{mnp} = \frac{dT_{mnp}}{d\tau}$$

$$T_{mnp} = e^{-\pi^2\left[\frac{m^2}{a^2}+\frac{n^2}{b^2}+\frac{p^2}{4L^2}\right]\tau}$$

Multiplying both sides of IX.18 by $\sin\frac{m'\pi}{a}x \sin\frac{n'\pi y}{b}\cos\frac{p'\pi z}{2L}$ and integrating over x,y,z at $\tau = 0$ gives

$$\int_0^a\int_0^b\int_{-L}^{L} q(\underline{r},0) \sin\frac{m'\pi x}{a}\sin\frac{n'\pi y}{b}\cos\frac{p'\pi z}{2L}\,dx\,dy\,dz = C_{m'n'p'}\,\frac{a}{2}\times\frac{b}{2}\times L$$

$$C_{mnp}\frac{abL}{4} = Q\sin\frac{m\pi}{2}\sin\frac{n\pi}{2} = -Q(-1)^{\frac{m+n}{2}}$$ using the initial conditions

$$q(\underline{r},\tau) = \frac{4Q}{abL}\sum_{\substack{m,n \\ \text{odd}}} -(-1)^{\frac{m+n}{2}}\sin\frac{m\pi x}{a}\sin\frac{n\pi y}{b}\, e^{-\pi^2\left[\frac{m^2}{a^2}+\frac{n^2}{b^2}\right]\tau}\sum_{\substack{p \\ \text{odd}}}\cos\frac{p\pi z}{2L}\, e^{-\frac{\pi^2 p^2}{4L^2}\tau}$$

Let $L\to\infty$ and $k \equiv \frac{p}{L}$;

$$\frac{1}{L}\sum_{\substack{p \\ \text{odd}}}\cos\frac{p\pi z}{2L}\, e^{-\frac{\pi^2p^2}{4L^2}\tau} \to \frac{1}{2}\int_0^\infty dk\,\cos\frac{\pi k z}{2}\, e^{-\frac{\pi^2k^2}{4}\tau} = \frac{1}{2\sqrt{\pi\tau}}\, e^{-\frac{z^2}{4\tau}}$$

$$\boxed{q(\underline{r},\tau) = \frac{2Q}{ab\sqrt{\pi\tau}}\, e^{-\frac{z^2}{4\tau}}\sum_{\substack{m,n \\ \text{odd}}} -(-1)^{\frac{m+n}{2}}\sin\frac{m\pi x}{a}\sin\frac{n\pi y}{b}\, e^{-\pi^2\left(\frac{m^2}{a^2}+\frac{n^2}{b^2}\right)\tau}}$$

In the preceding problem, the boundary condition assumed was that $n(\underline{r},\tau)$ is zero at the boundary. This seems fairly reasonable since $\lambda = \infty$ outside the medium and thus the free space outside appears as a perfect sink. However, a simple calculation* shows that the neutron density at a bounding surface behaves as tho $n(\underline{r},\tau)$ is a linear function of x, vanishing at $x = -2/3\,\lambda$. See FIG. 7. The coefficient 2/3 is not quite correct due to a simplifying assumption in the calculation. A more exact calculation gives $.7104\lambda$.

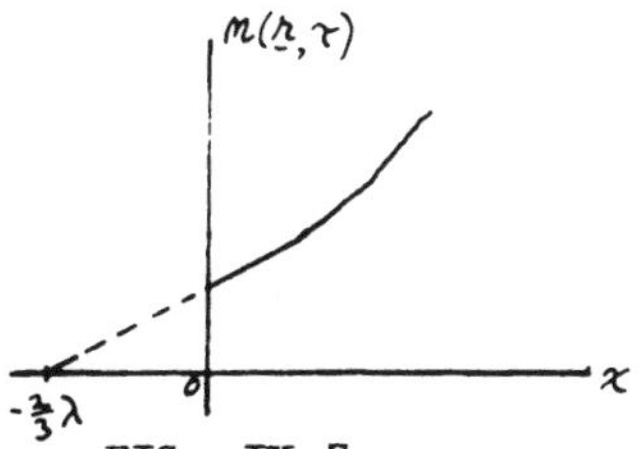


FIG. IX.7

This diffusion theory should be expected to give a neutron density which is too small for large distances from the neutron point source. This is easily seen from IX.11 (p. 185) where the distribution of neutrons which have had no collisions goes as $e^{-\frac{r}{\lambda}}$. The diffusion theory gives $e^{-\frac{r^2}{4\tau}}$ as the dependence. But there exists a large value of r where the $e^{-\frac{r}{\lambda}}$ term becomes larger than the $e^{-\frac{r^2}{4\tau}}$ term. In this region there will be more neutrons than the number given by the macroscopic diffusion theory.

A mono-energetic neutron source has a value $\tau$ associated with slowing down in a particular medium to a particular energy. The Ra-Be source can be approximated as giving neutrons of several ages. For a $\tau$ corresponding to E = 1.44 ev in graphite of density 1.6, the ages are given on the right. The solution to a diffusion problem with Ra-Be as a source would be the sum of the three separate solutions.

| $\tau$ | Relative Amount |
|---|---|
| 130 $cm^2$ | 15% |
| 340 | 69 |
| 815 | 16 |

Age of Ra-Be Source As Seen by 1.44 ev Neutrons in Graphite

2. Distribution of Thermal Neutrons

Let $n(\underline{r}) \equiv$ the number of thermal neutrons per $cm^3$

$D = \frac{\lambda v}{3}$ is the diffusion coefficient of thermal neutrons. ($\lambda$ is used rather than $\frac{\lambda}{1-\frac{2}{3\Lambda}}$ for the sake of simplicity

$\Lambda \equiv$ absorption mean free path

$T \equiv \frac{\Lambda}{v} =$ mean life of a thermal neutron. Since $\Lambda \propto \frac{1}{\sigma_a} \propto v$, T is, in most cases constant.

$q_T(\underline{r}) \equiv$ density of nascent neutrons, or the number of neutrons per $cm^3$ becoming thermal per second. It is obtained from $\nabla^2 q = \frac{\partial q}{\partial \tau}$ by $q_T(\underline{r}) \equiv q(\underline{r}, \tau_T)$.

$D\nabla^2 n$ is the increase in density per sec. due to diffusion.
$q_T$ is the increase in density per sec. due to slowing down.
$\frac{n}{T}$ is the decrease in density per sec. due to absorption.

* L.A. 255, Ch. VII, sec. 2. Also Goodman, p. 94.

Thus at equilibrium, $D\,\nabla^2 n(r) + q_T - \frac{n(r)}{T} = 0$

or

$$\boxed{\nabla^2 n(r) - \frac{3}{\lambda\Lambda} n(r) + \frac{3q_T}{\lambda V} = 0} \qquad \text{IX.19}$$

Problem: Consider an infinite plane source of fast neutrons of strength Q neutrons per cm$^2$ per second. Let the xy plane be the source. The problem is to determine n(z) in terms of the parameters $\lambda$, $\Lambda$, $\tau_{\text{thermal}} \equiv \tau_T$, and Q.

Let $a \equiv \sqrt{\frac{\lambda\Lambda}{3}}$. $a$ is called the diffusion length.

In terms of $a$, IX.19 becomes $\frac{d^2n}{dz^2} - \frac{1}{a^2} n = -\frac{3}{\lambda V} q_T(z)$.

Let n'(z - z') be the solution of the above equation where the right hand side is $-\frac{3}{\lambda V} q_T(z')\delta(z-z')$. Then the integral of n'(z-z') over z' will be the complete solution. Quantitatively,

$$n(z) = \int_{-\infty}^{\infty} n'(z-z')\,dz' \qquad \text{IX.20}$$

$$\frac{d^2n'}{dz^2} - \frac{1}{a^2} n' = -\frac{3}{\lambda V} q_T(z')\delta(z-z')$$

$n' = C\,e^{-\frac{1}{a}|z-z'|}$ is a solution of this equation.

C is to satisfy the condition that q (z') neutrons per cm$^2$ come off the plane z = z'.

Thus $-2D \left.\frac{dn'}{dz}\right]_{z=z'} = q_T(z')$ since $-D\frac{dn'}{dz}$ is the no. of neutrons passing thru a sq. cm per sec.

$$-2\frac{\lambda V}{3}\left(-\frac{1}{a}\right)C = q_T(z')$$

$$\therefore \quad n'(z-z') = \frac{3a\,q_T(z')}{2\lambda V} e^{-\frac{1}{a}|z-z'|} \qquad \text{IX.21}$$

Using IX.15, the plane source (at z=0) of high energy monoenergetic neutrons gives rise to the nascent neutron distribution:

$$q_T(z') = \frac{1}{8(\pi\tau_T)^{3/2}} \iiint Q\,\delta(z'')\,e^{-\frac{(r''-r')^2}{4\tau_T}}\,dx''\,dy''\,dz''$$

$$q_T(z') = \frac{Q}{2\sqrt{\pi\tau_T}} e^{-\frac{z'^2}{4\tau_T}}$$

Substituting into IX.20,

$$n(z) = \int_{-\infty}^{\infty} \frac{3a}{4\lambda V}\frac{Q}{\sqrt{\pi\tau_T}} e^{-\frac{z'^2}{4\tau_T}} e^{-\frac{|z-z'|}{a}}\,dz'$$

$$= \frac{3aQ}{4\lambda V\sqrt{\pi\tau_T}} e^{\frac{\tau_T}{a^2}}\left[e^{\frac{z}{a}}\int_{\left(\frac{z}{2\sqrt{\tau}}+\frac{\sqrt{\tau}}{a}\right)}^{\infty} e^{-u^2}\,2\sqrt{\tau}\,du + e^{-\frac{z}{a}}\int_{-\infty}^{\left(\frac{z}{2\sqrt{\tau}}-\frac{\sqrt{\tau}}{a}\right)} e^{-u^2}\,2\sqrt{\tau}\,du\right]$$

$$n(z) = \frac{3aQ}{4\lambda v} e^{\frac{\tau_T}{a^2}} \left\{ \left[1 - \mathrm{erf}\left(\frac{z}{2\sqrt{\tau}} + \frac{\sqrt{\tau}}{a}\right)\right] e^{\frac{z}{a}} + \left[1 + \mathrm{erf}\left(\frac{z}{2\sqrt{\tau}} - \frac{\sqrt{\tau}}{a}\right)\right] e^{-\frac{z}{a}} \right\}$$

where erf(x) is the error function which is defined as follows:

$$\mathrm{erf}(x) \equiv \frac{2}{\sqrt{\pi}} \int_0^x e^{-u^2} du$$

For large z the solution has the form

$$n(z) \approx \frac{3aQ}{2\lambda v} e^{\frac{\tau_T}{a^2}} e^{-\frac{z}{a}} = \frac{1}{2v}\left(\frac{3\Lambda}{\lambda}\right)^{\frac{1}{2}} Q\, e^{\frac{\tau_T}{a^2}} e^{-\frac{z}{a}}$$

Notice that the coefficient is the same as in IX.21 except for the factor $e^{\tau_T/a^2}$. This factor would be unity if the source were thermal.

The thermal column: The previous problem shows that if one has a plane source of high energy neutrons, the nascent thermal neutrons will drop off in a gaussian distribution, while the distribution of all the thermal neutrons is more spread out in the form $e^{-\frac{z}{a}}$. The diffusion length ($a = \sqrt{\frac{\lambda\Lambda}{3}}$) is about 50 cm for graphite. Also the root mean square path for the nascent neutrons of the gaussian distribution is about the same. ( $\overline{R^2} = 6\tau$ )

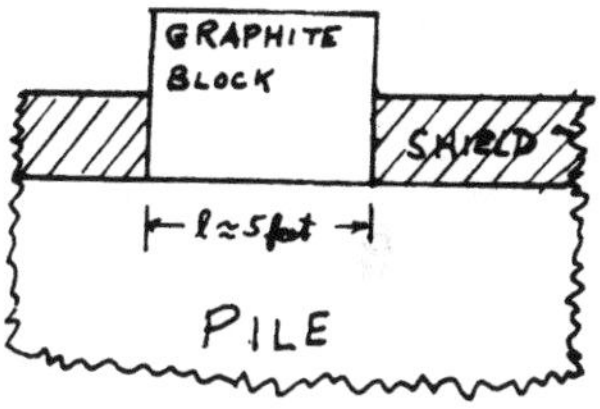


Practical use of this data is made in the design of thermal columns. A column of graphite with base of dimensions $l \times l$ is placed against a nuclear pile which is a source of fast neutrons. At distances greater than a few feet from the pile, the neutrons in the graphite will be thermal.

Assuming $q_\tau = 0$ except in this small layer near the pile and using IX.19, the equation for n(r) is

$$\nabla^2 n - \frac{n}{a^2} = 0$$

The approximate boundary conditions are

$$n = 0 \text{ at} \begin{cases} x = 0 \text{ and } l \\ y = 0 \text{ and } l \end{cases}$$

| Substance | $\rho$ | $a$ | $\lambda$ | $\Lambda = \frac{3a^2}{\lambda}$ |
|---|---|---|---|---|
| $H_2O$ | 1.0 | 2.85 | .43 | 57 |
| $D_2O$ | 1.1 | 170 | 2.4 | 36,500 |
| Be | 1.8 | 31 | 2.0 | 1,400 |
| C | 1.62 | 50 | 2.5 | 3,000 |

EXPERIMENTAL RESULTS

Let $n(\underline{r}) = \sum_{j,k=1}^{\infty} n_{jk}(z) \sin\frac{\pi j x}{l} \sin\frac{\pi k y}{l}$

$$\frac{d^2 n_{jk}}{dz^2} - \left[\frac{\pi^2}{l^2}(j^2 + k^2) + \frac{1}{a^2}\right] n_{jk} = 0$$

$$n_{jk}(z) = C e^{-\frac{1}{b_{jk}} z} \quad \text{where} \quad \frac{1}{b_{jk}} \equiv \sqrt{\frac{1}{a^2} + \frac{\pi^2}{l^2}(j^2 + k^2)}$$

$b_{jk}$ is maximum for the (1,1) mode.

$$\frac{1}{b_{11}} = \sqrt{\frac{1}{a^2} + \frac{2\pi^2}{l^2}}$$

For $l >> a$; $b_{\parallel} \approx a$

and for $l << a$; $b_{\parallel} \approx \frac{l}{\sqrt{2\pi^2}}$ is the effective diffusion length
Thus $b_{\parallel}$ can reach its maximum value (a) by increasing $l$.
The object is to get thermal neutrons out past the region of the nascent and higher energy neutrons which already reach out on the order of 50 cm ($\sqrt{R^2} \approx$ 50 cm). Thus $l > a$ is a condition for building a thermal column.

## D. SCATTERING OF NEUTRONS

### 1. Effect of Chemical Binding of Scatterer

When a neutron of energy much greater than the molecular binding energy hits a hydrogen nucleus in a molecule, it knocks this proton out of the molecule and on the average loses ½ of its energy to it.* However, if the energy of the neutron is less than the $h\nu$ of the molecular vibration, it cannot lose any energy to vibration or freeing of the hydrogen. Consequently the proton acts as if it has the mass of the molecule. This makes it hard for the slow neutron to lose energy. Thus there is a "slowing up of the slowing down" as the thermal region is approached. For energies below $h\nu$ the reduced mass approaches $\mu$ = M rather than $\mu$ = M/2.

Using the Born approximation, the differential scattering cross section is

$$\sigma(\theta) = \frac{\mu^2}{4\pi^2\hbar^4} \left| \int \psi^*_{final}\, U\, \psi_{in}\, d\tau \right|^2 \qquad \text{IX.22}$$

where $\psi_{final}$ and $\psi_{initial}$ are plane waves normalized for unit volume.

Thus $\sigma_{bound} = 4\,\sigma_{free}$ IX.22

The criterion for the applicability of the Born approximation is not fulfilled for thermal energies. However, it can be shown (Bethe B, p 122) that the Born approximation, with modifications, may be used here.

For the carbon-hydrogen bond in paraffin the longitudinal vibration is 3000 $cm^{-1}$ or 1/3 ev. The transverse vibration is 600 $cm^{-1}$ or 1/15 ev. The scattering cross section curve is shown in FIG. 8.

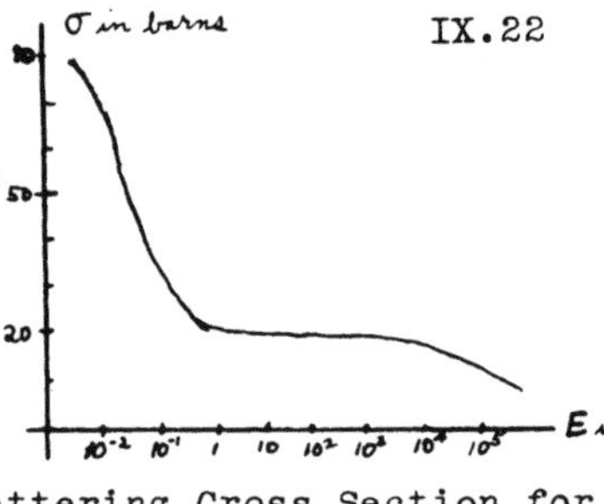


Scattering Cross Section for H (in paraffin)
FIG. IX.8

### 2. Low Energy Scattering

Since slow neutron wave lengths are on the order of interatomic distances ($\lambda_{thermal}$ = 1.81 angstroms), slow neutrons will play an analogous role to x-rays in analysis of crystalline substances.

Consider an interaction between a neutron and a nucleus. Expand the wave function in the asymptotic form:

$$\psi \rightarrow e^{ikz} + \sum_{l=0}^{\infty} \frac{c_l}{r} e^{ikr} P_l(\cos\theta)$$

In Schiff, p. 105, it is shown that

$$c_l = \frac{\hbar}{p}(2l+1)\, e^{i\beta_l} \sin\beta_l \qquad \text{for elastic collisions}$$

*Assuming elastic S-wave scattering (page 182)

where $\beta_l$ is the phase shift of the partial wave.
$\beta_l \approx 0$ for $pr_0 < lh$ (see p. 119)
Thus for slow neutrons all the $C_l$'s except $C_0$ may be neglected.

As shown on p. 119, $C_0 = -a$ where $a$ is the r axis intercept of the tangent of $r\psi$ at $r_0$.

Thus $\psi \approx e^{ikz} - \frac{a}{r} e^{ikr}$

The total scattering cross section is $\sigma_s = 4\pi a^2$ as shown on p. 119.

Since $-1 = e^{i\pi}$, $\psi \approx e^{ikz} + \frac{a}{r} e^{i(kr+\pi)}$ and there is a phase shift of 180° in the scattered wave for positive $a$. For negative $a$, there is no phase shift. In most cases $a$ is found to be positive. However, in the case of the singlet (N,P) interaction, $a$ was found to be rather large and negative.

---

Problem: Express both the absorption and the scattering cross sections in terms of $C_0$, which may be complex.

The scattering cross section, $\sigma_s$, is the outgoing current divided by the current of the plane wave.

$$\sigma_s = \frac{4\pi r^2 \times |\frac{C_0}{r} e^{ikr}|^2 v}{|e^{ikz}|^2 v} \qquad \sigma_s = 4\pi C_0^* C_0$$

The absorption cross section:

The incoming plane wave can be expanded in terms of orthogonal spherical waves (see Schiff p. 105):

$$e^{ikz} \to \frac{1}{kr}\sum_{l=0}^{\infty}(2l+1)\, i^l \sin(kr - l\tfrac{\pi}{2}) P_l(\cos\theta)$$ is the asymptotic form.

The $l = 0$ component of this expansion is

$$\frac{1}{kr}\frac{1}{2i}(e^{ikr} - e^{-ikr})$$

The complete zeroth component of $\psi$ is then

$$\frac{1}{r}\left(\frac{1}{2ik} + C_0\right) e^{ikr} - \frac{1}{r}\frac{1}{2ik} e^{-ikr}$$

This gives an incoming current of $4\pi r^2 \times \left|\frac{1}{2ikr} e^{-ikr}\right|^2 v$

and an outgoing current of $4\pi r^2 \times \left|\frac{1}{r}\left(\frac{1}{2ik} + C_0\right)\right|^2 v$

Absorption cross section, $\sigma_a$, is defined as follows:

$|e^{ikz}|^2 \sigma_a v$ = no. of particles lost per sec.
= (incoming current) - (outgoing current)

$$\sigma_a = 4\pi\frac{1}{4k^2} - 4\pi\left(\frac{1}{2ik} + C_0\right)\left(-\frac{1}{2ik} + C_0^*\right)$$

$$= \frac{\pi}{k^2} - 4\pi\left[\frac{1}{4k^2} + \frac{1}{2ik}(C_0^* - C_0) + C_0^* C_0\right]$$

$$\boxed{\sigma_a = \frac{4\pi}{k} \mathrm{Im}\{C_0\} - 4\pi|C_0|^2}$$ where Im stands for "the imaginary part of"

$$\sigma_a = \frac{4\pi}{k}\mathrm{Im}\{C_0\} - \sigma_s$$

Notice that this checks in the case of elastic collisions where $C_0 = \frac{1}{k} e^{i\beta_0}\sin\beta_0$ and $\mathrm{Im}(C_0) = \frac{1}{k}\sin^2\beta_0$. Then $\sigma_a = 0$.

## 3. Interference Phenomena

The following two paragraphs review the distinction between coherent and incoherent scattering and point out that in the case of neutron scattering there are two fundamentally different types of incoherent scattering.

The method of partial waves will be applied to N nuclei. Each nucleus will behave as a scattering center giving an S-wave (the higher order waves are very small for thermal neutrons). The scattered wave for the jth nucleus is

$$-\frac{a_j}{r_j} e^{ik(r_j + z_j')}$$

where $a_j$ is the scattering length
$r_j$ is the distance from the jth nucleus
$z_j'$ is the z coordinate of the jth nucleus

If multiple scattering and spin (see next paragraph) can be ignored, the total neutron scattered wave is

$$\sum_{j=1}^{N} \frac{-a_j}{r_j} e^{ik(r_j + z_j')}$$

The probability of observing a scattered neutron in any given position is given by the square modulus of the summation. This contains cross-product (or interference) terms. In the case of random positions of the nuclei, the interference terms tend to cancel, giving a scattered beam proportional to N. This is one of the two types of incoherent scattering which occurs for neutrons. If the nuclei are spaced in a regular lattice, there are directions (Bragg angles) where all the scattered waves are in phase and then the square of the summation gives a beam strength proportional to $N^2$; while at non-Bragg angles the intensity is much less than it would be for random-position scattering because of systematic cancellation of intensities. This type of scattering is commonly called coherent.

Due to physical considerations there is yet a third type of elastic scattering. In the case that the scattering nuclei have non-zero spin, there is a probability that a scattered neutron will have had its spin orientation changed. In this case the $I_z$ of one of the scattering nuclei must have changed by one in order to satisfy the law of conservation of total $I_z$.

It will next be shown that the probability of finding a neutron which has flipped its spin is given by the square modulus of only one partial wave; namely, the scattered wave from the nucleus which had its $I_z$ changed. The total scattered beam of neutrons which have flipped spin is given by the sum of the squares of such waves. Notice there are no interference terms at all in this case. This is the second type of incoherent scattering.

Let $\phi$ be the total initial wave function of the scatterer.

$\phi_j$ be the total wave function of the scatterer where the jth nucleus has changed its $I_z$

$c_j$ be the amplitude of the non-flip scattered wave from the jth nucleus

$c_j'$ be the amplitude of the spin flip scattered wave from the jth nucleus

Let the incoming wave be $e^{ikz}\binom{1}{0}$

Then the asymptotic solution to the entire Hamiltonian is

$$\psi \to e^{ikz}\binom{1}{0}\phi + \Sigma \frac{c_j}{r_j} e^{ik(r_j+z_j')}\binom{1}{0}\phi + \Sigma \frac{c_j'}{r_j} e^{ik(r_j+z_j')}\binom{0}{1}\phi_j$$

The probability of observing a neutron at (x,y,z) is the integral of $|\psi|^2$ over the $\phi$ configuration space.

The scattered part of this $= \left|\Sigma \frac{c_j}{r_j} e^{ik(r_j+z_j')}\right|^2 + \Sigma \frac{|c_j'|^2}{r_j^2}$

since the different $\phi_j$ must be orthogonal (by definition of a state).

The term $\left|\Sigma \frac{c_j}{r_j} e^{ik(r_j+z_j')}\right|^2$ is the probability of observing a a neutron which hasn't flipped, and contains interference terms. The term $\Sigma \frac{|c_j'|^2}{r_j^2}$ is the probability of observing a neutron which has flipped, and doesn't have interference terms.

The amount of spin flip incoherent scattering by protons will be calculated in the following problem. Then the general case of scatterer of spin I will be considered.

Problem: Find the probability that a neutron, scattered by protons, has flipped.

Before collision the following four spin combinations each have a probability of $\frac{1}{4}$: $\underset{N\,P}{\uparrow\uparrow}$ $\underset{N\,P}{\downarrow\downarrow}$ $\underset{N\,P}{\uparrow\downarrow}$ $\underset{N\,P}{\downarrow\uparrow}$

The problem may be reduced to finding the scattering of four randomly placed protons which give the above four interactions with the incoming neutron beam.

Since the first two cases are eigen functions of $\sigma_1 \cdot \sigma_2$ (the nuclear force has a $\sigma_1 \cdot \sigma_2$ term), they will each contribute a scattered amplitude of $-a_3$.

The third case ($\underset{N\,P}{\uparrow\downarrow}$) is a mixture of a triplet and a singlet state.

$$\uparrow\downarrow = \underset{(\text{TRIPLET})}{\frac{\uparrow\downarrow + \downarrow\uparrow}{2}} + \underset{(\text{SINGLET})}{\frac{\uparrow\downarrow - \downarrow\uparrow}{2}}$$

The scattered wave for this mixture of states is the mixture:

$$-\frac{a_3}{r} e^{ikr}\left(\frac{\uparrow\downarrow + \downarrow\uparrow}{2}\right) - \frac{a_1}{r} e^{ikr}\left(\frac{\uparrow\downarrow - \downarrow\uparrow}{2}\right) = \underset{(\text{NON-FLIP})}{-\frac{a_3+a_1}{2}\frac{1}{r}e^{ikr}(\uparrow\downarrow)} \underset{(\text{FLIP})}{-\frac{a_3-a_1}{2}\frac{1}{r}e^{ikr}(\downarrow\uparrow)}$$

Likewise $\downarrow\uparrow = \frac{\uparrow\downarrow + \downarrow\uparrow}{2} - \frac{\uparrow\downarrow - \downarrow\uparrow}{2}$ gives $\underset{(\text{NON-FLIP})}{-\frac{a_3+a_1}{2}\frac{1}{r}e^{ikr}(\downarrow\uparrow)} \underset{(\text{FLIP})}{-\frac{a_3-a_1}{2}\frac{1}{r}e^{ikr}(\uparrow\downarrow)}$

Thus each of these two cases contributes a scattering length $-\frac{a_3-a_1}{2}$ for neutron flipping and $-\frac{a_3+a_1}{2}$ for non-flipping.

In the case of flipping, the separate scattering lengths are to be squared before adding.

Thus, relative probability of flipping $= 2\left(\frac{a_3-a_1}{2}\right)^2$

In the case of no flip, the four scattered waves are to be added and then squared. In general there will be interference terms. However these interference terms drop out after integration over all directions ( provided no two protons are closer than several wave lengths). Thus only the squared terms are left and

relative prob. of scattering without flip $= a_3^2 + a_3^2 + 2\left(\frac{a_3+a_1}{2}\right)^2$

$$\text{Probability of flipping} = \frac{(a_3-a_1)^2}{4a_3^2 + (a_3+a_1)^2 + (a_3-a_1)^2}$$

It has been shown (p. 120) that $a_3 = .589 \times 10^{-12}$ cm
$a_1 = -2.32 \times 10^{-12}$ cm

Thus, percentage of incoherent scattering = 66%

If the nuclei all have spin I, the possible total angular momentum of a neutron and a scatterer is either $I + \frac{1}{2}$ or $I - \frac{1}{2}$. In general, the interaction forces will be somewhat different in these two cases, leading to two different scattering lengths. Let $a_+$ be the length for the $I + \frac{1}{2}$ interaction and $a_-$ for the $I - \frac{1}{2}$ interaction.

The probability is $\frac{I+1}{2I+1}$ that the resultant spin of the neutron plus any one of the nuclei $= I + \frac{1}{2}$. This is because there are $2I + 2$ possible orientations of $I + \frac{1}{2}$ and $2I$ possible orientations of $I - \frac{1}{2}$.

In the case of a Bragg reflection from a perfect crystal, the phases of each scattered wave could be the same, and it can be shown that the nuclei would appear to have the effective scattering length, $a$, defined as follows:

$$a = \frac{I+1}{2I+1}a_+ + \frac{I}{2I+1}a_- \qquad \text{for coherent scattering} \qquad \text{IX.23}$$

The corresponding intensity for a crystal of N nuclei at a distance R is

$$I_{coh} = \frac{1}{R^2}N^2a^2$$ neutrons per unit volume for an incident beam of one neutron per $cm^3$

However $a_+$ and $a_-$ interactions are randomly distributed among the regular scattering centers (in the ratio $\frac{I+1}{I}$ ). This gives a background of random phase incoherent scattering length. It can be shown* that this incoherent scattering has the following average intensity in the non-Bragg directions:

$$I_{incoh} = \frac{I(I+1)}{(2I+1)^2}\frac{N}{R^2}(a_+ - a_-)^2 \qquad \text{for random phase scattering}$$

In addition there is a second type of incoherent scattering: the spin flip scattering. It can be shown* that this has the following isotropic contribution to the intensity:

$$I_{flip} = \frac{2}{3}\frac{I(I+1)}{(2I+1)^2}\frac{N}{R^2}(a_+ - a_-)^2 \qquad \text{for spin flip scattering}$$

* These expressions (not given by Fermi) were derived rather hurriedly and may possibly be incorrect.

In the case of an isotope mixture or random positioning of several different types of nuclei, the situation is quite similar. Ignoring spin, the effective coherent scattering length is

$$a = \Sigma w_j a_j$$

where $w_j$ is the abundance of the jth nucleus. In addition there will be an average isotropic background of random phase (or position) scattering, the same as before.

It will next be shown that for heavy nuclei the scattering length is about the same as the radius of the nucleus. The radius of the potential well is about five times $r_o$ (radius of the N-P well) and the depth is ~30 Mev (see p160). Thermal neutrons will have several nodes inside such a large well. The phase of the sine wave at the edge of different wells will be more or less random. This makes it unlikely that a will be negative. See FIG. 9 for two slightly different well sizes. It is easily seen that a ≈ R for most cases. In addition to hydrogen, only two cases are known of a negative scattering length. These are lithium and manganese.

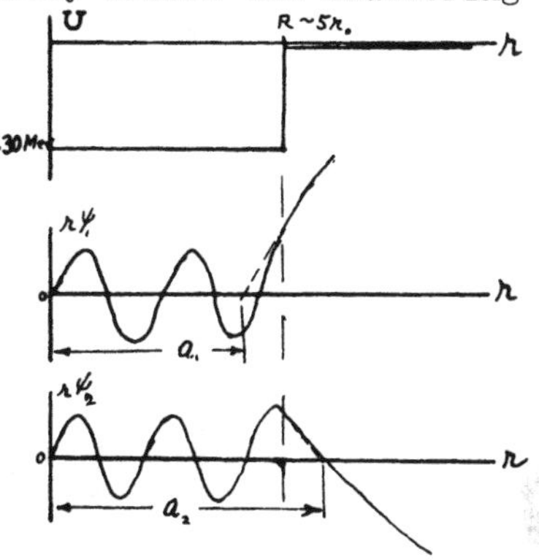


FIG. IX.9

## 4. Scattering in Ortho and Para-hydrogen

(N,P) scattering results give $a_3$ and $a_1$ (knowing the BE of the deuteron)*, but do not tell us whether $a_1$ is + or -. Such knowledge is necessary in order to determine whether the singlet state of the deuteron can exist, or is virtual.

In 1937 Schwinger and Teller (Phys. Rev. 52, 286) worked out the cold neutron scattering cross sections for both ortho and para-hydrogen and showed that if experimental results could be obtained, that the sign of $a_1$ would be determined. Experiments were then performed which show that the singlet state of deuterium is virtual.

For ortho-hydrogen the proton spins are parallel (symmetric spin function) and the spatial part must have odd parity (i.e., odd angular momentum $\ell$). For para-hydrogen spin is anti-parallel (anti-symmetric spin state) and the space part must have even $\ell$. Para-hydrogen can be obtained by keeping $H_2$ at a low temperature in the presence of a catalyst which speeds up the very low reaction rate between ortho and para.

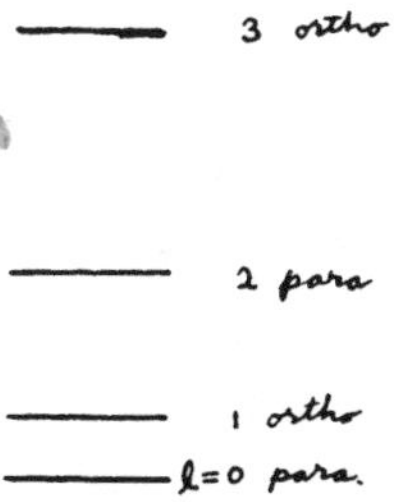


Rotational Levels for $H_2$

The neutron beam must be cold enough so that the neutron wavelength is much greater than the separation of the two hydrogen nuclei. An experimental method of accomplishing this is discussed on p. 203. The effective scattering length for para-hydrogen is then $2a$ where $a$ is the same as in IX.23.

* Ch. VI. p 120.

Thus $a = \frac{3}{4}a_3 + \frac{1}{4}a_1$ $\qquad a_3 = .589 \times 10^{-12}\ cm$, $\quad a_1 = -2.32 \times 10^{-12}\ cm$.

$$\sigma_{para} = 4\pi(.276)^2 \times 10^{-24}\ cm^2$$

which is fairly small for $a_3$ and $a_1$ of opposite signs. The calculation of $\sigma_{ortho}$ is more complicated (see Bethe D. p 50-53). The result is $\sigma_{ortho} = 6.29[(3a_3+a_1)^2 + 2(a_3-a_1)^2] + 1.45(a_1-a_3)^2$

$$\frac{\sigma_{ortho}}{\sigma_{para}} = \begin{cases} 35 & \text{for } a_1 \text{ and } a_3 \text{ of opp. sign} \\ 1.4 & \text{" " " " same "} \end{cases}$$

Recent experimental results are given by Sutton and others in Phys. Rev., 72, 1147 (1947) which show that $a_1$ is negative.

5. Crystalline Diffraction

In many ways neutron and x-ray diffraction are quite similar. Both are quite useful in determining crystalline structure. X-ray diffraction is due to scattering by orbital electrons, while neutron diffraction is due to scattering by nuclei. Thus x-rays are useful in gaining knowledge of electronic structure, while neutrons are useful in determining molecular structure or the positions of nuclei. Thus neutron diffraction can give some knowledge which cannot be obtained by x-rays. The disadvantages of neutron diffraction are mainly monetary and political, since the only reasonable sources of neutrons are nuclear piles which are at present not freely availiable for research because of security regulations.

Molecular structure can be studied by observing the various molecular form factors associated with the different diffraction angles and orders of diffraction. Consider a particular diffracted beam at an angle of incidence θ and order n where θ and n satisfy the Bragg condition

$$n\lambda = 2d \cos\theta \qquad \text{IX.24}$$

where d is the separation of Bragg planes of identical nuclei.

The molecular form factor is defined as $\sum_{s=1}^{N} a_s e^{2\pi i n \frac{d_s}{d}}$ over a unit cell of N nuclei. $d_s$ is the distance from the initial plane to the plane containing the sth nucleus. See FIG. 10.

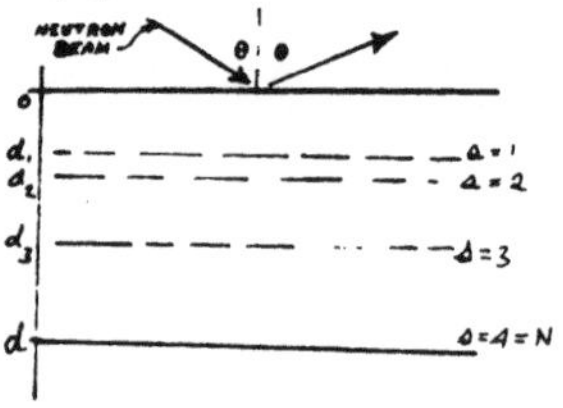


FIG. IX.10

In the case of the (1,1,1) planes for NaCl:

$$\text{form factor} = a_{Na} + a_{Cl}e^{in\pi}$$

$$= a_{Na} + a_{Cl} \qquad \text{for n even}$$

$$= a_{Na} - a_{Cl} \qquad \text{for n odd}$$

The table to the right shows the relative form factors depending whether the a's are of the same or opp. sign.

| | same sign | opposite sign |
|---|---|---|
| n odd | small | large |
| n even | large | small |

Relative Size of (1,1,1) Form Factor (NaCl)

Thus the relative intensities for different orders of diffraction are as follows:

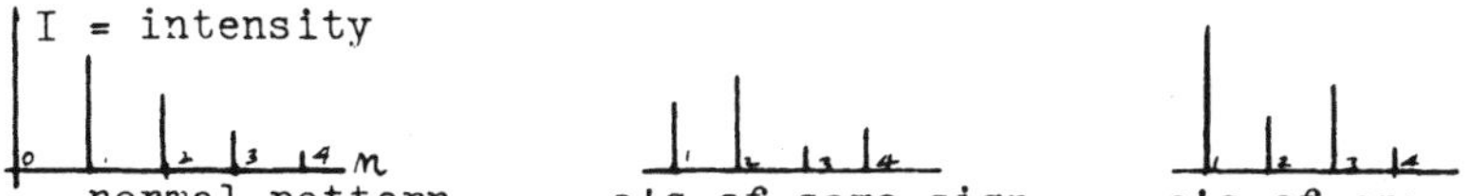


normal pattern a's of same sign a's of opp. sign

FIG. IX.11

The experimental results (for the (1,1,1) plane) for PbS which has the same structure as NaCl are

| n | 1 | 2 | 3 | 4 |
|---|---|---|---|---|
| I | 7230 | 10700 | 808 | 750 |

The results for the (1,1,1) plane of LiF are

| n | 1 | 2 | 3 |
|---|---|---|---|
| I | 10,000 | 0 | 300 |

This indicates that Pb and S have a's of the same sign, while Li and F have a's of opposite sign which are about equal in magnitude. For more results and the various techniques used see the following Physical Review articles: 71, 589; 71, 636; and 71, 752.

## 6. Index of Refraction

In the case of random position scattering, it so happens that the separate scattered waves interfere in such a way that if the total scattered wave is added to the initial wave, the resultant wave appears to have a different wave length in the medium. This is the same as giving the medium an index of refraction, n. This is analogous to the situation in optics, where the index of refraction of a medium can be computed by taking the sum of the initial beam amplitude and all the scattering amplitudes. One difference is that neutron scattering is isotropic while the electromagnetic scattering is dipole reradiation by electrons.

In the following problem the slow neutron index of refraction of a medium will be calculated in terms of the neutron wave length, the density of particles, and the scattering length.

Problem Consider an infinite slab of thickness $w \ll \lambda_0$ where $\lambda_0$ is the incident neutron wave length. There are N identical nuclei of scattering length $a$ per cm$^3$. Show that

$$1-n = \frac{N\lambda_0^2 a}{2\pi}$$

where n is the index of refraction.

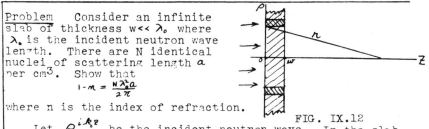

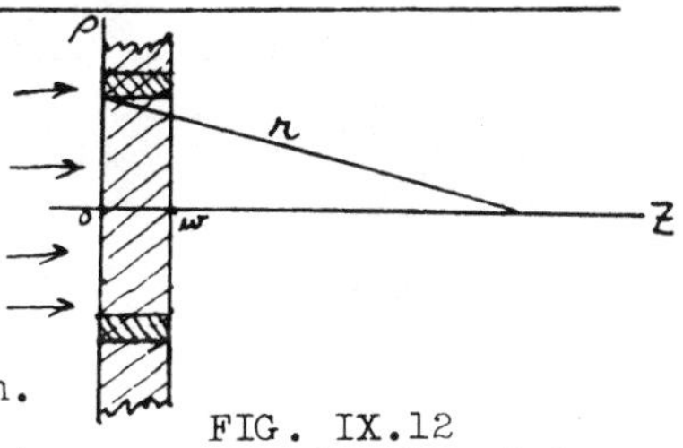


FIG. IX.12

Let $e^{ik_0 z}$ be the incident neutron wave. In the slab the propagation constant will appear as $nk_0$. $k = \frac{2\pi}{\lambda}$

Thus $e^{i[mk_0 w + k_0(z-w)]}$ is the wave after passing thru the slab. The slab is sufficiently thin so that the attenuation is negligible. The wave after passing thru the slab is also equal to the sum of the incident amplitude and scattered amplitudes: $e^{ik_0 z} + \int_0^\infty \frac{c_0}{r} e^{ik_0 r} N w \, 2\pi \rho \, d\rho$

Equating the two expressions and using $\rho \, d\rho = r \, dr$ gives

$$e^{ik_0 z} e^{ik_0 w(m-1)} \approx e^{ik_0 z} + 2\pi c_0 N w \int_z^\infty e^{ik_0 r} \, dr$$

The difficulty presents itself that the $\int_z^\infty e^{ik_0 r} dr$ is indeterminant at the upper limit. The usual remedy is to include in the integrand a function which slowly decreases from unity at large r. This is physically reasonable, since there always exist external factors which cause the far distant contribution to be weakened. Thus the factor $e^{-b^2 r}$ will be included in the integrand. After integrating, the limit will be taken as $b^2 \to 0$.

$$\int_z^\infty e^{-b^2 r} e^{ik_0 r} dr = \frac{1}{ik_0 - b^2}\left[e^{(ik_0 - b^2)r}\right]_z^\infty = \frac{-e^{(ik_0 - b^2)z}}{ik_0 - b^2}$$

$$\lim_{b^2 \to 0} \int_z^\infty e^{-b^2 r} e^{ik_0 r} dr = \frac{i}{k_0} e^{ik_0 z}$$

$$\therefore e^{ik_0 z}\left[1 + i k_0 w(m-1)\right] \approx e^{ik_0 z} + \frac{i}{k_0} 2\pi c_0 N w \, e^{ik_0 z}$$

Equating the imaginary parts,

$$k_0 w(m-1) \approx \frac{2\pi c_0 N w}{k_0}$$

$$m - 1 \approx \frac{2\pi N c_0}{k_0^2}$$

$$\boxed{1 - m \approx \frac{\lambda_0^2}{2\pi} N a} \qquad \text{since} \quad c_0 = -a \ \text{and} \ k_0 = \frac{2\pi}{\lambda_0}$$

Thus it is seen that $n < 1$. In this case total reflection from a surface is possible. From Snell's law the condition for total reflection is that $\cos\varepsilon \geq n$ or for small $\varepsilon$ :

$$1 - \frac{\varepsilon_0^2}{2} \approx n$$

where $\varepsilon_0$ is the max. angle for total reflection

$$\varepsilon_0 = \sqrt{2(1-n)}$$

$$= \sqrt{\frac{N\lambda^2 a}{\pi}}$$

$$= \lambda \sqrt{\frac{N}{\pi}} \left(\frac{\sigma}{4\pi}\right)^{\frac{1}{4}} \qquad \text{since } \sigma = 4\pi a^2$$

7. Microcrystalline Scattering

If a continuous spectrum neutron beam hits a microcrystal at an angle $\theta$, only certain values of n and $\lambda$ will satisfy the condition that $n\lambda = 2d \sin\theta$. These few sharp wave lengths will be reflected strongly and will be taken out of the rest of the beam after about $10^4$ Bragg planes. Assuming a perfect crystal, the rest of the beam experiences no scattering except for the contributions to the index of refraction. This is because of systematic cancelation in non-Bragg directions and is to be distinguished from random position scattering which, in addition to the index of refraction contribution, also gives isotropic scattering.

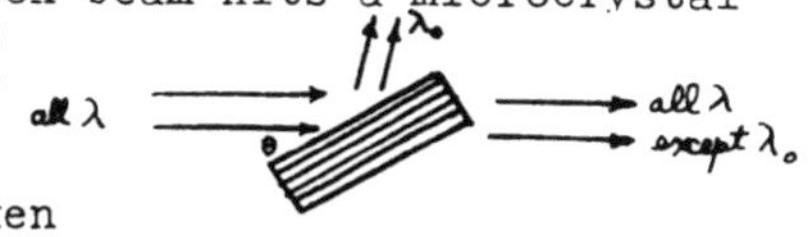


For a substance made up of many microcrystals of random orientations, the Bragg condition for any given wave length will be fulfilled sooner or later. The part of the beam which is scattered away by Bragg reflections is expressed in terms of an apparent atomic scattering cross section, $\sigma_{app}$. Fermi, Sturm, and Sachs (in Phys. Rev. 71, 589) show that

$$\sigma_{app} = \sigma \sum_{b_s < \frac{2}{\lambda}} \frac{N\lambda^2}{8\pi b_s} \qquad \text{IX.25}$$

where $\sigma$ is the atomic scattering cross section of each nucleus
N is the number of atoms per $cm^3$
$a_s$ is the separation between Bragg planes
$b_s \equiv \frac{n}{a_s}$ where n is the order of diffraction

The summation may be easily visualized by making use of the reciprocal lattice. Then the sum includes all the points within a sphere of radius $= \frac{2}{\lambda}$

For $\lambda$ large enough such that $\frac{2}{\lambda} <$ any $b_s$, $\sigma_{app} = 0$.

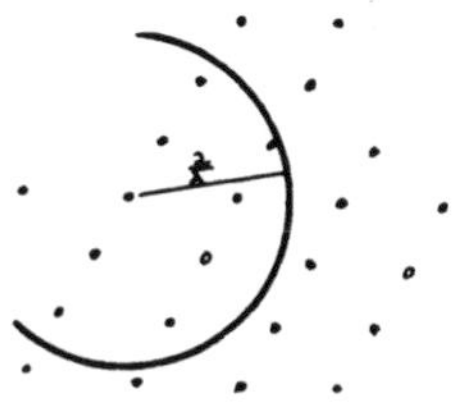


Reciprocal Lattice

Plotting IX.25 against V gives FIG. 13. $\sigma_{app}$ goes as $\lambda^2$ or $1/V^2$ between each set of points. As the velocity (or radius of the circle) increases, the spacing between jumps decreases and the number of points included approaches the volume of the sphere times the density of points. Thus the curve becomes flat.

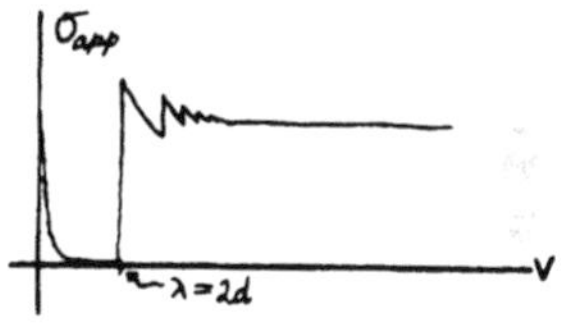


FIG. IX. 13

In the region near the origin there is a small 1/V compoment due to absorption and imperfections. For graphite the sharp edge occurs at $\lambda = 6.69$ Å which is an energy of 0.0018 ev or 1/14 room temperature. It is possible to obtain a very cold neutron beam by sending thermal neutrons thru a polycrystalline graphite filter, since only those neutrons of the distribution will get thru which have E < 0.0018 ev (see shaded part of FIG. 14).

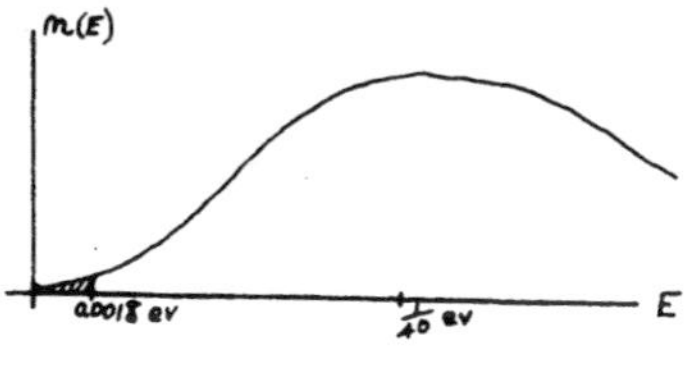


Effect of Graphite Filter
FIG. IX. 14

## 8. Polarization of Neutron Beams

Since the scattering lengths $a_+$ and $a_-$ are not generally equal, a crystal with all of its nuclei oriented in the same direction will have different cross sections for the two directions of the spin of the incident neutrons. If $\sigma_+ > \sigma_-$, more of the neutrons with spin up will be scattered away, leaving a transmitted beam which is predominantly spin down. The percentage of polarization would approach 100% as the path thru the crystal was increased. However this method is not feasible, since temperatures on the order of 0.01°K would be necessary to line up the nuclear spins.

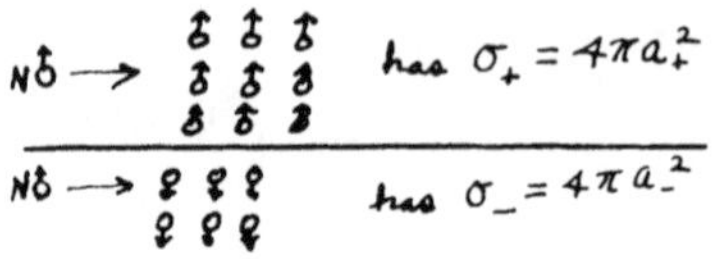


In ferromagnetic materials, the atomic magnetic moments can be lined up. The Born approximation will now be used to see how this can give different cross sections for the two possible neutron orientations.

Let $U = b\delta(\underline{r}) \mp \mu_N \mathcal{H}_z$ represent the interaction potential between the neutron and the Fe atom. $b\delta(\underline{r})$ represents the nuclear potential well. $\mu_N$ is the neutron magnetic moment, and $\mathcal{H}_z$ is the average magnetic field of the Fe atom in its lattice position. Using the Born approximation one obtains

$$\sigma(\theta) = \frac{M^2}{4\pi^2\hbar^4}\left|\int U\, e^{\frac{i}{\hbar}(\underline{k}'-\underline{k})\cdot\underline{r}}\, d^3\underline{r}\right|^2$$

$$= \frac{M^2}{4\pi^2\hbar^4}\left|b \mp \mu_N \int \mathcal{H}_z(\underline{r})\, e^{\frac{i}{\hbar}(\underline{k}'-\underline{k})\cdot\underline{r}}\, d^3\underline{r}\right|^2$$

where the region of integration is that part of the lattice belonging to one Fe atom. For slow neutrons $\lambda$ is on the order of the lattice distance and the exponential is $\approx 1$ over the region of integration.

Thus $$\sigma \approx \frac{M^2}{\pi\hbar^4}\left(b \mp \mu_N \int \mathcal{H}_z\, d^3\underline{r}\right)^2$$

To obtain an idea of the order of magnitude of the magnetic field contribution, b will be taken as zero in the following calculation.

Let $b = 0$

$\lambda = \infty$

$\mathcal{H}_z = 23{,}000$ gauss (saturated Fe)

$\int_{\text{unit cell}} \mathcal{H}_z\, d^3r = 2.7 \times 10^{-19}$

then $\sigma = 4.9 \times 10^{-24}$

The experimental result for thermal neutrons on Fe is $\sigma = 12 \pm 3.15$ barns

The polarized beam will be that orientation which has $\sigma$ = 9 barns.

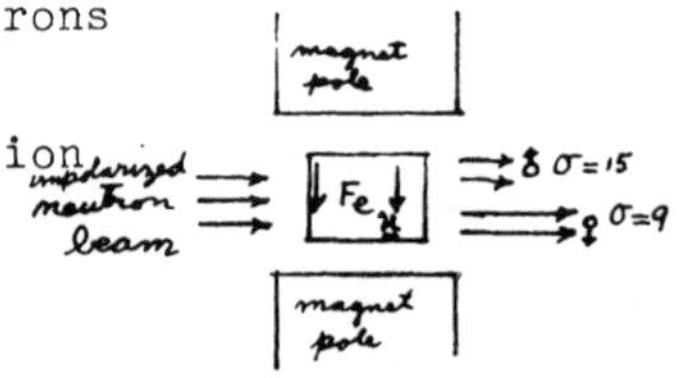


A disadvantage of this method is that it happens to be very sensitive to a lack of complete polarization of the Fe. Fe is completely polarized per domain, but in those small number of domains which are not properly oriented, the neutrons will precess about $\mathcal{H}$ and

*Phys. Rev. 73, 1277 (1948)

undo most of the work done by the previous "good" domains. This effect is calculated in the following problem.

Problem Consider a block of Fe with f the ratio of "good" domains (those completely lined up) to the total number of domains. Assume the "bad" domains are at right angles to $\mathcal{H}$ and that all domains are $10^{-4}$ cm. thick. Find the greatest amount of polarization obtainable for an unpolarized thermal neutron beam of unlimited intensity.

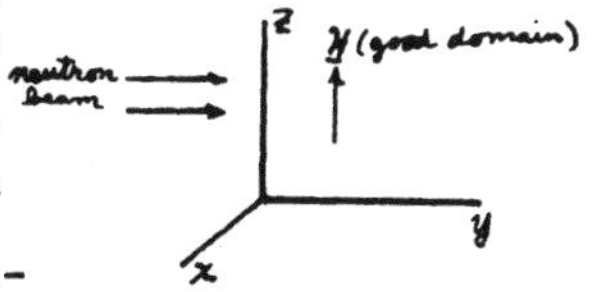


Let $n_+$ be the number of neutrons whose wave functions are $e^{iky}\binom{1}{0}$

$n_-$ be the number of neutrons whose wave functions are $e^{iky}\binom{0}{1}$

$x \equiv \frac{n_+}{n_-}$ be the ratio of neutrons with spin up to those with spin down.

Initially x = 1. The Fe block will be considered as unit sections of one bad domain ($10^{-4}$cm) next to $\frac{f}{1-f}$ good domains. In general x will be changed after passing thru this unit section. We will make use of the criterion on the final x that it be unchanged in passing thru a unit section.

In the good domains:

The $\binom{1}{0}$ beam is reduced by $n_+ N\sigma_+ d$

The $\binom{0}{1}$ beam is reduced by $n_- N\sigma_- d$

where N is the number of Fe atoms per $cm^3$ and $d \equiv \frac{f}{1-f}\times 10^{-4}cm$

is the distance thru the good domains.

For convenience we will use the following matrix notation:

$$\begin{pmatrix} n_+ \\ n_- \end{pmatrix}_{final} = \begin{pmatrix} 1-c\sigma_+ & 0 \\ 0 & 1-c\sigma_- \end{pmatrix} \begin{pmatrix} n_+ \\ n_- \end{pmatrix}_{initial} \qquad \text{IX.26}$$

where $c \equiv N\frac{f}{1-f}\times 10^{-4}$

In the bad domains:

Time dependent perturbation theory will be used to determine the probability that a neutron change its spin during the time of transit, $\Delta t$, thru a bad domain. $\Delta t = \frac{10^{-4}}{v} = 3.82\times 10^{-10}$ sec.

The use of perturbation theory is justified if the probability of changing spin state is small (as will be seen shortly). If $a_+$ and $a_-$ are the probability amplitudes that a neutron be $\binom{1}{0}$ or $\binom{0}{1}$, then the general formula

$$\dot{a}_m = -\frac{i}{\hbar}\sum_n H_{mn} a_n e^{\frac{i}{\hbar}(E_m - E_n)t}$$

gives

$$\dot{a}_+ = -\frac{i}{\hbar}(H_{++}a_+ + H_{+-}a_-)$$

Consider only one of the $n_-$ neutrons separately. Initially $a_- = 1$ and $a_+ = 0$.

Then $$a_+ = -\frac{i}{\hbar} H_{+-} \Delta t$$

$$|a_+|^2 = |H_{+-}|^2 \left(\frac{\Delta t}{\hbar}\right)^2$$

$$H_{+-} = (1\ 0)\, \mu_N\, \underline{\sigma} \cdot \underline{\mathcal{H}} \begin{pmatrix}0\\1\end{pmatrix}$$

where $\underline{\sigma} \cdot \underline{\mathcal{H}} = \begin{pmatrix}0 & 1\\ 1 & 0\end{pmatrix}\mathcal{H}_x + \begin{pmatrix}0 & -i\\ i & 0\end{pmatrix}\mathcal{H}_y$ since $\mathcal{H}_z = 0$

$$H_{+-} = \mu_N(\mathcal{H}_x - i\mathcal{H}_y) = \mu_N \mathcal{H} e^{i\phi}$$

$$|H_{+-}|^2 = |H_{-+}|^2 = \mu_N^2 \mathcal{H}^2$$

$$|a_+|^2 = \left(\frac{\mu_N \mathcal{H} \Delta t}{\hbar}\right)^2$$

Let $$w \equiv \left(\frac{\mu_N \mathcal{H} \Delta t}{\hbar}\right)^2$$

Thus the contribution to $n_+$ from $n_-$ is $wn_-^\circ$, and $n_-$ is decreased in $\Delta t$ to $(1 - w)n_-^\circ$ where $n_-^\circ$ is the initial number of neutrons with spin down.

For a neutron initially with spin up we have the following:

$a_+(0) = 1$

$a_-(0) = 0$

$$|a_-|^2 = \left(\frac{\mu_N \mathcal{H} \Delta t}{\hbar}\right)^2 = w$$ also, since $|H_{+-}|^2 = |H_{-+}|^2$

This gives an additional contribution of $wn_+^\circ$ to $n_-$. The total contributions to both $n_+$ and $n_-$ can be written in the previous matrix form as

$$\begin{pmatrix}n_+\\ n_-\end{pmatrix} = \begin{pmatrix}1-w & w\\ w & 1-w\end{pmatrix}\begin{pmatrix}n_+^\circ\\ n_-^\circ\end{pmatrix}$$

The total effect on $n_+$ and $n_-$ in the unit section will be the product of this matrix times the matrix in IX.26. Let $\underset{\sim}{A}$ be this matrix product:

$$\underset{\sim}{A} = \begin{pmatrix}1-c\sigma_+ & 0\\ 0 & 1-c\sigma_-\end{pmatrix}\begin{pmatrix}1-w & w\\ w & 1-w\end{pmatrix} = \begin{pmatrix}(1-c\sigma_+)(1-w) & (1-c\sigma_+)w\\ (1-c\sigma_-)w & (1-c\sigma_-)(1-w)\end{pmatrix}$$

The problem is to find the eigenvector $\begin{pmatrix}n_+\\ n_-\end{pmatrix}$ such that $\underset{\sim}{A}\begin{pmatrix}n_+\\ n_-\end{pmatrix} = a\begin{pmatrix}n_+\\ n_-\end{pmatrix}$

Using $n_+ = \chi n_-$,

$$(1-c\sigma_+)(1-w)\chi n_- + (1-c\sigma_+)w n_- = a\chi n_-$$

$$(1-c\sigma_-)w\chi n_- + (1-c\sigma_-)(1-w)n_- = a n_-$$

Dividing,

$$\frac{(1-c\sigma_+)(1-w)x + (1-c\sigma_+)w}{(1-c\sigma_-)wx + (1-c\sigma_-)(1-w)} = x$$

$$x = \frac{-c(\sigma_+ - \sigma_-)(1-w) \pm \sqrt{c^2(\sigma_+ - \sigma_-)^2(1-w)^2 + 4(1-c\sigma_-)(1-c\sigma_+)w^2}}{2(1-c\sigma_-)w} \qquad \text{IX.27}$$

The $\pm$ sign gives the two mathematical eigenvalue solutions. Since a choice of the minus sign makes x negative, the solution which we are seeking must be that corresponding to the plus sign.

$$w = \left(\frac{\mu_n \mathcal{H} \Delta t}{\hbar}\right)^2 = 6.50 \times 10^{-3} \quad \text{for } \mathcal{H} = 23{,}000 \text{ gauss}$$

$$c = \frac{f}{1-f} N \times 10^{-4} = \frac{f}{1-f} \times 8.50 \times 10^{18}$$

For f = .99 (99 good domains per bad domain), x = .67 from IX.27.

Since the percentage of polarization is $\frac{m_+ - m_-}{m_+ + m_-}$, this is 20% polarization.

The solutions for three values of f are given in the following table.

| f | x | Percentage Polarization |
|---|---|---|
| .99 (1 bad per 99 good) | .67 | 20% |
| .995 (1 bad per 199 good) | .476 | 35.5% |
| .999 (1 bad per 999 good) | .115 | 79.5% |

## E. THEORY OF CHAIN REACTIONS

For security reasons Dr. Fermi followed fairly closely the article reprinted below. To avoid clearance delays, we reprint it directly, by courtesy of *Science*.

# Elementary Theory of the Chain-reacting Pile

(Reprinted by permission from Science Jan. 10, 1947)

**Enrico Ferm**
***Institute for Nuclear Studies, University of Chicag***

THE RESULTS AND THE METHODS DISCUSSED in the following outline of the theory of a chain-reacting pile working with natural uranium and graphite have been obtained partly independently and partly in collaboration by many people who participated in the early development work on the chain reaction. Very important contributions to the theoretical ideas were given by Szilard and Wigner. Many physicists contributed experimental results that helped to lead the way, among them, H. L. Anderson and W. H. Zinn, first at Columbia University and later at the Metallurgical Laboratory of the University of Chicago; R. R. Wilson and E. Creutz, at Princeton; and Allison, Whitaker, and V. C. Wilson, at the University of Chicago. The production of the chain reaction was finally achieved in the Metallurgical Laboratory directed by A. H. Compton.

## Absorption and Production of Neutrons in a Pile

We consider a mass, "the pile," containing uranium spread in some suitable arrangement throughout a block of graphite. Whenever a fission takes place in this system, an average number ($\nu$) of neutrons is emitted with a continuous distribution of energy of the order of magnitude of 1,000,000 EV. After a neutron is emitted, its energy decreases by elastic collisions with the atoms of carbon and to some extent also by inelastic collisions with the uranium atoms. In the majority of cases the neutrons will be slowed down to thermal energies. This process requires about 100 collisions with carbon atoms. After the energy of the neutron is reduced to thermal value, the neutron keeps on diffusing until it is finally absorbed. In several cases, however, it will happen that the neutron is absorbed before the slowing-down process is completed.

The neutron may be absorbed by either the carbon or the uranium. The absorption cross-section of carbon for neutrons of thermal energy is quite small, its value being approximately .005 $\times$ $10^{-24}$ cm.[1] For graphite of density 1.6, this corresponds to a mean free path for absorption of about 25 m. It is believed that the absorption cross-section follows the 1/v law, and consequently the absorption cross-section, which is already quite small at thermal energies, becomes practically negligible for neutrons of higher energy. It is therefore a sufficiently good approximation to assume that absorption by carbo during the slowing-down process can be neglected.

**This paper, presented at the Chicago meeting of the American Physical Society, June 21, 1946, is based on work performed under Contract No. W-7401-eng-37 with the Manhattan District at the Metallurgical Laboratory, University of Chicago.**

The absorption of a neutron by uranium may lea either to fission or to absorption by a (n, $\gamma$) process. W shall refer to this last possibility as the process of reso nance absorption. The relative importance of fission an resonance absorption in the different energy intervals i not the same. In this respect we can consider roughl three intervals:

(1) Neutrons with energy above the fission threshol of $U^{238}$—We can call these conventionally "fast neu trons." For fast neutrons the most important absorp tion process is fission, which normally takes place in th abundant isotope $U^{238}$. Resonance absorption is smalle but not negligible.

(2) Neutrons of energy below the fission threshold c $U^{238}$ and above thermal energy—We shall refer to thes neutrons as "epithermal neutrons." For epitherma neutrons the most important absorption process is th resonance capture. The cross-section for this proces as a function of energy is quite irregular and presents large number of resonance maxima that can be fairly wel represented by the Breit-Wigner theory. In practica cases the resonance absorption becomes important fo neutron energy below about 10,000 EV and increases a the energy of the neutrons decreases.

(3) Neutrons having thermal agitation energy c "thermal neutrons"—For thermal neutrons both the reso nance and fission absorption processes are important In this energy range both cross-sections follow approxi mately the 1/v law, and therefore their relative impor tance becomes practically independent of the energy Let $\sigma_f$ and $\sigma_r$ be the cross-sections for fission and reso nance absorption for neutrons of energy kT, and $\eta$ be th average number of neutrons emitted when a therma neutron is absorbed by uranium. Then $\eta$ differs from $\nu$ since only the fraction $\sigma_f/(\sigma_r + \sigma_f)$ of all the therma neutrons absorbed by uranium produces a fission. It is therefore,

$$\eta = \nu\sigma_f/(\sigma_f + \sigma_r). \qquad (1$$

The preceding discussion leads one to conclude tha only a fraction of the original fast neutrons produced wil end up by producing a fission process. For systems c finite size, further losses of neutrons will be expected b leakage outside the pile.

Limiting ourselves for the present to systems of prac tically infinite dimensions, we shall call $P$ the probabilit that a fast neutron ultimately is absorbed by the fissio process. The average number of neutrons produced in the "second generation" by the first neutron will then be

$$k = P\nu. \qquad (2)$$

Usually, $k$ is called the "reproduction factor" of the system. A self-sustaining chain reaction evidently is possible only when $k > 1$. If this is the case, the reaction actually will take place provided the leakage loss of neutrons is sufficiently small. This, of course, can always be achieved if the size of the pile is large enough.

## Life History of a Neutron

When a fast neutron is first emitted in our pile, the following events may take place:

(1) There is a small probability that the neutron will be absorbed by uranium before its energy has been appreciably decreased. If this is the case, the absorption leads often to fission of $U^{238}$. The probability of such fast fissions, however, is usually only a few per cent. Indeed, if the system contains little uranium and a large amount of carbon, the elastic collisions with carbon tend to reduce the energy very rapidly to a value below the fission threshold of $U^{238}$. If, on the other hand, the system is very rich in uranium, the inelastic collision processes become very probable and rapidly reduce the energy of the original fast neutron to a fairly low value before it has a chance to produce a fission in $U^{238}$.

(2) In the large majority of the cases, therefore, the neutron is not absorbed as a fast neutron and rapidly loses its energy, mostly due to collision against the carbon atoms. One can prove in an elementary way that it takes about 6.3 collisions against the carbon atoms to reduce the energy by an average factor of $e$. Consequently, it will take about 14.6 collisions in order to reduce the energy by a factor of 10, and about 110 collisions to reduce the energy from 1,000,000 EV to the thermal energy value of 1/40 v. While this slowing-down process is in progress, the neutron may be absorbed by the resonance process in uranium. We shall call $p$ the probability that a neutron is not absorbed before reaching thermal energy. One of the most important factors in designing a pile consists in trying to minimize the probability that neutrons are removed from the system by resonance absorption during the slowing down.

(3) If the neutron is not absorbed during the slowing-down process, it eventually reaches thermal energy and ultimately will be absorbed by either uranium or carbon. If uranium and carbon were mixed uniformly, the probability for these two events would be in the ratio of the absorption cross-sections of uranium and carbon for thermal neutrons multiplied by the atomic concentrations of the two elements. Since actually the mixture is not uniform, this is only approximately true. We shall call $f$ the probability that a thermal neutron is absorbed by uranium. In designing a chain-reacting pile one will normally try to adjust things so as to have both $f$ and $p$ as large as possible. Unfortunately, the two requirements are contradictory, because in order to make $f$ large, one shall try to build a system very rich in uranium in order to reduce the probability of absorption of thermal neutrons by carbon. On the other hand, in a system containing a relatively small amount of carbon the slowing-down process will be relatively slow, and consequently the probability of resonance absorption during the slowing down will be large.

It is clear, therefore, that one shall have to conciliate two opposite requirements by finding an optimum value for the ratio of uranium to carbon.

In a homogeneous mixture of uranium and carbon the values of $f$ and $p$ depend only on the relative concentrations of the two elements. If we do not restrict ourselves, however, to homogeneous mixtures only, one can try to obtain a more favorable situation by proper arrangement of the geometrical distribution of the two components. This actually is possible to a considerable extent, because of the following circumstances. The resonance absorption which is responsible for the loss of neutrons during the slowing down has very sharp cross-section maxima of the Breit-Wigner type. Therefore, if the uranium, instead of being spread through the graphite mass, is concentrated in rather sizable lumps, we will expect that the uranium in the interior of a lump will be shielded by a thin surface layer from the action of neutrons with energy close to a resonance maximum. Therefore, the resonance absorption of a uranium atom inside the lump will be much less than it would be for an isolated atom. Of course, self-absorption in a lump reduces not only the resonance absorption but also the thermal absorption of uranium. One can expect theoretically, however, and experiment has confirmed, that at least up to a certain size of lumps the gain obtained by reducing the resonance loss of neutrons overbalances by a considerable amount the loss due to a lesser absorption of thermal neutrons.

The typical structure of a pile is a lattice of uranium lumps embedded in a matrix of graphite. The lattice may be, for example, a cubic lattice of lumps or a lattice of rods of uranium. This latter arrangement is slightly less efficient from the point of view of the neutron absorption balance but often presents some practical advantages, since it makes easier the removal of the heat produced by the pile. In the present discussion we shall consider only lattices of lumps.

It is useful to give some typical figures for the probabilities of the various absorption processes. These probabilities, of course, are not constant but depend on the details of the structure of the lattice. Average figures for a good lattice will be given as an example. When a neutron is first produced by a fission taking place in a lump of uranium, it may have a probability of the order of 3 per cent of being absorbed, giving rise to fission before loosing any appreciable amount of energy. In 97 per cent of the cases when this does not happen the neutron will initiate its slowing-down process, and it may either be absorbed by the resonance process during the slowing down or reach thermal energy. The probability of resonance absorption during the slowing down may be of the order of 10 per cent, so that 87 per cent of the original neutrons will be slowed down to thermal energies. Of these, perhaps 10 per cent may be absorbed by carbon and the remaining 77 per cent by uranium. If we assume for the purpose of example that $\nu = 2$, we shall have in one generation the processes summarized in Table 1. For the example given, the reproduction factor will be, therefore,

$$k = .06 + .77\eta. \tag{3}$$

Consequently, a lattice of the type described would have a reproduction factor larger than 1, provided $\eta$ is larger than 1.22.

In order to evaluate the reproduction factor one must

TABLE 1

| Probability (%) | Type of process | Neutrons produced per neutron absorbed | Neutrons per generation by one neutron |
|---|---|---|---|
| 3 | Fast fission | 2 | .06 |
| 10 | Resonance absorption | 0 | 0 |
| 10 | Absorption by carbon | 0 | 0 |
| 77 | Absorption by uranium at thermal energies | $\eta$ | .77 $\eta$ |

be able to calculate the probabilities for the various processes mentioned. Some points of view which may be used in the practical calculation will be indicated briefly.

### Probability of Fission Before Slowing Down

The value of this quantity is very easily calculable for a very small lump of uranium. In this case it is obviously given by

$$P_F = \sigma_F nd, \tag{4}$$

where $\sigma_F$ is the average value of the fission cross-section for fission neutrons; $n$ is the concentration of uranium atoms in the lump; and $d$ is the average value of the distance that the neutron produced in the lump must travel before reaching the surface of the lump. The case of a lump of larger size is more complicated, since then multiple collision processes become important and both elastic and inelastic scattering play a considerable role. In particular, the last process for a lump of large size effectively slows down the neutrons before the fission threshold of $U^{238}$ and brings them down to an energy level in which they are readily absorbed by the resonan[ce] process.

### Resonance Absorption

If we had a single atom of uranium in a graphite m[e]dium where fast neutrons are produced and slowed dow[n] to thermal energy, the probability per unit time of resonance absorption process of neutrons with energ[y] larger than thermal energy would be given by the follo[w]ing expression:

$$\frac{q\lambda}{.158}\int \sigma(E)\,\frac{dE}{E}, \qquad (5)$$

where $q$ is the number of fast neutrons entering the sy[s]tem per unit time and unit volume, $\lambda$ is the mean fr[ee] path, and $\sigma(E)$ is the resonance absorption cross-secti[on] at energy E. The integral must be taken between a lo[w] limit just above thermal energy and an upper limit equ[al] to the average energy of the fission neutrons. One w[ould] expect that the largest contribution to the integral w[ill] be due to the Breit-Wigner peaks of $\sigma(E)$.

The above formula would be very much in error in t[he] case of a lattice of lumps. As already indicated, this [is] due to the fact that inside a lump there is an importa[nt] self-screening effect that reduces very considerably t[he] density of neutrons having energy close to a resonan[ce] maximum.

The best approach to a practical solution to the pro[b]lem is therefore a direct measurement of the number [of] neutrons absorbed by resonance in lumps of uraniu[m] of various sizes.

Measurements of this type have been performed fi[rst] at Princeton University, and the results have been su[m]marized in practical formulas that are used in the ca[l]culations.

### Probability of Absorption at Thermal Energi[es]

If uranium and carbon were uniformly mixed, a the[r]mal neutron would have a probability

$$\frac{N_U\sigma_U}{N_C\sigma_C + N_U\sigma_U} \qquad (6)$$

to be absorbed by uranium. In this formula $N_C$ and $N_U$ represent the numbers of atoms of carbon and of uraniu[m] per unit volume, and $\sigma_C$ and $\sigma_U$ represent the cross-se[c]tions of carbon and uranium for thermal neutrons.

More complicated is the case of a lattice distribution [of] lumps of uranium in graphite, since the density of the[r]mal neutrons throughout the system is not uniform b[ut] is large at the places far from the uranium lumps a[nd] smaller near and inside the uranium lumps, due to t[he] fact that the absorption of thermal neutrons is mu[ch] greater in uranium than in graphite. Let $\overline{n_C}$ and $\overline{n_U}$ [be] the average densities of thermal neutrons in the graph[ite] and in the uranium lumps. The number of thermal neutrons absorbed by uranium and by carbon will be proportional to $N_U \sigma_U \bar{n}_U$ and $N_C \sigma_C \bar{n}_C$, and we will have, therefore, instead of Equation (6), the corrected formula,

$$f = \frac{N_U \sigma_U \bar{n}_U}{N_U \sigma_U \bar{n}_U + N_C \sigma_C \bar{n}_C}. \tag{7}$$

For practical purposes it is usually sufficiently accurate to calculate $\bar{n}_C$ and $\bar{n}_U$, using the diffusion theory. The approximation is made to substitute the lattice cell by a spherical cell having volume equal to that of the actual cell, with the boundary condition that the radial derivative of the density of neutrons vanishes at the surface of the sphere. It is also assumed that the number of neutrons that are slowed down to thermal energies per unit time and unit volume is constant throughout the graphite part of the cell. This approximation is fairly correct, provided the dimensions of the cell are not too large. With these assumptions one finds the following formula for the probability, $f$, that thermal neutrons be absorbed by uranium:

$$f = \frac{3\alpha^2}{\alpha^3 - \beta^3} \frac{(1-\alpha)(1+\beta)e^{-\beta+\alpha} - (1+\alpha)(1-\beta)e^{\beta-\alpha}}{(\alpha + s - s\alpha)(1+\beta)e^{-\beta+\alpha} - (\alpha + s + s\alpha)e^{\beta-\alpha}} \tag{8}$$

where $\alpha$ and $\beta$ represent the radius of the lump and the radius of the cell expressed taking the diffusion length in graphite, $l = \sqrt{\lambda\Lambda/3}$, as unit of length. It is further

$$s = \frac{\lambda}{\sqrt{3}} \frac{1+\gamma}{1-\gamma}, \tag{9}$$

where $\gamma$ is the reflection coefficient of the lump for thermal neutrons.

### Lattice Containing a Large Number of Cells

The density of neutrons of any given energy in a lattice containing a large number of cells is a function of the position in the lattice. One can arrive at a simple mathematical description of the behavior of such a system by neglecting in first approximation the local variation of such functions due to the periodic structure of the lattice and substituting for the actually inhomogeneous system an equivalent homogeneous system. In this section we shall accordingly simplify the problem by substituting for all densities of neutrons values obtained by averaging the actual values over the volume of the cell. The densities will then be represented by smooth functions such as one would expect in a homogeneous uranium-graphite mixture.

Let $Q(x,y,z)$ be the number of fast neutrons produced per unit time and unit volume at each position in the lattice. These neutrons diffuse through the mass and are slowed down. During this process some of the neutrons are absorbed at resonance. Let $q(x,y,z)$ be the number of neutrons per unit time and unit volume which become thermal at the position $x, y, z$—$q$ is called the "density of the nascent thermal neutrons."

We shall assume that if an original fast neutron is generated at a point, 0, the probability that it becomes thermal at a given place has a Gaussian distribution around 0. This assumption may be justified by considering that the diffusion process of slowing down consists of very many free paths. Experimentally one finds that the distribution curve of the nascent thermal neutrons around a point source of fast neutrons is represented only approximately by a Gaussian distribution, and formulas have been used in which the actual distribution is described as a superposition of two or three Gaussian curves with different ranges. For the purpose of the present discussion, however, we shall take only one. For each fast neutron produced only $p$ neutrons reach thermal energy. The distribution of nascent thermal neutrons produced by a source of strength 1, placed at the origin of the coordinate, shall then be represented by

$$q_1 = \frac{p}{\pi^{3/2} r_0^3} e^{-r^2/r_0^2}. \tag{10}$$

For graphite of density 1.6 the range, $r_0$, is of the order of 35 cm. The density of nascent thermal neutrons at point P can be expressed in terms of Q by adding up the contribution of all the infinitesimal sources, $Q(P')d\tau'$ ($d\tau'$ represents the volume element around the point, P′). We obtain in this way

$$q(P) = \frac{p}{\pi^{3/2} r_0^3} \int Q(P') e^{-((P'-P)^2/r_0^2)} d\tau'. \tag{11}$$

The density, $n(x,y,z)$, of the thermal neutrons is connected to $q$ by the differential equation,

$$\frac{\lambda v}{3} \Delta n - \frac{v}{\Lambda} n + q = 0, \tag{12}$$

where $\lambda$ is the collision mean free path of thermal neutrons, $v$ is their velocity, and $\Lambda$ is the mean free path for absorption of a thermal neutron. Equation (12) is obtained by expressing a local balancing of all processes whereby the number of thermal neutrons at each place tends to increase or decrease. The first term represents the increase in number of neutrons due to diffusion ($\lambda v/3$ is the diffusion coefficient of thermal neutrons); the second, the loss of neutrons due to absorption; and the third, the effect of the nascent thermal neutrons.

It should be noted that the absorption mean free path $\Lambda$ in Equation (12) is much shorter than the corresponding quantity, $\Lambda_0$, in pure graphite. Indeed, the absorption in a lattice is due mostly to the uranium. In first approximation $\Lambda$ is given by

$$\Lambda = (1 - f)\Lambda_0. \tag{13}$$

In practical cases $\Lambda$ may be of the order of magnitude of 300 cm., whereas $\Lambda_0$ in graphite without uranium is about 2,500 cm.

When a thermal neutron is absorbed by uranium, $\eta$ new neutrons are produced by fission. This number should be increased by a few per cent in order to take into account the effect of the small probability of fast fission. Let $\epsilon\eta$ be the total number of fast neutrons so corrected.

The number of thermal neutrons absorbed per unit volume and unit time is $\frac{vn}{\Lambda}$. Of these, the fraction $f$ is absorbed by uranium. We have, therefore,

$$Q = f\eta\epsilon\frac{v}{\Lambda}n + Q_0, \tag{14}$$

where $f\eta\epsilon\frac{v}{\Lambda}$ represents the number of fast neutrons produced in the chain reaction process, and $Q_0$ represents the number of fast neutrons produced by an outside source if one is present. In most cases, of course, $Q_0$ will be equal to 0. From Equations (11), (12), and (14) we can eliminate all unknowns except $n$, and we find

$$\frac{3}{\lambda\Lambda}n - \Delta n = \frac{3p\epsilon\eta f}{\pi^{3/2}r_0^3\Lambda\lambda}\int n(P')e^{-(P'-P)^2/r_0^2}d\tau' + \frac{3p}{\pi^{3/2}r_0^3\lambda v}\int Q_0(P')e^{-(P'-P)^2/r_0^2}d\tau'. \tag{15}$$

A solution of this equation is obtained readily by developing both $Q_0$ and $n$ in a Fourier series. The general term of this development, corresponding to $Q_0$ of the form $Q_0 \sin\omega_1 x \sin\omega_2 y \sin\omega_3 z$, is:

$$n = \frac{(\Lambda p Q_0/v)\sin\omega_1 x\sin\omega_2 y\sin\omega_3 z}{\left(1+\frac{\lambda\Lambda}{3}\omega^2\right)e^{\omega^2 r_0^2/4} - \epsilon p f\eta} \tag{16}$$

where $\omega^2 = \omega_1^2 + \omega_2^2 + \omega_3^2$.

When the dimensions of the pile are finite but very large compared with the mean free path, the boundary condition is that all densities must vanish at the surface. If the pile, for example, is a cube of side $a$ and the origin of the coordinates is taken in one of the corners, it is:

$$\omega_1 = \frac{\pi n_1}{a}; \quad \omega_2 = \frac{\pi n_2}{a}; \quad \omega_3 = \frac{\pi n_3}{a}, \tag{17}$$

where $n_1$, $n_2$, $n_3$ are positive integral numbers that define the various Fourier components. The critical dimensions of the system are such that the denominator of Equation (16) vanishes for the 1,1,1 harmonic, since in this case the density of the neutrons becomes infinitely large. The critical condition can be expressed, therefore by the equation:

$$\left(1+\frac{3\pi^2}{a^2}\frac{\lambda\Lambda}{3}\right)e^{3\pi^2 r_0^2/4a^2} = \epsilon p f\eta \tag{18}$$

The right-hand side in this formula is the reproduction factor, $k$, for a system of infinite size. We can therefore write the critical condition as follows:

$$k = \left(1+\frac{3\pi^2}{a^2}\frac{\lambda\Lambda}{3}\right)e^{3\pi^2 r_0^2/4a^2}. \tag{19}$$

In most cases both the exponent of $e$ and the term added to 1 in the parentheses are small compared with 1, and so the previous expression can be simplified to:

$$k = 1 + \frac{3\pi^2}{a^2}\left(\frac{\lambda\Lambda}{3} + \frac{r_0^2}{4}\right). \tag{20}$$

This formula can be used in order to calculate the critical side of a pile of cubical shape. If, for example, we assume for a special lattice numerical values of $\lambda$ = 2.6 cm., $\Lambda$ = 350 cm., $r_0^2$ = 1,200 cm.², and k = 1.06, we find for the critical side of a cubical pile, a = 584 cm. Naturally these constants are merely hypothetical, and though included within the possible range, are in practical cases strongly dependent on the details of the lattice structure.

It is useful to derive an approximate relationship between the power produced by a pile and the intensity of thermal neutrons inside it. Roughly 50 per cent of the thermal neutrons absorbed in a pile give rise to fission and the energy released per fission is of the order of 200 MEV. This corresponds to about $1.6 \times 10^{-4}$ erg per thermal neutron absorbed. Since the number of thermal neutrons absorbed per unit volume is $vn/\Lambda$ the energy produced is approximately

$$\frac{vn}{\Lambda}1.6 \times 10^{-4} \cong 4.6 \times 10^{-7}\ vn \text{ ergs/cm.}^3\text{ sec.} \tag{21}$$

Naturally, the power is not produced uniformly throughout the pile because $n$ is a maximum at the center and decreases to 0 at the edge of the pile. For a cubical pile $n$ is represented approximately by

$$n = n_0 \sin\frac{\pi x}{a}\sin\frac{\pi y}{a}\sin\frac{\pi z}{a}, \tag{22}$$

where $n_0$ is the density of neutrons at the center of the pile. Integrating the previous expression (21) over all the volume of the pile, one obtains the following formula for the power:

$$W = \frac{8}{\pi^3}4.6 \times 10^{-7}\ nv a^3 = 1.2 \times 10^{-7}\ n_0 v a^3. \tag{23}$$

If, again, we take as an example a pile with a side of 584 cm., we find $W = 24 n_0 v$ ergs/sec. When the pile is operating at a power of 1 kw., the flux of thermal neutrons at the center is therefore about $n_0 v = 4 \times 10^8$ neutrons/cm.$^2$ sec.

## Description of a Graphite Pile at Argonne Laboratory

The first pile was erected under the West Stands on the campus of the University of Chicago at the end of 1942. After having been operated there for a few months it was moved to the Argonne Laboratory, near Chicago, where it has been used until now for various research purposes.

The lattice of that pile is not the same throughout the structure. Since only a small amount of uranium metal was available at that time, metal has been used in the central portion of the pile and uranium oxide in the outer portion.

The intensity of operation of the pile is recorded by a number of $BF_3$ ionization chambers connected to amplifiers or to galvanometers.

Since this pile has no cooling devices built into it, the power produced is limited by the necessity of avoiding an excessive temperature rise. The pile could be operated indefinitely at a power of 2 kw. and is often operated for periods of the order of one or two hours up to about 100 kw.

One feature that is often used for neutron research work is the thermal column, a column of graphite having sides of about 5 x 5 feet, which is built on the center of the top of the pile and goes through the top shield. The neutrons that diffuse from the pile into this column are rapidly reduced to thermal energy so that the neutrons inside the column a few feet above the top of the pile are practically pure thermal neutrons.

The pile is also equipped with a number of holes in the shield and removable stringers of graphite that make it possible to explore phenomena inside the pile or to introduce samples for neutron irradiation.

When the pile is operated at 100 kw., the flux of thermal neutrons at the center is about $4 \times 10^{10}$ neutrons/cm.$^2$ sec.

NOTE ADDED IN PROOF CONCERNING EQUATION VIII.2:

According to page 142, the matrix element in "Golden Rule #2", eqn. VIII. 2, is a "suitable average" for various final states. To indicate more explicitly what is meant by "suitable average", consider a transition in which state A consists of particles A and a, and these may transform to particles B and b:

$$A + a \longrightarrow B + b$$

The matrix element will be the sum of all matrix elements between initial and final states of all possible values of vector angular momentum (total momentum and its projection), divided by the number of terms in the sum, the number of terms being $(2I_A + 1)(2I_a + 1)(2I_B + 1)(2I_b + 1)$, where $I_A$, for example, is the spin of particle A. Now only matrix elements for which vector angular momentum is conserved are nonzero. It is this fact that makes the matrix element depend on the final state in this particular example. (In other problems, the matrix element may depend on additional variables, such as direction of emission of particle a.) Writing out this average,

$$\overline{|\mathcal{H}_{A\to B}|^2} = \frac{1}{N}\sum_{J_A}\sum_{J_{A_z}}\sum_{J_B}\sum_{J_{B_z}} |\mathcal{H}_{J_A J_{A_z}\to J_B, J_{B_z}}|^2 \quad \text{where} \quad J_A = I_A + I_a\text{, etc.}$$

$$N = (2I_A+1)(2I_a+1)(2I_B+1)(2I_b+1)$$

$$= \frac{1}{N}\sum_{J_A}\Big(\sum_{J_{A_z}}\sum_{J_{B_z}} |\mathcal{H}_{J_A, J_{A_z}\to J_B, J_{B_z}}|^2\Big) = \frac{1}{N}\sum_{J_A} |\mathcal{H}_{J_A\to J_B}|^2 (2J_A+1)$$

since matrix elements are zero unless $J_A = J_B$ and $J_{A_z} = J_{B_z}$.

and this is the "suitable average" to be inserted, in this particular case. Note that for the inverse reaction, we get in a similar manner:

$$\overline{|\mathcal{H}_{B\to A}|^2} = \frac{1}{N}\sum_{J_A} |\mathcal{H}_{J_B\to J_A}|^2 (2J_A+1) = \overline{|\mathcal{H}_{A\to B}|^2}$$

in accordance with the statement on page 145.

# CHAPTER X. COSMIC RAYS

This field is expanding very rapidly. Many facts are known, but the present theories to explain them are mostly tentative. For brevity we shall take many liberties and talk as though both fact and theory were better established than is the case.

First let us set down some facts about the pressure and equivalent height of the atmosphere. We shall define the quantity <u>atmospheric depth</u> x measured in g/cm$^2$, which is numerically equal to the pressure measured in grams-force/cm$^2$. At sea level $x = 1030$ g/cm$^2$, and the density of the air is $1.225 \times 10^{-3}$ g/cc. If the air maintained this density all the way from sea level to the "top" of the atmosphere, the equivalent height of the atmosphere would be about 8 km. If the density were at all heights proportional to the pressure, then it would follow that the mean height of the atmosphere is 8 km and that

$$x = 1030\, e^{-z/8} \quad \text{g/cm}^2$$

where z is the altitude in km. This is only very roughly true.

The first three sections of this chapter will attempt to give a bird's eye view of the field of cosmic radiation. We shall first describe the primary radiation incident upon the top of the atmosphere, and then discuss the secondary radiation produced when the primary particles collide with the nuclei of the air. Since we shall try to follow a particle and its secondaries from the top to the bottom of the atmosphere, we shall postpone discussion of latitude effects until the end of the chapter.

## A. PRIMARY RADIATION

A charged cosmic ray particle approaching the earth from outer space comes under an appreciable influence of the earth's magnetic field at a distance of several thousand km from the earth. If the energy of the particle is low, it may be caused to curve so much that it never gets to the earth. If, however, the low energy particle is moving along one of the earth's lines of force, then it will eventually end up near one of the earth's magnetic poles.

We shall show in section D. that the lowest momentum* with which a particle can arrive at any point on the earth's surface is determined by both the magnetic latitude, $\lambda$, and by the angle between the particle's velocity and a vector pointed west. For the moment we shall simply give some idea of these dependences by quoting X.15 for the special case where the particle is coming in along the meridian plane (the plane including north, south, and the zenith).

Then

$$p_{min} = 15\ \frac{\text{Bev}}{c}\ \cos^4 \lambda. \qquad \text{X.1}$$

---

*Since magnetic deflection determines the momentum, not the energy, of a particle ($\mathcal{H}\rho = pc/Ze$), we often speak of the momentum of a cosmic ray particle. For example, the momentum of a particle expressed in ev/c is numerically equal to its energy if it is extremely relativistic and to 2c/v times its energy if it is non-relativistic.

Observations of the Zeeman effect in the spectrum of solar radiation indicate that the sun may have a magnetic moment that produces an $\mathcal{H}$ of the order of 10 gauss at the surface of the sun. This Zeeman effect is very weak; no actual splitting of lines, but merely polarization effects, are observed. To add to the uncertainties, older data indicate that this field may have been stronger 10 years ago than it is now.

If the sun does have a magnetic moment, most low-energy cosmic rays will be deflected from their orbits and never get to the earth*. The data are inconclusive, but Pomerantz has sent up balloons at the earth's north magnetic pole and established the incidence of primary particles with energy as low as around $10^8$ ev.

Information on primary radiation comes primarily from balloon-carried equipment**. A successful flight may last for say 8 hr at 100,000 ft ($x \approx 10$ g/cm$^2$). Rockets have transported equipment up to 150 Km ($x \approx 10^{-4}$ g/cm$^2$), but the data are harder to interpret because of the short flights ($\approx$ 5 min.), scattering from the rocket, and other difficulties***

At the top of the atmosphere it appears that primary particles, mostly protons, arrive isotropically**** from all directions of outer space at a rate that apparently does not vary with time by more than a few tenths of a percent for particles with $E > 10$ Bev.

Because of the latitude effect it is important to specify $\lambda$ when giving intensities. TABLE I gives the total vertical intensity of particles of all energies, near the top of the atmosphere.

| Magnetic Latitude, | $\lambda = 55^\circ$ N | $\lambda = 30^\circ$ N |
|---|---|---|
| Protons | 0.15 (?) | 0.06 |
| Alphas | 0.05 (?) | 0.009 0.003 |
| $6 \leq Z \leq 9$ | $(1.1 \pm 0.2) \times 10^{-3}$ | $(3.0 \pm 1.0) \times 10^{-4}$ |
| $10 \leq Z$ | $(3.5 \pm 0.6) \times 10^{-4}$ | $(1.0 \pm 0.3) \times 10^{-4}$ |

TABLE X.I Vertical intensity of primary radiation in particles sec$^{-1}$ cm$^{-2}$ steradian$^{-1}$ (i.e. per unit solid angle)*****

*A list of references is given at the end of the chapter (p 236). Papers on the effects of these stellar fields are given in section 112 of reference 7. See also Pomerantz and Vallarta, Phys. Rev. 76, 1889 ('49).

**Schein et al., Phys. Rev. 59, 615 ('41)
Hulsizer and Rossi, Phys. Rev. 73, 1402 ('48)

***Van Allen, Sky and Telescope 7, 171 ('48); Phys. Rev. 75, 57 ('49).

****Where this is energetically possible as discussed in section D.

*****The proton data are given by Schein. The heavier particles were first reported by Frier, Lofgren, Ney, Oppenheimer, and Bradt and Peters (Phys. Rev. 74, 213 ('47)) and the data in the table were given by Bradt and Peters, Phys. Rev. 77, 62 ('50). (continued on next page)

The intensity of electrons (or high-energy photons) is less than $10^{-3}$ for all $\lambda$, and possibly zero.*** Unless they come from the sun, there can be no neutrons, and there can surely be no known mesons in the primary flux, since these particles are unstable.

The energy spectrum (FIG. X.2) is discussed by Montgomery (page 147 in reference 1 at the back of this chapter). The spectrum of Fig. X.2 could not be observed at any point on earth other than at the magnetic poles. For other latitudes and for a given direction of incidence, only that part of the spectrum to the right of the cut-off momentum (the top scale of the Fig. gives a special case of this -- the general equation is X.14, p. 230) arrives at the top of the atmosphere.

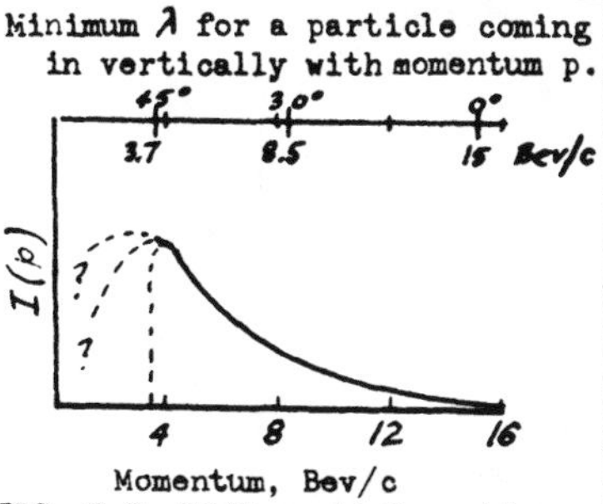

FIG. X.2 Differential spectrum of primary protons arriving at a magnetic pole.

## B. SECONDARY RADIATION

Upon entering the atmosphere, the primary particles collide with nuclei, freeing or creating all the known fundamental particles (and maybe more!).

The heavy primary nuclei are "absorbed" (broken up) more rapidly than the protons. Thus nuclei with Z > 12 disappear* as $e^{-x/20}$ whereas the sort of protons that produce small stars disappear more nearly as $e^{-x/150}$.

Of the interesting heavy-nuclear collisions in the top few percent of the atmosphere, we shall say only that they ultimately produce protons, neutrons, mesons, etc., very rapidly**.

Next we shall discuss the sequence of events when a high-energy nucleon (say a 5 Bev proton) collides with a nucleus of nitrogen or oxygen and produces a large star.

First, presumably after about $10^{-23}$ sec, a few particles, 5 pions ($\pi$ mesons) plus one or two nucleons, for example, are seen (or in the case of neutrons, supposed) to come off nearly

---

footnote continued from last page:

In none of the high altitude measurements from which the primary intensity is deduced is it possible to tell whether a particle was counted while it was going up or down. As originally suggested by Wheeler and calculated by Rossi, it could well be that one particle out of five is going up. However, Biehl et al. (Phys. Rev. 76, 914, ('49)) find experimentally that the upwards fraction is negligibly small.

*Bradt and Peters, Phys. Rev. 75, 1779 ('49)

**Marshak, Phys. Rev. 76, 1736 ('49); Leprince-Ringuet, Phys. Rev. 76, 1273 ('49).

***Hulsizer, Phys. Rev. 73, 1253 ('48)

straight forward*. There are also created, by means not well understood,** gammas, or perhaps electrons, which then multiply into cascade showers (p. 49).

Apart from creating pions and chipping off local nucleons, the high-energy particle apparently "heats" larger portions of the nucleus. Part of this energy may eventually be dispersed

FIG. X.4a A nuclear interaction occurring in a photographic emulsion exposed in the stratosphere showing the paths of all charged particles (neutral particles, of course, do not leave any track). The light track A of developed Ag grains is produced by a proton of several Bev kinetic energy and is interpreted as the incident particle which collides with an atom of C, N, or O in the emulsion. The three heavy tracks are due to low-energy evaporation nucleons. Nine light tracks are produced by particles emitted in the general direction of travel of the incident particle and are all relativistic.

These particles may be identified by examining their small-angle scattering. In similar cases it has been shown that, say, 3/4 of the charged relativistic particles are pions, the rest protons. Photograph by J.J. Lord and Marcel Schein.

FIG. X.4b (This star is described by Lord, Schein, and Vidale in Phys. Rev. 76, 321 ('49). The statements made here are only the most probable conclusions). The star took place in emulsion exposed in the stratosphere.

The incident particle is a primary α near minimum ionization. Its total kinetic energy must be at least 3 Bev (0.75 Bev per nucleon) and is probably much greater. The α initiates a star and continues on, still near minimum ionization. One of the prongs of the star is a low-energy $\pi^+$ which finally decays into a $\mu^+$ at the end of a trajectory that never leaves the emulsion.

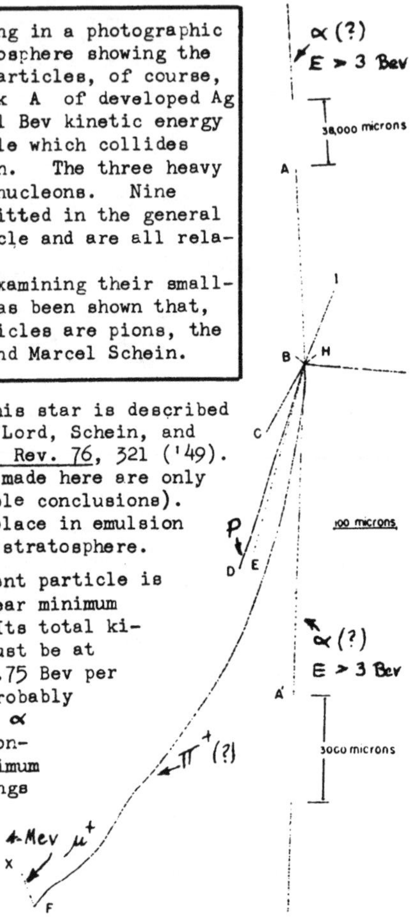


*There is considerable interest in determining the relative importance of "multiple" and "plural" pion formation. A multiple collision is one where several pions are formed simultaneously; a plural collision describes the successive encounter of the high-energy particle with different nucleons as it passes through the nucleus, one pion being created per encounter. Recent data favor multiple production with low "multiplicity" (Lord, Fainberg, and Schein, Phys. Rev. 80, 970 ('51); Fermi, Phys. Rev. 81, 1683 ('51)).

Proton-induced stars lead to the ejection of more positive than negative pions. Owen and Wilson (Proc. Phys. Soc. Lond. 64, 410 (Apr. '51)) report a "positive excess" of 15% to 25% depending upon the proton energy.

**See footnote pp. 221a-b.

over the entire nucleus, so that within about $10^{-21}$ sec neutrons (theoretically with E ≈ 10 to 20 Mev) and protons (E ≈ 10 to 30 Mev) evaporate isotropically*** as discussed on pp 162-3. Some heavier particles may boil off too.

FIG. X.4a shows such a star (initiated by a proton of several Bev kinetic energy).

Of course this isotropic evaporation need not follow a high-energy forward ejection of particles but may occur alone. For example, when a 100 Mev neutron hits a nucleus it will probably cause a small star, i.e. two to four nucleons may evaporate isotropically.

A given collision may have characteristics somewhere between the extremes described above*. Notice that the new nucleons may be sufficiently energetic to produce secondary stars, the cumulative result being a cascade nuclear shower**

While in the midst of definitions, we might as well explain some other terms that the reader will encounter in the literature:

A burst is a sudden local increase in ionization, observed in an ionization chamber. Even a small star may register as a burst.

In a pure penetrating shower, a number of particles having a common origin traverse on the order of 10 cm Pb or more without any (electronic) cascade multiplication. A pure penetrating shower is of nuclear origin; however gammas are created in some of these high-energy (Bev range) showers, so there may be an accompanying cascade. Even a small portion of the large cascade nuclear showers mentioned above may be called a penetrating shower.

$\delta$-rays. When a heavy ionizing particle, particularly a slow one, ionizes an atom, it may give the electron an impulse sufficient to allow it to produce secondary ionization. Thus we see tracks ($\delta$-rays) which point away from the path of a non-relativistic heavy particle but have a component along this path if the heavy particle is relativistic.

So far in our chronological survey of what happens to our representative 5-Bev proton, we are still near the top of the atmosphere surrounded by all sorts of fundamental particles. We shall now discuss the fate of the most important of these, illustrating our remarks with the curves on pp 222-3, where the radiation is conveniently analysed into "components." The reader should be warned, however, that the following analysis of the cosmic radiation into a few "separate" components is really quite idealized. Thus, the curve of FIG. X.6, p 222, marked P (for Protons) really has a very heterogeneous constitution -- namely all heavy particles capable of causing nuclear events. Rossi calls it the N-component (N for "nuclear"), and about all that can be said about it is that it does not contain muons ($\mu$-mesons) or any of the "electronic component" (see p. 221a).

*For a statistical analysis of 100-to 700-Mev stars, see Harding, Lattimore, and Perkins, Proc. Roy. Soc. A196, 325 ('49).

**Fretter, Phys. Rev. 76, 933 ('49).

***In the cm. system of the nucleus, but this is practically the lab. system, since, as can be seen from the photographs, the nucleus is not observed to move.

As drawn by Rossi on Fig. X.6, p. 222, the intensity of the N-component decreases exponentially with atmospheric depth x with an "apparent absorption thickness" (or just absorption thickness, for short) of 125 $g/cm^2$. The experimental rate of absorption may be considerably less than the collision rate, since there are collisions such that only the energy of the nucleons changes or such that the number of N-component particles actually increases.

If we define a geometrical collision thickness

$$T_{coll.} = \frac{1}{\sigma} AM_1 \qquad g/cm^2$$

where $M_1$ is one atomic mass unit and $\sigma$ is the cross-sectional area of an "air" nucleus, $= \pi(1.5\ A^{1/3} \times 10^{-13})^2$; we find $T_{coll.} = 57\ g/cm^2$, about one half of the absorption thickness shown in Fig. X.6.

For future reference we also calculate $T_{coll.}$ for nucleons passing through Pb. It is 140 $g/cm^2$.

1. Protons Both primary and secondary protons will produce stars if their energy is > 0.1 Bev. Below this energy the rate of ionization increases rapidly (see Fig. II.4, p. 33) and the proton is usually brought to rest.

2. Neutrons These will cause nuclear events until their energy drops to the kev range. After this they will continue to lose energy by scattering, until they get down to about 1 ev, where they will be absorbed by nitrogen. These neutrons produce $C^{14}$ through an (n,p) reaction. Since ${}_6C^{14} \xrightarrow{\beta^-} {}_7N^{14}$ in 5100 y, the $C^{14}/C^{12}$ ratio in living matter (which is in equilibrium with the air) will be slightly higher than the ratio in dead organic matter*. A determination of this ratio gives a good indication of the age of an archeological specimen.

Recent data on the neutron component are presented in FIG. X.3, upon which we shall make a few comments:

1. The variation of slope (absorption thickness) with latitude is to be expected, and would also have shown up in Fig. X.6 (p. 222) if equatorial intensities had been plotted. This variation results simply because the higher average-energy primary protons at small latitudes produce longer chains of nuclear events, hence are absorbed less rapidly.
2. In an apocrypha at the end of the chapter we discuss how the experimental left-hand scale is calibrated to give the absolute scale at the right, and we point out the connection between these neutron production rates and the proton intensities of Fig. X.6, p. 222.

Slow neutron energy density After the energy of a neutron has been reduced so much that it causes few nuclear events, then the energy distribution of the slow neutrons will be determined by IX.10, p. 184. If we include an altitude-dependent amplitude factor, the energy-distribution will apply to a $cm^3$ of atmosphere and nV will be given explicitly by IX.10a.

Experimentally, for these low energies:

$\sigma_{scat.} = 0.36\ cm^2\ g^{-1}$. Thus the mean free path, $L = 1/\sigma =$

*Libby, Science 110, 678 ('49).

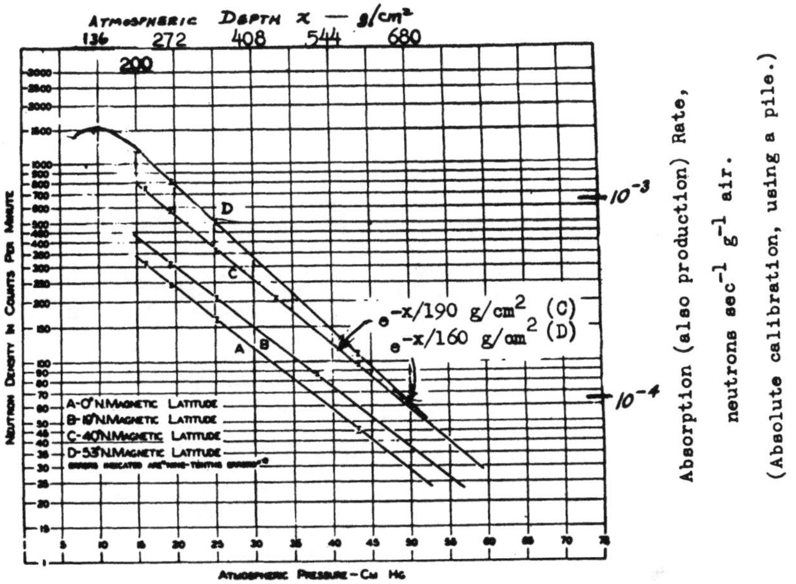


FIG. X.3 Apparent absorption of neutron-producing radiation at various magnetic latitudes*. See text, p. 220.

---

$2.78\ g/cm^2$. To express this in terms of a length use $\rho_{air} = 1.225 \times 10^{-3}\ g/cm^3$. Then L = 2280 cm.

So neutrons do not diffuse very far through the atmosphere, at least at sea level.

$\xi_{air}$ = 0.135, so a neutron's energy is reduced to 1/e in about 7 collisions.

$\sigma_{absorb.} = .00915\ E^{-1/2}\ cm^2\ g^{-1}$, in air. E is expressed in ev.

Putting these values into IX.10a, we find

$$n(E)V = A(x)\,\frac{1}{E}\,e^{-\frac{0.379}{\sqrt{E}}} \quad volt^{-1}\,cm^{-3}$$

where E is in ev. The form of this dependence is illustrated in the sketch at left.

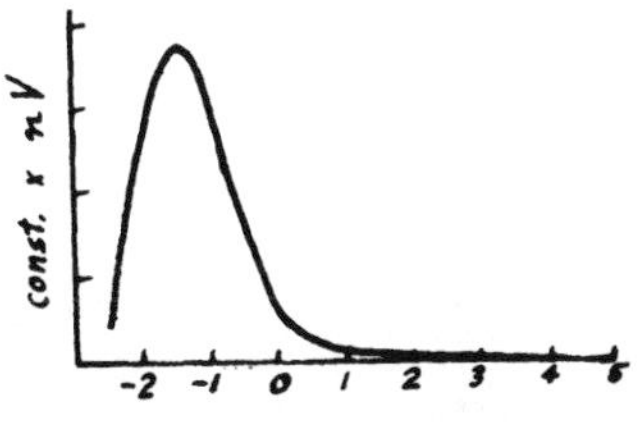


FIG. X.4 Distribution of the quantity nV for low energy neutrons.

A more complete discussion is given by Bethe, Korff, and Placzek, Phys. Rev. 57, 573 ('40).

---

*Simpson, Phys. Rev. 73, 1389 ('48), and Yuan, Phys. Rev. 74, 504 ('49).

**3. Mesons** Pions may be produced in any star caused by an incident particle with E > 3.5 to 4.0 Bev. But, as stated in Ch. VII,

$$\pi \rightarrow \mu + \nu \quad (?)$$

with a mean life $\tau = 10^{-8}$ sec in the rest system of the pion.

Hence $\tau_{lab.} = \gamma \tau_{rest} = \frac{W}{M_\pi c^2} \tau_{rest} \sim 10^{-7}$ sec. for $W \approx M_\pi \gamma c^2 \sim 1$ Bev

In this time the pion can travel only about 30 m, which is negligible compared with the "height" of the atmosphere.

The muon formed when the pion decays has $\tau_\mu = 2.15 \times 10^{-6}$ sec, so it will travel a distance of 600 meters x $W/M_\mu c^2$. Thus a fair fraction of the muons will reach the surface of the earth and some have been observed 1000 m of water equivalent below the surface of the earth.

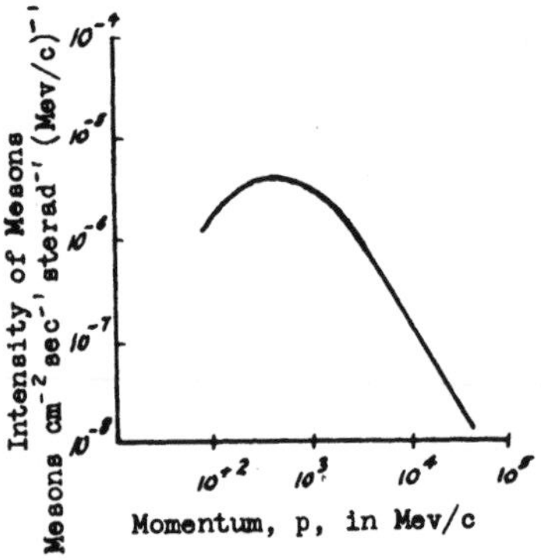


FIG. X.5 Differential momentum spectrum of $\mu$'s at sea level**

Relativistic muons are much more penetrating than relativistic protons. The main reason is that the muon-nucleon collision cross section is small ( $\sim 10^{-27}$ cm$^2$) compared with the proton-nucleon cross section. In addition the ionization loss -dE/dx stays close to the relativistic minimum down to energies lower than a Bev or so where a proton starts to slow down and consequently to ionize intensely. The great difference in the penetrating power of muons and protons is illustrated by Fig. X.6, p. 222, particularly for large values of x.

FIG. X.5 gives the experimental momentum spectrum for muons at sea level.

**4. Electrons and Gammas (the "electronic component")**

We do not distinguish between electrons and gammas when either has an energy in the Mev range or higher, since electrons give rise to photons in a radiation length of 360 m (II.58, p. 47) and these photons create electron pairs in a comparable pair-production length if the energy $> \approx 100$ Mev or else they free Compton electrons if the energy is low. This component loses its energy mainly after multiplication (p. 49).

Sources of the electronic component are "knock-on" electrons and the disintegration of muons:

$$\mu^{\pm} \xrightarrow{2.15\ \mu sec} e^{\pm} + 2\nu \quad (?)$$

as discussed at the beginning of Ch. VII.

These sources of electrons will account for the intensity of the electronic component near sea level, but will not explain the much greater intensity at atmospheric depths of a few hundred g/cm$^2$ (see Fig. X.7, p. 223). Apparently photons are created during the very high energy processes taking place in the upper one quarter of the atmosphere

*However, while traversing this path the pion may collide with a nucleon. The collision thickness seems to be about the geometrical one - see Camerini, Fowler, Lock, and Muirhead, Phil. Mag. 41, 413 ('50).

**J.G. Wilson, Nature 158, 415 ('46).

***As will be mentioned on p. 237, when 350 Mev protons from the Berkeley cyclotron are incident upon a target, gammas are emitted (cont. on next page)

## C. ANALYSIS INTO HARD AND SOFT COMPONENTS

To quote Rossi: "It is well known that the curve which represents the coincidence rate between two or more g-m tubes arranged in a straight line as a function of the thickness of a lead absorber placed between them shows a rather sudden change in slope at a thickness near 10 cm,* being steeper for small thicknesses. This fact is due to the presence in the cosmic radiation of electrons which are easily absorbable. Independently of its interpretation, it leads to an empirical separation of [charged] particles into a hard component, which includes all particles capable of traversing a given lead thickness, and a soft component, which includes all particles which are capable of traversing the counter walls but which are stopped by the given lead thickness. The choice of the critical thickness is, of course, somewhat arbitrary and the subdivision of cosmic-ray particles into a hard and a soft component has a meaning only because the relative intensities of the two components depends only slightly on this choice."

Rossi defines the critical thickness of lead as 167 g/cm$^2$ (14.7 cm). When the lead is removed, the soft particles have to traverse the equivalent of 5 g/cm$^2$ of brass in going between the sensitive volumes of the two counters. We shall now consider how much energy various particles must possess in order that they (or their secondaries) may penetrate 167 g/cm$^2$ Pb and 5 g/cm$^2$ of brass.

Experimentally, the situation is shown in TABLE X.II.

---

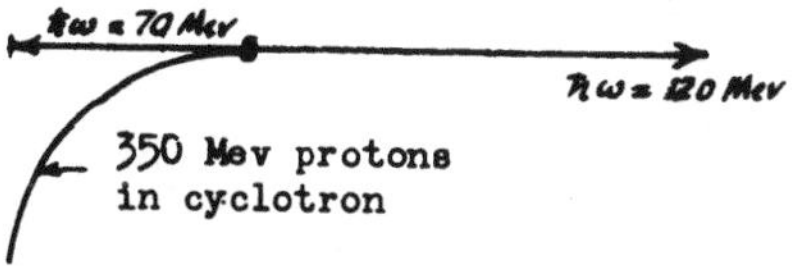


(footnote continued from last page) mostly in the forward and backward directions. For 300 Mev protons, $\sigma = 0.5 \times 10^{-27}$ cm$^2$, but $\sigma$ rises by by a factor of 3, to $1.5 \times 10^{-27}$, at 350 Mev. If the cross-section continues to rise this rapidly with increasing energy, it will account for some of the electronic component in the upper atmosphere.

This process of emission could be explained with the help of a neutral meson, $\pi^0$ (These mesons can be thought of as explaining non-exchange forces in the same way that $\pi^+$ and $\pi^-$ explain exchange forces.)

A $\pi^0$ most probably has a mass of about 300 m, so that it is energetically impossible for it to decay into a proton, P, and an antiproton, P$^-$. However, the P and P$^-$ may act as a virtual state for the reaction

$$\pi^0 \rightarrow P + P^- \rightarrow 2\gamma$$

where the second reaction is the inverse to pair production. The mean life for this process should be between $10^{-12}$ and $10^{-14}$ sec, and theory leads to a $\sigma$ (P,γ) comparable with that observed at 350 Mev (Lewis, Oppenheimer, et al, Phys. Rev. 73, 127 ('48)). The 50 Mev difference in energy between the forward and backward radiated beams can be interpreted as Doppler effect.

An entirely different process that would lead to the emission of photons during high-energy nuclear events has been proposed by Hayakawa (Phys. Rev. 75, 1759 ('49)). He points out that a high energy proton appears, electrically, like a plane wave of incident energy. If this proton exchanges with a neutron during a nuclear event, there will be an extremely rapid deceleration of charge and we can say, very crudely, that the plane wave continues along its path without the proton.

*At low altitudes

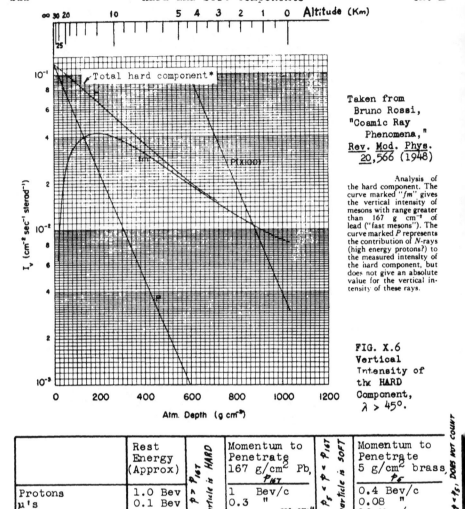

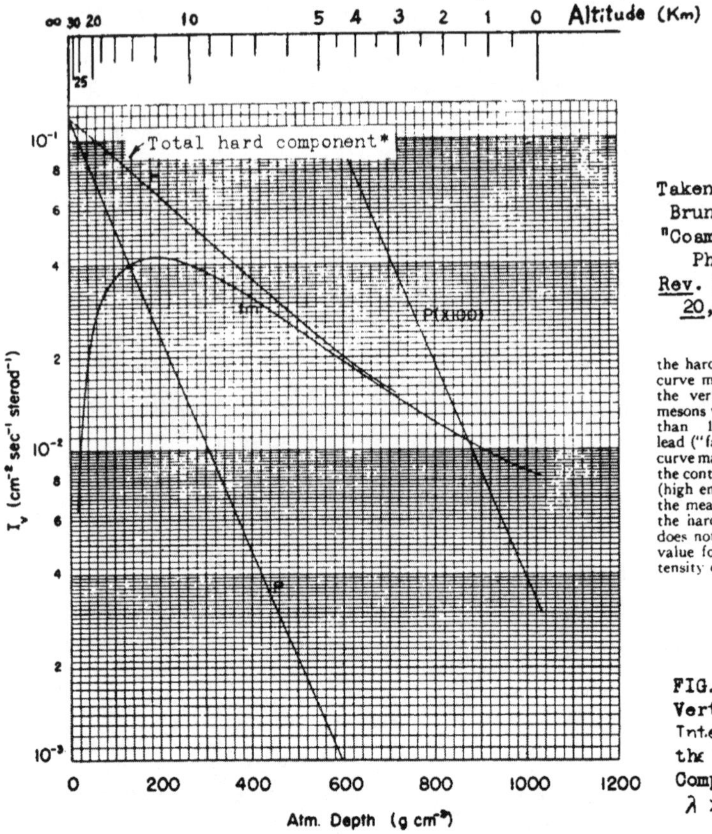


Taken from Bruno Rossi, "Cosmic Ray Phenomena," Rev. Mod. Phys. 20, 566 (1948)

Analysis of the hard component. The curve marked "*fm*" gives the vertical intensity of mesons with range greater than 167 g cm$^{-2}$ of lead ("fast mesons"). The curve marked *P* represents the contribution of *N*-rays (high energy protons?) to the measured intensity of the hard component, but does not give an absolute value for the vertical intensity of these rays.

FIG. X.6 Vertical Intensity of the HARD Component, $\lambda > 45°$.

| | Rest Energy (Approx) | If $p > p_{167}$ particle is HARD | Momentum to Penetrate 167 g/cm$^2$ Pb, $p_{167}$ | If $p_5 < p < p_{167}$ particle is SOFT | Momentum to Penetrate 5 g/cm$^2$ brass, $p_5$ | If $p < p_5$, DOES NOT COUNT |
|---|---|---|---|---|---|---|
| Protons | 1.0 Bev | | 1 Bev/c | | 0.4 Bev/c | |
| μ's | 0.1 Bev | | 0.3 " | | 0.08 " | |
| Electronic Comp. | 0.5 Mev | | 10.0 " see small type | | 10 Mev/c | |

TABLE X.II Momentum of Hard and Soft Components

From the TABLE it will be seen that the hard component is essentially relativistic charged nuclei and mesons, whereas the soft component involves mainly non-relativistic heavy charged particles plus practically all the electronic component.

**Qualitative remarks on values of TABLE II:**

The momenta necessary to enable heavy charged particles to penetrate 167 g/cm$^2$ Pb comes from relativistic ionization loss calculations.

*Measured by Gill, Schein, and Yngve, Phys. Rev. 72, 733 ('47).

Taken from
Bruno Rossi,
"Cosmic Ray
Phenomena,"
Rev. Mod. Phys.
20, 567 (1948)

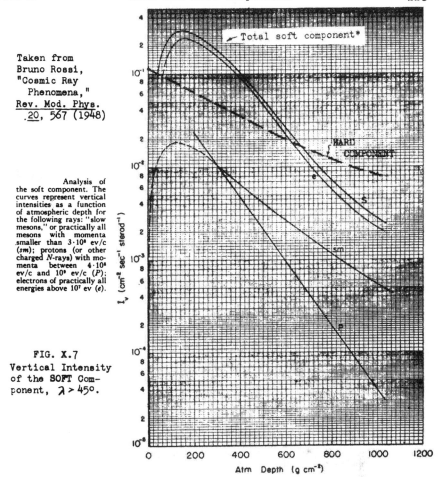

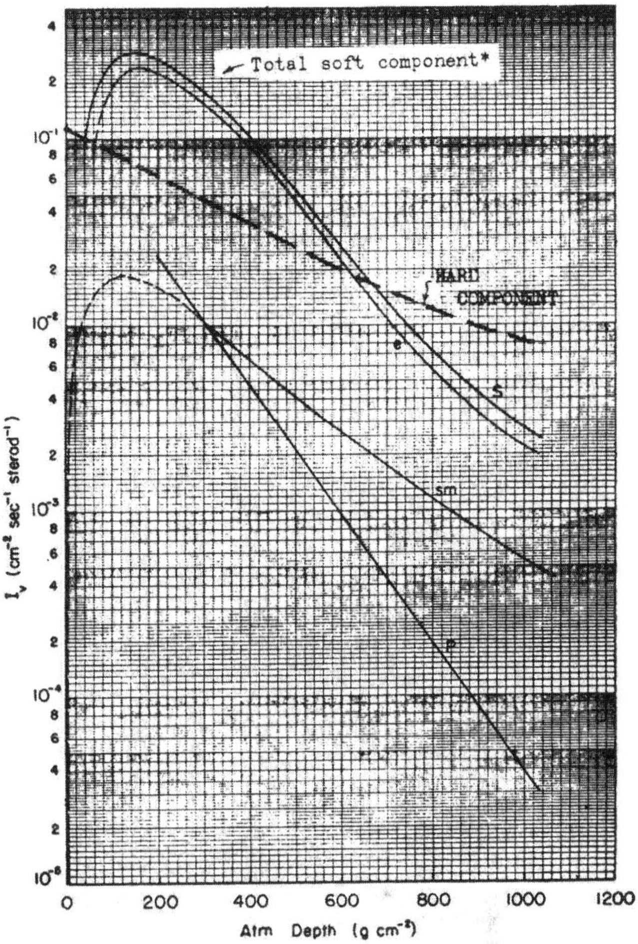


Analysis of the soft component. The curves represent vertical intensities as a function of atmospheric depth for the following rays: "slow mesons," or practically all mesons with momenta smaller than $3 \cdot 10^8$ ev/c (*sm*); protons (or other charged *N*-rays) with momenta between $4 \cdot 10^8$ ev/c and $10^9$ ev/c (*P*); electrons of practically all energies above $10^7$ ev (*e*).

FIG. X.7
Vertical Intensity of the SOFT Component, $\lambda > 45°$.

Curves of the results of these calculations may be found at the back of Montgomery's book. Of course there may well be a nuclear collision as the pion and particularly as the proton traverses the Pb, since we showed on p. 220 that the "geometric" collision thickness for protons through Pb is 140 g/cm$^2$, which is very close to the thickness of Pb under consideration. The experimental absorption thickness at these energies is 300-400 g/cm$^2$.

The figure of $10^{10}$ ev necessary for the electronic component to penetrate 167 g/cm$^2$ Pb is due to Greisen (Phys. Rev. 75, 1071 ('49)). Electronic component of such a high energy will multiply into a cascade shower rapidly, since its radiation length is only 0.6 cm; but the problem is complicated because one must then take into account the higher penetrating power of gammas at lower energies (Fig. II.21, p. 50).

*Measured by Millikan, Neher, and Pickering, Phys. Rev. 63, 234 ('43).

The momentum required by a (soft) charged particle to penetrate 5 g/cm$^2$ brass can be predicted merely by use of the range-energy equations based on ionization loss (II.16, p. 32; Feather's rule, on the same page, works fine for electrons).

While discussing how much energy an electron would have to have to be "hard", we might point out that Pomeranchuk (J. Phys. USSR 2, 65 ('40)) has considered the radiation by an electron travelling in the $\mathcal{H}$ of the earth and concludes that an electron can never reach the earth with an energy $> 10^{17}$ ev. Also he would not expect to observe photons with $E > 10^{19}$ ev.

We are now able to summarize all the discussion of Sec. B and of this section by presenting FIGs. X.6 and X.7, which are taken from Rossi's paper, p. 566. The dashed curve at the top of the soft-component figure is simply the hard component line of Fig. X.6 redrawn for comparison.

These curves apply for $\lambda > 45°$ and for radiation from the zenith. For the hard component at sea level, the radiation from a direction making an angle $\theta$ with the zenith varies roughly as $\cos^2 \theta$.

A few points worth noticing:

At sea level the radiation is due mainly to fast muons; the total intensity is roughly 1/10 of the primary intensity, and about 1/5 of this total is "soft".
At 100 to 200 g/cm$^2$, the intensity of the electronic component is 2 to 3 times greater than that of the primary radiation.

Upon closer examination of these curves, the reader will discover many other important facts.

(During this discussion Dr. Fermi had no time to mention the interesting topic of extensive showers, large nuclear showers involving up to $10^6$ particles and total energies of $10^{15}$ ev. The reader is referred to Montgomery or to the symposium of the Interuniversity Cosmic Ray Laboratories, Echo Lake, June 1949, which is to be published soon.)

## D. MOTION OF CHARGED PARTICLES IN THE EARTH'S MAGNETIC FIELD

Since this chapter is only 24 pages long, the 8 pages devoted to parts 1., 2., and 3. of this section are not justified by their small relative importance in cosmic rays. We were unable, however, to give a satisfactory derivation of X.14 more concisely. The reader who is willing to accept this equation may omit parts 1. and 2., reading only the results at the top of page 230. The notation used there is as follows:

$\lambda$ is the geomagnetic latitude;
$\gamma$ is the angle that the trajectory of a cosmic ray particle makes with a vector pointed west;
Z is the charge of the primary particle;
p is the momentum of the primary particle.

## 1. Trajectories

The magnetic field of the earth may be approximately considered as due to a magnetic dipole of moment $\mu = 8.1 \times 10^{25}$ erg/gauss located several hundred miles from the center of the earth and pointing roughly south.* North is here defined as the direction towards the earth's North Magnetic Pole, located in the Canadian arctic region.

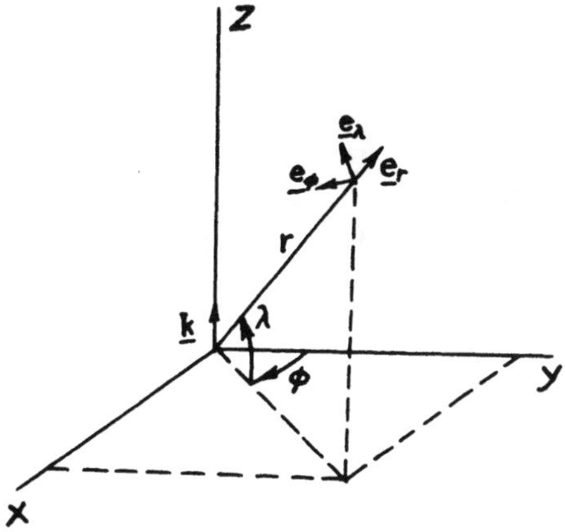


$$\underline{e}_\phi = \underline{e}_r \times \underline{e}_\lambda$$
$$\underline{k} = \underline{e}_r \sin\lambda + \underline{e}_\lambda \cos\lambda$$

FIG. X.8

FIG. X.8 shows the coordinate system we shall use; note that the z-axis lies along the earth's magnetic axis, that the angles are defined somewhat differently from those conventionally used in physics, and that for a point $(r, \lambda, \phi)$ on the earth's surface ($r = r_e = 6.4 \times 10^8$ cm), $\lambda$ and $\phi$ are respectively the geomagnetic latitude and west longitude. The vectors $\underline{e}_r$, $\underline{e}_\lambda$, and $\underline{e}_\phi$ point, respectively, to the zenith, to the north, and to the west. Note also that the velocity is

$$\underline{V} = \dot{r}\,\underline{e}_r + r\dot{\lambda}\,\underline{e}_\lambda + r\cos\lambda\,\dot{\phi}\,\underline{e}_\phi \qquad \text{X.5}$$

The vector potential of the earth's magnetic field is

$$\begin{aligned} \underline{A} &= -\underline{\mu} \times \underline{\nabla}\frac{1}{r} \\ &= \mu\underline{k} \times \underline{\nabla}\frac{1}{r} \\ &= \frac{\mu\cos\lambda}{r^2}\,\underline{e}_\phi \end{aligned} \qquad \text{X.6}$$

while the magnetic field itself is

$$\underline{\mathcal{H}} = \underline{\nabla}\times\underline{A} = \mu\,\frac{\cos\lambda\,\underline{e}_\lambda - 2\sin\lambda\,\underline{e}_r}{r^3} \qquad \text{X.7}$$

$\mathcal{H}$ decreases as $r^{-3}$, so at an altitude equal to $r_e$, $\mathcal{H}$ is only 1/8 as strong as at the surface. (At the poles, $\mathcal{H} = 0.61$ gauss; at the equator, 0.31). A particle may be considered as relatively unaffected by $\mathcal{H}$ if the radius of curvature, $\rho$, of its orbit is much larger than $r_e$, the range within which the field is appreciable. For motion perpendicular to a uniform $\mathcal{H}$, we know that the "magnetic rigidity," $P = pc/Ze$, is equal to $\mathcal{H}\rho$**. (P has the dimensions of energy per charge.)

---

*Because of local variations, this approximation is poor near the surface of the earth, but it improves at high altitudes.

** The equation of motion is: $\dot{\underline{p}} = \frac{Ze}{c}\underline{V}\times\underline{\mathcal{H}}$.
If V is perpendicular to a uniform $\mathcal{H}$, then

$$\dot{p} = p\frac{V}{\rho} = \frac{Ze}{c}V\mathcal{H},$$

$$P \equiv \frac{pc}{Ze} = \mathcal{H}\rho$$

Thus we can find the minimum order of magnitude of P such that the particle will be relatively unaffected, from the condition

$$P > \mathcal{H} r_e = 0.31 \times 6.4 \times 10^8 = 2 \times 10^8 \text{ gauss cm}.$$

For Z = 1, this corresponds to an energy of about 60 Bev.

Since a static magnetic field cannot perform work, a charged particle under its influence moves so that E,V,p, and P are all constant. The Lagrangian of such a particle is*

$$L = -Mc^2\sqrt{1-\beta^2} + \frac{ze}{c}\underline{A}\cdot\underline{V}$$

Using X.5 and X.6 this becomes

$$L = -Mc^2\sqrt{1-\beta^2} + \frac{ze}{c}\frac{\mu\cos\lambda}{r}\cos\lambda\,\dot\phi$$

This expression is independent of $\phi$, so we may conclude from the Lagrangian equation of motion

$$\frac{d}{dt}\frac{\partial L}{\partial\dot\phi} = \frac{\partial L}{\partial\phi}.$$

that $p_\phi = \frac{\partial L}{\partial\dot\phi}$ is a constant of the motion.

$$p_\phi = \frac{\partial L}{\partial\dot\phi} = -M\left(c^2\frac{\partial}{\partial V}\sqrt{1-\beta^2}\right)\frac{\partial V}{\partial\dot\phi} + \frac{ze}{c}\mu\frac{\cos^2\lambda}{r};$$

$$\text{From X.5,}\quad V^2 = \dot r^2 + r^2\dot\lambda^2 + r^2\cos^2\lambda\,\dot\phi^2,$$

$$\text{So}\quad p_\phi = M\left(\frac{V}{\sqrt{1-\beta^2}}\right)r^2\frac{\cos^2\lambda}{V}\dot\phi + \frac{ze}{c}\mu\frac{\cos^2\lambda}{r} = p\left\{\frac{r^2\cos^2\lambda}{V}\dot\phi + \frac{\mu}{P}\frac{\cos^2\lambda}{r}\right\}$$

We now define $\gamma$ as the angle between $\underline{V}$ and $\underline{e}_\phi$, i.e. the angle that the trajectory makes with a vector pointing west. Then

$$\cos\gamma = \frac{V_\phi}{V} = \frac{r\cos\lambda\,\dot\phi}{V} \qquad \text{X.8}$$

$$\therefore \quad \boxed{\frac{p_\phi}{p} = r\cos\lambda\cos\gamma + \frac{\mu}{P}\frac{\cos^2\lambda}{r} = \text{const} \equiv b} \qquad \text{X.9}$$

This integral of the motion is known as Störmer's theorem**. The significance of b becomes clear when we examine the motion in the equatorial plane (which we shall treat more fully later). Here |b| is the impact parameter, as may be seen from FIG. X.9, in which the particle is presumed to be at such a great distance from the earth that the magnetic term in X.9 is negligible.

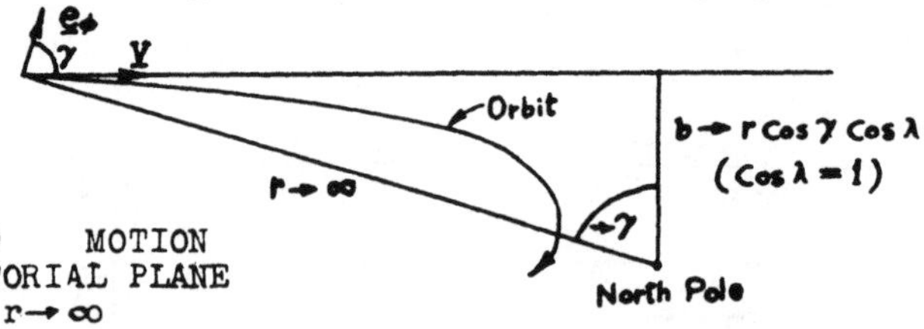


FIG. X.9 MOTION IN EQUATORIAL PLANE as $r \to \infty$

*Bergmann, "An Introduction to the Theory of Relativity," eqn. 7.36

**Störmer was the first to work out the mathematics of the motions of charged particles in the field of the earth. Z. Astrophys. 1, 237 ('30) Originally he hoped to use these equations to explain the northern lights.

If we now make the substitution $r = \sqrt{\frac{\mu}{|P|}}R$, then not only X.9 but all our equations will become dimensionless. The length $\sqrt{\frac{\mu}{|P|}}$ is called a Störmer: $\sqrt{\frac{\mu}{|P|}} = \sqrt{\frac{\mu|z|e}{pc}} = 4.9 \times 10^{9}\sqrt{\frac{|z|}{pc\,(\text{in Bev})}}$ cm .

(In the extreme relativistic case, pc is just the particle energy.)

The radius of the earth, $r_e = 6.378 \times 10^{8}$ cm $= 0.13\sqrt{\frac{pc(\text{Bev})}{|z|}}$ Störmers

We shall use a capital letter to express a length, velocity, etc., in terms of Störmers: e.g.

$$R_e = 0.13\sqrt{\frac{pc\,(\text{Bev})}{|z|}}$$

so, for future reference,

$$pc\ (\text{Bev}) = \frac{|z|}{(0.13)^2}R_e^2 = 60\,z\,R_e^2 \qquad \text{X.10}$$

Frequently we speak of a particle as having a momentum of n Störmers. By this we simply mean that the real value of pc is obtained by substituting $R_e = n$ in X.10.

X.9 now becomes

| (For positive particles) | (For negative particles) | |
|---|---|---|
| $R\cos\lambda\cos\gamma + \frac{\cos^2\lambda}{R} = B$ | $R\cos\lambda\cos\gamma - \frac{\cos^2\lambda}{R} = B$ | X.11 |

Solving for $\cos\gamma$,

| | | |
|---|---|---|
| $\cos\gamma = \frac{B}{R\cos\lambda} - \frac{\cos\lambda}{R^2}$ | $\cos\gamma = \frac{B}{R\cos\lambda} + \frac{\cos\lambda}{R^2}$ | X.12 |

The plane through the moving particle and the z-axis will be called the meridian plane. For any given value of B, certain areas in this plane, where $|\cos\gamma|>1$, will be forbidden to the particle. FIG. X.10 illustrates these forbidden regions for a particle of positive charge. The same figures will apply for a negative particle if the sign of B is changed and the cross-hatchings are interchanged.

For the next few pages we shall be talking about allowed and forbidden areas. The forbidden areas are really forbidden, but the reader should not assume that the particle can necessarily be anywhere in the allowed areas, since there are other restrictions that we shall talk about later, such as the conditions that the particle may not pass through the earth or through a forbidden region in order to get into an allowed one.

The sketches of Fig. X.10 apply for the whole trajectory of the particle, but we shall now consider the end of a trajectory and inquire whether a particle of a given energy, E, can reach a given latitude, $\lambda$, at the surface of the earth, from a given direction making an angle $\gamma$ with the west. When we specify E, this merely determines the radius of the earth in Störmers, and thus fixes the point in Fig. X.10 where the trajectory ends.

First put the given values for $R_e$, $\lambda$, and $\gamma$ into X.11 and solve for B. Draw a sketch such as those in Fig. X.10 corresponding to this value of B, and mark the point $R_e, \lambda$. This point will, of course, lie in an <u>allowed</u> area, since for any direction we have

$B = -0.4$

As B becomes more negative, forbidden region increases, and hump disappears.

I

$B = -0.06$

Equator

I

Axis of earth (z-axis)

$B = 0$

This is not a circle.

1　　2

Scale is in dimensionless units (see p. 227)

I

$B = 1$

$(R, \lambda)$

R

$\lambda$

1　　2

I

Asymptote crosses equator at 18I.

$B = 1.94$

1　　2

I

$B = 2$

1　　2

II　　I

$B = 2.032$

Radius of the earth ($R_e$) for two energies.

$R \propto \sqrt{\rho}$

1　　2

I

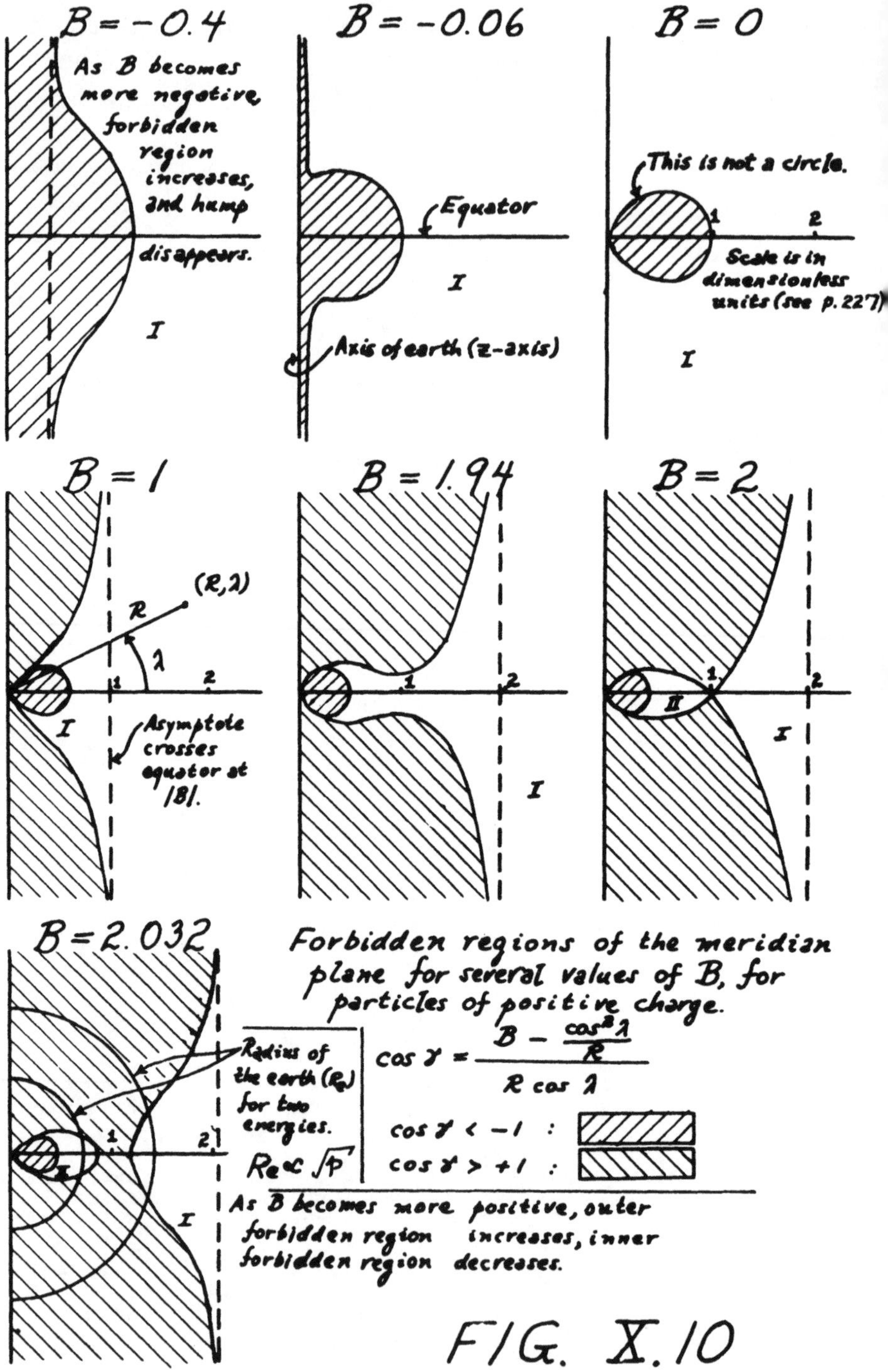

Forbidden regions of the meridian plane for several values of B, for particles of positive charge.

$$\cos \gamma = \frac{B - \dfrac{\cos^2 \lambda}{R}}{R \cos \lambda}$$

$\cos \gamma < -1$ :

$\cos \gamma > +1$ :

As B becomes more positive, outer forbidden region increases, inner forbidden region decreases.

FIG. X.10

specified, $|\cos \gamma| \leq 1$.

But there are two types of allowed areas: type I extends to infinity and is available to cosmic ray particles; type II is insulated from infinity by forbidden areas, and the only particles that could circulate in a region corresponding to a type II area would be those shot from a high energy machine on the earth. It is crucial to notice that these type II areas never occur unless $B > 2$ (for positive particles) and never lie outside the semicircle of radius one.

So long as we choose $R_e \geq 1$, it is not even necessary to compute B; we know that our mark must fall in a type I area. This is equivalent to saying that, so long as $R_e > 1$ ($E > 60$ Bev), $\gamma$ is arbitrary (the particle can come in from any direction it pleases).

If $R_e < 1$, however, we must calculate B before we can come to any decision. If our choice of $\gamma$ has made $B < 2$ (for positive particles) the jaws of the forbidden region will not be closed, and the particle will still be observable. But if $\gamma$ is too small B will be $> 2$, and the particle cannot get to the earth.

Let us find the **observable** values of $\gamma$; inserting in X.12 the condition

| $B < 2$ (for $Z > 0$), | $B > -2$ (for $Z < 0$), |
|---|---|
| $\cos\gamma < \frac{2}{R_e \cos\lambda} - \frac{\cos\lambda}{R_e^2}$ | $\cos\gamma > -\frac{2}{R_e \cos\lambda} + \frac{\cos\lambda}{R_e^2}$ |

or, stated differently,

| $\gamma > \gamma_{0+}$ | $\gamma < \gamma_{0-}$ |
|---|---|
| $\cos\gamma_{0+} \equiv \frac{2}{R_e \cos\lambda} - \frac{\cos\lambda}{R_e^2}$ | $\cos\gamma_{0-} \equiv \frac{-2}{R_e \cos\lambda} + \frac{\cos\lambda}{R_e^2}$ |

Considering only the condition for positive particles, we may say that the minimum angle that the velocity makes with a vector pointing west is $\gamma_{0+}$. FIG. X. 11 illustrates the allowed cone of observable trajectories for the case $\gamma_{0+} = 150^{\circ}$.

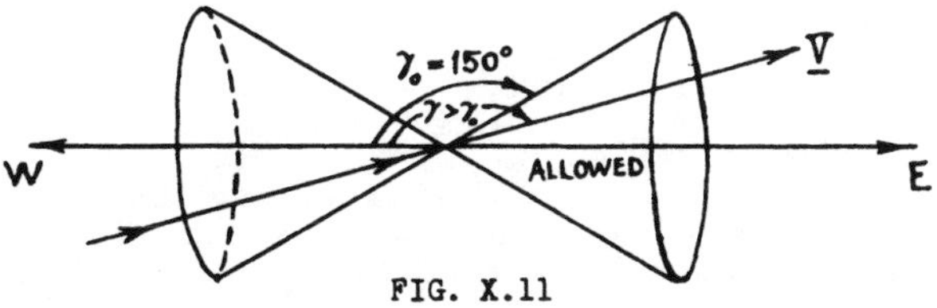


FIG. X.11

The inequality above may be solved for $R_e$ to yield the following condition for observable momenta for a given latitude and direction:

| $R_e > \frac{1 - \sqrt{1 - \cos\gamma \cos^3\lambda}}{\cos\gamma \cos\lambda}$ | $R_e > \frac{1 - \sqrt{1 - \cos(\pi - \gamma) \cos^3\lambda}}{\cos(\pi - \gamma) \cos\lambda}$ |
|---|---|

Note that the condition for negative particles is the same as for positive, except that $\pi - \gamma$ is substituted for $\gamma$, or east for west, which seems reasonable.

Using X.10 we can rewrite this equation in terms of Bev/c.

$$\underset{z>0}{p\left(\frac{Bev}{c}\right)} > 60z\left[\frac{1-\sqrt{1-\cos\gamma\cos^3\lambda}}{\cos\gamma\cos\lambda}\right]^2 \quad \Bigg| \quad \underset{z<0}{p\left(\frac{Bev}{c}\right)} > -60z\left[\frac{1-\sqrt{1-\cos(\pi-\gamma)\cos^3\lambda}}{\cos(\pi-\gamma)\cos\lambda}\right]^2 \qquad \text{X.14}$$

For particles arriving in the meridian plane, cos $\gamma$ = 0, and X.14 simplifies to

$$p > 15(\cos^4\lambda)|z| \text{ Bev/c} \qquad \text{X.15}$$

TABLE X.III presents the critical momenta for particles arriving from the west (cos $\gamma$ = -1), from the zenith (cos $\gamma$ = 0), and from the east (cos $\gamma$ = 1).

| Direction from which particle arrives | Critical momenta in Bev/c | | | | |
|---|---|---|---|---|---|
| | $\lambda = 0°$ | $\lambda = 30°$ | $\lambda = 45°$ | $\lambda = 60°$ | $\lambda = 90°$ |
| West | 10 | 6.4 | 3.1 | 0.9 | 0 |
| Zenith | 15 | 8.5 | 3.7 | 0.93 | 0 |
| East | 60 | 13.4 | 4.6 | 1.0 | 0 |

TABLE X.III

## 2. Illustration in the Equatorial Plane --- Shadow Effect.

In the next two pages we shall derive no new equations, but merely illustrate our results by considering in detail a simple case - motion in the equatorial plane. The reader who is not interested in an example may skip to the last paragraph of this numbered section. In this paragraph the shadow effect of the earth is introduced.

If a particle is moving in the equatorial plane, then, by symmetry, it must continue to move in this plane. For this case, the Störmer integral, X.11, takes the simple form

$$\underset{z>0}{R\cos\gamma + \frac{1}{R} = B} \quad \Bigg| \quad \underset{z<0}{R\cos\gamma - \frac{1}{R} = B} \qquad \text{X.17}$$

We shall consider only positive particles. FIG. X.12 illustrates orbits for three typical values of B.

The first illustration is that of B = 0, ie., the particle is headed for the center of the earth, originally. The magnetic field deflects it to the east. At the perigee (point closest to the earth) it is headed due east (cos $\gamma$= -1). From X.17 we find that R = 1 at the perigee. By drawing in all the possible trajectories of particles with B = 0, we see that the whole plane is filled except the region within the circle R = 1. This corresponds to the forbidden region of the B = 0 diagram of Fig. X.10. It is, in fact, the area in the equatorial plane generated by rotating the B = 0 diagram of Fig. X.10 about the z-axis. Getting back to Fig. X.12, if we now draw in the earth (a circle with radius $R_e$) we see two cases: first, if the earth lies within the forbidden region, particles with B = 0 obviously do not reach the earth, which is another way of saying that there is no conceivable direction of observation, or that $\gamma$ is imaginary, or that $|\cos\gamma|>1$; secondly, if the surface of the earth is outside the forbidden region, all the particles are intercepted at an angle found from the figure or from X.17, with R = $R_e$. Since

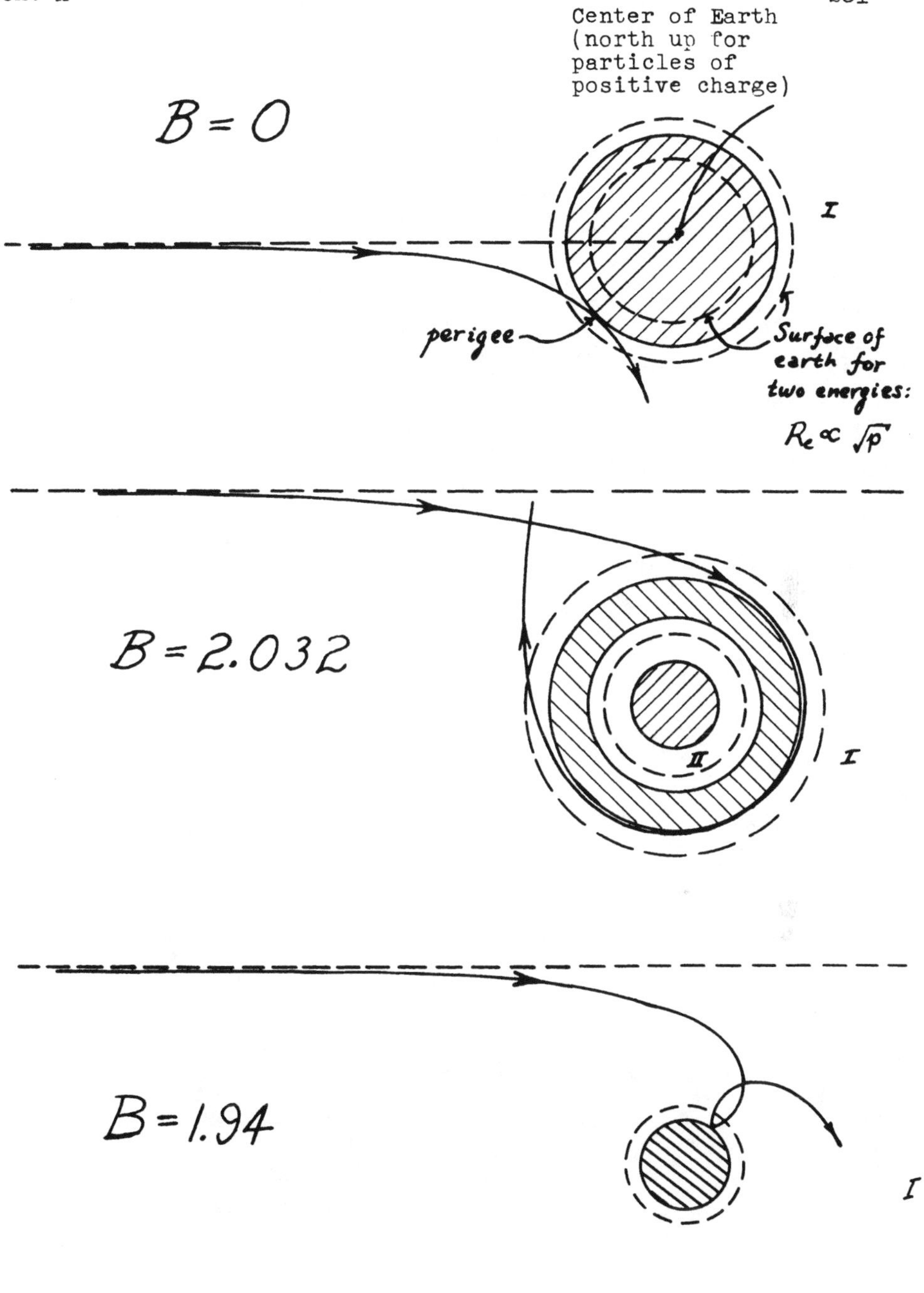


TRAJECTORIES
FIG. X. 12

$R_e$ is a function of energy, the angle of observation is a function of energy for a given B. The trajectory cuts the surface of the earth at two points, at only one of which is the particle observable; for at the other, observation has been forbidden by the shadow of the earth (this case is trivial, since it corresponds to observation of a direction below the horizon).

We next consider the case B = 2.032. Here there are two allowed regions, I and II. The particle from infinity remains in region I and arrives at its perigee now heading west, since the forbidden region that it grazes corresponds to $\cos\gamma > 1$ (the opposite of the previous case). If the surface of the earth lies in region I, the result is similar to that of the previous case. However, if the surface of the earth lies in region II, it will not intercept particles from infinity, even though there is a real $\gamma$ corresponding to such a position. There exist trajectories of motion confined to region II, but these are not realized by cosmic ray particles. Thus all solutions $(R_e, \gamma)$ of X.17 for B = 2.032 corresponding to region II represent observations forbidden because the particles cannot enter the region.

Finally, for B in the neighborhood of 1.94, the orbit looks something like that shown at the bottom of Fig. X.12. Here the shadow effect comes into its own. For although the trajectory cuts the surface of the earth **two** times from above the horizon, only the first intersection is observable. This effect has been considered by Lemaitre and Vallarta in their more rigorous treatment.*

## 3. Intensity of Allowed Radiation.

In Fig. X.11 we illustrated our conclusion that, at a given latitude, particles of a certain energy would be confined to a cone of incident directions. We now ask "What will be the angular distribution within this cone of particles in a given differential energy range?" One might think that there would be some tendency for the particles to come in along the axis of this cone, but this is not the case. The distribution is isotropic within the confines of the cone (excluding shadow effect). We shall now prove a theorem even more general, namely that the intensity, in a given differential energy range, of cosmic radiation reaching the earth in an allowed direction is the same as its intensity at "infinity."

We shall use the Liouville theorem, which states that the density of particles in phase space is a constant of the motion. The element of volume in phase space enclosing dN particles therefore remains constant. It equals

$$dx\, dy\, dz\, d\pi_x d\pi_y d\pi_z = \text{const.}$$

The conjugate momenta, $d\pi_i$ are defined as follows:

$$\pi_i = \frac{\partial L}{\partial \dot{x}_i} = \frac{MV_i}{\sqrt{1-\beta^2}} + \frac{e}{c} A_i = p_i + \frac{e}{c} A_i(x, y, z)$$

If we now transform from the $x, y, z, \pi_x, \pi_y, \pi_z$ space to $x, y, z, p_x, p_y, p_z$ space, the constant element of volume becomes

*Vallarta, "On the Allowed Cone of Cosmic Radiation," (1938) Phys. Rev. 74, 1837 ('49)

$$J\left(\frac{\pi_x, \pi_y, \pi_z}{p_x, p_y, p_z}\right) dx\, dy\, dz\, dp_x\, dp_y\, dp_z$$

where $J = \begin{vmatrix} \frac{\partial \pi_x}{\partial p_x} & \frac{\partial \pi_x}{\partial p_y} & \cdots \\ \cdots & \cdots & \end{vmatrix} = \begin{vmatrix} 1 & 0 & 0 \\ 0 & 1 & 0 \\ 0 & 0 & 1 \end{vmatrix} = 1$ is the Jacobian.

Thus also the element of volume

$$dx\, dy\, dz\, dp_x\, dp_y\, dp_z = \text{const.}$$

Finally, the density $\frac{dN}{dx\, dy\, dz\, dp_x\, dp_y\, dp_z}$ in quasi-phase space is therefore a constant of the motion.

If we take the z-axis along the velocity vector, we can write $(dx\, dy\, dz)(dp_x dp_y dp_z) = (V\, dt\, dA)(p^2\, dp\, d\omega)$, where dA is an element of area perpendicular to the velocity, and $d\omega$ is an element of solid angle defining the direction of motion of the particles. Since V and p are constant for motion in a static magnetic field, we find that $I \equiv dN/(dA\, dt\, dp\, d\omega)$ is a constant of the motion. But I is the intensity, so we have proved what we set out to prove.

Next consider two groups of cosmic ray particles, both with momenta between p and p + dp, both confined to the same sized differential of solid angle, but arriving at the same point on the earth from different directions. Follow these two groups in phase space to two far distant points. If we now assume that the distribution of particles in phase space is uniform spacewise, and is isotropic, then these two far distant points are equally probably populated. The reverse of this reasoning shows that the intensity of a given momentum interval at the earth is the same for all directions that are allowed.

## 4. Charge of Primary Radiation.

The existence of the allowed cone, developed in part 1. of this section, shows that the total intensity from the west will be greater than that from the east for positive particles; vice-versa for negative particles.

According to the actual value found for the excess of high-energy radiation coming from the west over that from the east at altitudes as high as 35,000 ft, it must be concluded that the majority of the primary radiation is positive*

## 5. Latitude Effect and the Primary Spectrum.

From Eq. X.15, p. 230, for example, we see that the critical momentum decreases as we move northward from the equator, so that we should expect the

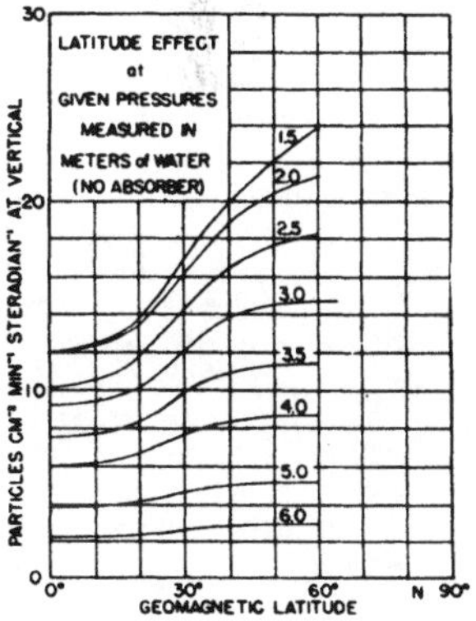


Fig. X.13 Latitude Effect for the Soft Component**

*Schein et al., Phys. Rev. 73, 928 ('48)

**From Biehl, Neher, and Roesch, Phys. Rev. 76, 927 ('49). These data are from vertical, unshielded counter telescopes, but the effect is substantially the same for omnidirectional observation and for the hard component.

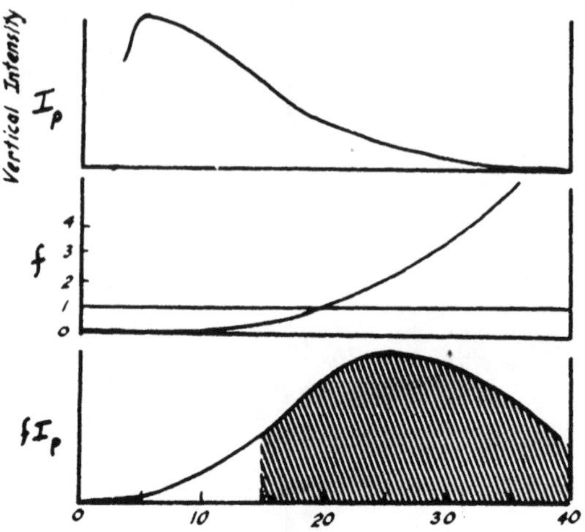


Primary Momentum, Bev/c

FIG. X.14

total intensity of cosmic radiation to increase. FIG. X.13, p. 233, illustrates this latitude effect and shows that it is quite small at sea level*. This will be made plausible with the help of FIG. X. 14.

This figure is schematic only, not to scale. The top curve, $I_p$, represents the probable differential primary spectrum (particles $cm^{-2} sec^{-1} sterad^{-1}$ per unit momentum range).

A quantitative relation is unknown, but the second curve, f, represents qualitatively the probability of observing a (primary or secondary) particle of any momentum at sea level, resulting from one primary particle of momentum p. It indicates that below 15-20 Bev primary energy or momentum, unit intensity at high altitudes produces less than unit intensity at sea level, and at high momenta the reverse is true.

The third curve is the product of the first two. Thus a cosmic ray telescope at sea level, looking straight up from the equator, sees the (shaded) intensity to the right of the 15 Bev/c abscissa (the cut-off momentum 15 Bev/c comes from Eq. X.15). Since this is a large fraction of the whole area (which would be seen from the magnetic poles), the latitude effect at sea level is small. This is the result that we wanted.

We shall now restate this last argument in a slightly more quantitative way since it also serves to summarize some of the ideas presented in this chapter.

At sea level we see mainly muons (and their secondary electrons). Fig. X. 5, p. 221a, shows that the average energy of these muons is about 1 Bev. From the curves on p. 355 of Montgomery's book we see that a muon at minimum ionization will lose about 2.5 Bev in penetrating the atmosphere, so that our muon which arrives at sea level with an energy of about one Bev must have started down from the top of the atmosphere with an energy of about 3 Bev. But a 3 Bev muon is most probably one of 5 or more such muons leaving a star initiated by a heavy particle with an energy of 20 Bev or more. In summary, the cosmic radiation observed at sea level is the end-result of a chain of events started by a primary particle with an energy in the tens of Bevs. Such particles will clearly not exhibit much latitude effect.

*It was first discovered by J. Clay

## Appendix: DISCUSSION OF FIG. X.3, p. 221

To get from the experimental counting rates given by the left-hand scale to the absolute production rates given by the right-hand one, the counter is calibrated in a pile where "nV" is known ("nV" is what we would call a flux if we were dealing with a beam rather than with isotropic velocities). If we neglect the high-energy collimated neutrons in the atmosphere (an assumption which might be justified for $x >> 57$ g/cm$^2$) we can then call the atmospheric neutron velocities isotropic, so that the calibration is justified. Once we know nV we can calculate the rate of neutron absorption per gram of air, and we can certainly replace "absorbed" by "produced" since there is an equilibrium condition.

----------

It might be heartening to point out that the order of magnitude of these production rates can be predicted by elementary considerations, starting from the proton intensity given in Fig. X.6, p. 222.

We shall have to make three arbitrary assumptions:

1. For $x > 57$ g/cm$^2$ the density of neutrons and protons is very roughly the same. (At these depths most nucleons are secondaries, and the number of secondary neutrons and protons should not be too different.)
2. A neutron is "absorbed" as soon as it hits its first "air" nucleus.
3. It hits an air nucleus after travelling 57 g/cm$^2$, i.e. after a distance $L = 57/\rho$ cm ($\rho$ is the density of air).

Also remember the assumptions of the calibration: a. Isotropic velocities, b. Equilibrium.

With these assumptions, the production rate = the collision rate ≡ R;

$$R = \frac{nV}{L} \qquad \text{collisions sec}^{-1}\text{cm}^{-3} \qquad \text{X.20}$$

where V is the average velocity, L is the mean free path.

All that remains is to state nV in terms of $I_{vert.}$, and this may be done almost immediately from the sketch. The number of particles effusing normally through the unit-area hole in the box is, by definition,

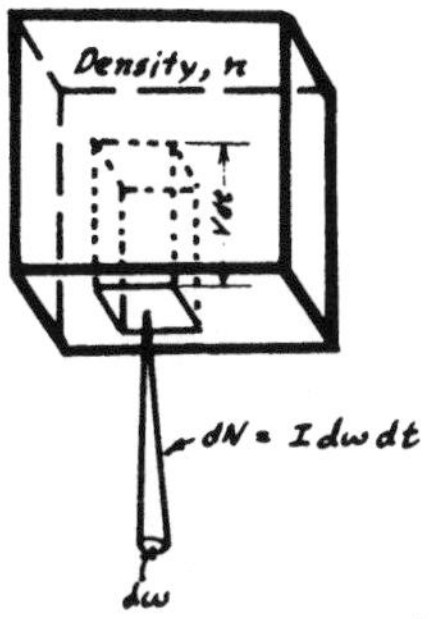


$$dN = I_{vert.}\, dt\, d\omega$$

but, as discussed in every book on kinetic theory, it should follow from the sketch that

$$dN = nV\, dt\, \frac{d\omega}{4\pi}$$

so

$$nV = 4\pi I_{vert.}$$

Thus X.20 becomes

$$R = \frac{4\pi I_{vert.}}{L} = \frac{4\pi I_{vert.}}{57/\rho} \qquad \text{collisions sec}^{-1}\ \text{cm}^{-3}$$

However the production rate of Fig. X.3 is expressed per gram rather than per cm$^3$, so our expression must still be divided by $\rho$. The $\rho$'s cancel, and we have

$$R' = \frac{4\pi I_{vert.}}{57} \qquad \text{collisions sec}^{-1}\ \text{g}^{-1}$$

If from Fig. X.6, p. 222, we pick a value such as $I_{vert.} = 5 \times 10^{-2}$ at $x = 400$, we find

$$R' \sim 10^{-3} \quad \text{at } x = 400 \ g/cm^2$$

Comparison with Fig. X.3, p. 221, shows that the agreement, fortuitously, is quite good.

---

## REFERENCES FOR COSMIC RAYS

1. Montgomery, D.J.X., "Cosmic Ray Physics," 1949
2. Jánossy, L., "Cosmic Rays," 1948
3. Rossi. B., "Interpretation of Cosmic Ray Phenomena," Rev. Mod. Phys. 20, 537 ('48).
4. Heisenberg, W., "Cosmic Radiation," 1943
5. Cosmic Ray Symposium, Rev. Mod. Phys. 21, ('49)
6. Jauch, J.M., "Cosmic Rays," Nucleonics 4, (Apr and May '49)
7. Tiomno and Wheeler, "Guide to the Literature of Elementary Particle Physics, Including Cosmic Rays," American Scientist 37 (Apr. and July '49) Also bound separately.
8. Papers delivered at the International Conference on Cosmic Rays, Como, Italy; Sept. 1949 (to be published)
9. Symposium of the Interuniversity Cosmic Ray Laboratories, Echo Lake, June 1949. (To be published early in 1950 by the U.S. Government Printing Office.)

## PROBLEMS

1. If you were travelling in an interplanetary space ship, what sort of apparatus would you need to measure the total intensity of cosmic rays?

2. Assume that your laboratory is at the top of the atmosphere. Design an experiment to set an upper limit on the intensity of primary electrons in cosmic rays.

3. Plot the semi-vertex angle of the allowed cones ($\chi_0$) as a function of p at $\lambda = 50^{\circ}$.

NOTES ON MESON TABLE, P. 133

This table may not be complete. There have been reports of charged and neutral "V-particles" - mesons with a rest mass of perhaps 2200m. See Rochester and Butler, Nature 160, 855 ('47); Seriff, Leighton, Hsiao, Cowan, and Anderson, Phys. Rev. 78, 290 ('50); Armenteros, Barker, Butler, Cachon, and Chapman, Nature 167, 501 ('51).

The MASSES of the charged mesons are due to Barkas, Smith, and Gardener, Phys. Rev. 82, 103 ('51). The mass of the $\pi^0$ is discussed below.

The MEAN LIFE of the $\pi^-$ is given by Lederman, Bernardini, Booth, and Tinlot, Phys. Rev. 82, 335 ('51). The mean life of the $\pi^0$ is discussed below.

THE NEUTRAL $\pi$-MESON

We might first mention two general review articles on work done with the Berkeley 350-Mev synchrocyclotron- Chew and Moyer, Am. Jour. Phys. 18, 125 ('50) and 19, 17 and 203 ('51); and a review of experiments performed inside the vacuum tank, by Gardner et al., Science 111, 191 ('50).

Next we mention briefly three experiments which indicate the existence of a $\pi^0$ which decays into two gammas, probably in less than $10^{-13}$ sec.

First, $\gamma$'s have been observed to come from various targets when bombarded with high-energy protons*. These $\gamma$'s have a production cross-section whose dependence upon proton energy is much like that of charged pion cross-sections. The $\gamma$-energy is roughly 70 Mev on the average (half of the energy of a $\pi^\pm$) and their energy spread is in agreement with the Doppler shift due to the velocity of the parent mesons. After examining the position of the shadow of a lead screen (see the schematic sketch) and calculating the distances traveled by the $\pi^0$'s before they decay (allowing for the relativistic time dilatation), the experimenters have been able to state that the mean life of the $\pi^0$'s is $< 10^{-11}$ sec. In this time the pion travels only a few mm.

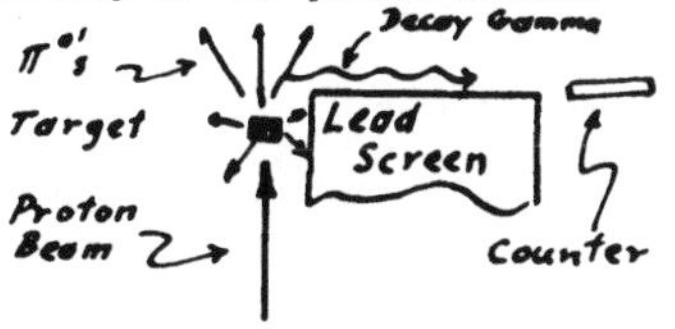


The upper limit on this mean life has been reduced to $5 \times 10^{-14}$ sec by examining (indirectly) the point of origin of gammas produced by the decay of $\pi^0$'s produced in cosmic ray stars, observed in emulsions****. These gammas create electron pairs, and the bisector of the angle between the two tracks is extrapolated back towards the star. It is found to pass very close to the star.

In a second experiment, $\gamma$'s of about 70 Mev ($\frac{M_\pi c^2}{2} = 68$ Mev (?)) of nuclear origin have been observed during the bombardment of nuclei by 330 Mev x-rays from the Berkeley synchrotron**. Coincidence and angular correlation measurements show that these $\gamma$'s are emitted in pairs, most probably from particles traveling with speeds up to about 0.8c. This speed is calculated by considering the aberration of two $\gamma$'s emitted by a relativistic particle: seen from the rest system of the emitter, the two photons must leave in opposite directions, but seen from the lab, they must come off in a forward cone.

Apparently the reaction is

$$\gamma + N \longrightarrow N + \pi^0; \qquad \pi^0 \xrightarrow{<10^{-11}\,\text{sec}} 2\gamma$$

where N stands for a Nucleon.

---

*Bjorklund, Crandall, Moyer, and York, Phys. Rev. 77, 213 ('50).
**Steinberger, Panofsky, and Steller, Phys. Rev. 78, 802 ('50).
***It is assumed that the reader has already read the footnote, p. 221a
****Carlson, Hooper, and King, Phil. Mag. 41, 701 ('50)

As an example, suppose that a 300 Mev photon is absorbed by a nucleon and that one $\pi^{\circ}$ of rest mass 135 Mev is created. Moreover, consider the special case in the c-m system where the pion comes off forward, while the nucleon flies back. In this case the reader can easily check that in the lab the nucleon has only about $\frac{1}{2}$ Mev kinetic energy, leaving the pion with kinetic energy slighly greater than its rest mass. Now it is trivial to show that a particle with KE = $Mc^2$ has a speed of 0.866c. Thus the $\gamma$-emitters, determined by the aberration measurements to have speeds up to about 0.8c, are probably $\pi^{\circ}$'s.

It has been shown* that an isolated $\pi^{\circ}$ with spin $\frac{1}{2}$ or 1 cannot decay into two $\gamma$'s. Unless one wants to include the possibility of spin $> 1$, it looks as if a $\pi^{\circ}$ has spin 0. Since the data seem to show a fair amount of similarity between charged and neutral pions, there may even be tentative indications that all pions have spin 0.

A third experiment which indicates the mass of the $\pi^{\circ}$ consists in allowing $\pi^-$'s (made in the Berkeley cyclotron) to come to rest in a tank of high-pressure hydrogen**. We have already discussed on p. 133 the capture into Bohr orbits of mesons by a nucleus. A pion which is in a state of low angular momentum will probably interact with the nucleus within the $10^{-8}$ sec mean life of the pion. In case the nucleus is nothing but a single proton, the result is not a star, but probably one of the two following reactions

$$P + \pi^- \rightarrow N(9\text{ Mev}) + \gamma(132\text{ Mev})$$

or $$P + \pi^- \rightarrow N(\text{almost at rest}) + \pi^{\circ}(\text{few Mev})$$
followed by $$\pi^{\circ} \rightarrow 2\gamma\ (\text{each of } \sim 65\text{ Mev}).$$

Both of these processes seem to occur.

The $\pi^-$'s are created inside the cyclotron where it is hard to run electronic counting equipment, so it has not been possible to determine the speed of the $\pi^{\circ}$'s by measuring the angular correlation of the coincident $\gamma$'s, but one can still examine the Doppler effect, looking at the $\gamma$'s with a pair spectrometer***. From this Doppler data it is claimed that $1.3\text{ Mev} < (M_{\pi^-} - M_{\pi^{\circ}})c^2 < 4.7\text{ Mev}$.

## SCINTILLATION COUNTERS

It has not been mentioned in this book that when charged particles (and also- in particular- when $\gamma$'s) pass through many materials (solids, liquids, and gases) these materials may give off visible light. The light may then be detected with a photomultiplier tube. This is the principle of the scintillation counter, which is becoming of increasing importance. These counters can be made with pulse widths as little as $10^{-9}$ sec, and have other characteristic advantages: their response depends fairly linearly upon the energy which the scintillator absorbs, so that they may be used as spectrographs****; they are sensitive; and a scintillating crystal does not have to be surrounded with some more or less transparent envelope, as do other counters.

For review articles see "Fluorescence of Liquids under $\gamma$-Bombardment," Kallman and Furst, Nucleonics 7, 69 (July 1950); and a review by Bell and Jordan, Nucleonics, 5, 30 (Oct. 1949).

---

*Yang, Phys. Rev. 77, 243 ('50)
**Panofsky, Aamodt, and York, Phys. Rev. 79, 825 ('50)
***i.e. one converts some of the gammas to pairs in a thin foil, measures the total energy of the pair by determining the electron orbits in a magnetic field, using counters in coincidence.
****Bell and Cassidy, Phys. Rev. 79, 173 ('50); Hofstader and McIntyre, Phys. Rev. 79, 389 ('50)

## COSMIC RAYS: Fluctuations with Time

Diurnal Effect. For a discussion of this daily variation in intensity see Elliot and Dolber, Proc. Roy. Soc. Lond A63, 137, ('50).

Solar Effect. These variations follow solar activity (solar flares, radio interference, etc.). See Forbush, Stinchcombe, and Schein, Phys. Rev. 79, 501 ('50); Adams, Phil. Mag. 41, 503 ('50); Simpson, Phys. Rev. 81, 895 ('51).

## GENERAL REFERENCES FOR NUCLEAR PHYSICS

Bethe, H.A., et al. "Nuclear Physics" Rev. Mod. Phys. 8, 82 (1936), 9, 69 and 245 (1937). Referred to in this book as Bethe A, B, and C.
Bethe, H.A. "Elementary Nuclear Theory," Wiley, 1947; called Bethe D.
Fermi, E. "Elementary Particles," Yale Press, 1951.
Frisch, O. "Progress in Nuclear Physics, 1" Academic Press, 1950.
Gamow, G., and C. Critchfield "Theory of Atomic Nucleus and Nuclear Energy Sources," Oxford, 1949.
Goodman, C. "Science and Engineering of Nuclear Power," Addison-Wesley, 1947.
Heitler, W. "Quantum Theory of Radiation" Oxford 1944.
LA 255 (AECD 2664); Fermi, E., "Neutron Physics"
NBS 499 "Nuclear Data"(National Bureau of Standards), Superintendent of Documents, 1951.
Rasetti, E. "Elements of Nuclear Physics" Prentice-Hall, 1936.
Rosenfeld, L., "Nuclear Forces" Interscience 1948.
Schiff, L.I. "Quantum Mechanics" McGraw-Hill 1949.

More specialized references are found at the end of the individual chapters, and in the footnotes.

The following bibliographies may also prove useful:

Beyer, R.T. "Foundations of Nuclear Physics" Dover 1949.
Tiommo and Wheeler, "Guide to the Literature of Elemtary Particle Physics, Including Cosmic Rays" Am. Scientist 37, 202 (1949); also bound separately.

For biological applications, instruments, see

Siri, W.E. "Isotopic Tracers and Nuclear Radiations" McGraw-Hill 1949 (one of the Nuclear Energy Series).

Note added Feb., 1953

Some of the major developments since the writing of these notes are covered in the following books:

Marshak, R.E. "Meson Physics" McGraw Hill, 1952,
Thorndike, Alan "Mesons - A Summary of Experimental Fact" McGraw Hill, 1952.

LePrince Ringuet, L. "Cosmic Rays" Prentice-Hall, 1950,
Rossi, Bruno "High-Energy Particles" Prentice-Hall, 1952,
Wilson, J.G, et al. "Progress in Cosmic Ray Physics - A Summary of the Copenhagen Conference" Interscience, 1952.

The ideas of charge-independence (isotopic spin) and their applications to mesons and to light nuclei are discussed by M. Gell-Mann and R. Hildebrand in a forthcoming issue of the American Journal of Physics.

Many new developments are covered in two volumes of notes by R.P. Feynman (1951-52) available from California Institute of Technology, Pasadena, Calif.

Reference should also be made to a very complete new text describing nuclear phenomena involving energies below about 50 Mev - "Theoretical Nuclear Physics" by J.M. Blatt and V.F. Weisskopf, Wiley, 1952.

The following list gives most of the symbols used. Symbols that are quite standard are not listed. Where a symbol has a special meaning in only one chapter, the chapter number is stated. If a symbol has a meaning that needs fuller explanation, the page number following it gives the place in the text where the definition is found. Numerical values are on p. 240.

| | |
|---|---|
| $a_0$ | Bohr radius, 5.29 x $10^{-9}$ cm |
| $a_1, a_3$ | Singlet and triplet scattering length, VI and IX |
| amu | Atomic mass unit, = $M_1$ |
| A | Atomic mass number, N + Z |
| $\underline{A}$ | Vector potential |
| b | Collision parameter, 27, 226 |
| BE | Binding energy |
| $c_l$ | VI and IX, coeff. of $l$th partial wave, 117 |
| D | II, thickness |
| e | electronic charge = -4.805 x $10^{-10}$ esu |
| $\underline{e}_r$, $\underline{e}_\theta$, etc., | unit vectors |
| E | Energy eigenvalue, T + U |
| $\mathcal{E}$ | Electric field strength, $-\nabla\phi$ |
| f(θ) | Angular dependence of wave |
| $\underline{F}$ | Total angular momentum of atom, $\underline{I} + \underline{J}$ |
| g | Coupling constant, Fermi constant |
| $\hbar$ | $h/2\pi$ |
| $\mathcal{H}$ | Magnetic Field |
| $\underline{I}$ | Total nuclear ang. momentum |
| $\underline{I}$ | Intensity, erg $cm^{-2} sec^{-1}$ |
| I | II, average ionization potential, 30 |
| $\underline{J}$ | Total ang. momentum of extra-nuclear atom |
| k | Propagation constant, $2\pi/\lambda$ |
| $l_R$ | Radiation length, 49 |
| $l_p$ | Mean free path for pair production, 49 |
| m | REST mass of electron |
| M | Rest mass of a particle |
| n | Index of refraction |
| N | a Neutron |
| N | Atoms or nuclei per $cm^3$ II, electrons $cm^{-3}$, 28 |
| P | Magnetic rigidity, $(pc/ze) = \mathcal{H}\rho$ |
| $q(\underline{r},\varepsilon)$ or $q(\underline{r},\tau)$ | IX, slowing down density, 187 |
| Q | IX, rate of production of neutrons, 184 |
| Q | Energy of reaction, exothermic |
| Q | Quadrupole moment |
| $r_e$ | Radius of electron and of proton |
| $r_\oplus$ | X, radius of earth |
| $r_0$ | Radius of potential well |
| | Range |
| Ry | Rydberg energy, -13.52 ev |
| s, S | Spin angular momentum |
| T | Kinetic energy, half-life, temperature. |
| U | Potential energy |
| $v$ | VIII, velocity |
| $\underline{V}$ | Velocity |
| W | Total relativistic energy, 5 |
| x | Atmospheric depth, g $cm^{-2}$, 215 |
| Z | Charge of particle in units of electronic charge. |
| $z$ | Charge of incident particle |
| $\alpha$ | II, $h\nu/mc^2$, energy parameter |
| $\alpha$ | VI, $\frac{3ze^2}{\hbar V}$, 126 |
| $\beta$ | V/c |
| $\beta_l$ | phase shift of $l$th partial wave |
| $\gamma$ | $(1-\beta^2)^{-1/2}$ |
| $\gamma$ | X, angle between $\underline{V}$ and west, 226 |
| $\Gamma$ | VIII, energy width of resonance at ½ max., 154 |
| $\varepsilon$ | IX, ln neutron energy |
| $\theta_L$ | IX, neutron scattering angle in lab. system, 182 |
| $\Theta$ | Total scattering angle, 36 |
| $\kappa$ | Range of meson field, 135 |
| $\lambda$ | Mean free path, 184 |
| $\lambda$ | Radioactive decay const, 1 |
| $\lambda$ | X, magnetic latitude, 225 |
| $\lambda\!\!\!^-_c$ | Compton wavelength/$2\pi$ |
| $\Lambda$ | IX, absorption mean free path, 184 |
| $\mu$ | reduced mass |
| $\mu_N$ | Nuclear magneton |
| $\underline{\mu}$ | Magnetic moment |
| $\nu$ | V, $h\nu/mc^2$ |
| $\xi$ | IX, reduction in ln(neutron energy) per collision, 183 |
| $\underline{\sigma}$ | Pauli spin operator, 112 |
| $\sigma(\theta)$ | Differential cross section |
| $\sigma$ | Total cross section |
| $\sigma_c$ | Klein-Nishina scattering cross section(Compton), 41 |
| $\sigma_p$ | II, $\sigma$ for photoelectric effect |
| $\sigma_T$ | II, Thomson cross section |

| | |
|---|---|
| $\xi$ | II, Thickness in $g/cm^2$ |
| $\tau$ | Mean life |
| $\tau(\epsilon)$ | IX, neutron "age" |
| $\omega$ | angular frequency, solid angle |
| $\Omega$ | Volume of box for normalization |
| $*$ | Denotes excited nucleus |
| $\sim$ | Order of magnitude |
| $\approx$ | Approximately |

# TABLE OF PHYSICAL CONSTANTS

General Physical Constants:

$a_0 = \hbar^2/me^2 = 0.529 \times 10^{-8}$ cm. Bohr Radius

$\alpha = e^2/\hbar c = 1/137.03$ Fine structure constant

$$\frac{r_e}{\lambda_c} = \frac{\lambda_c}{a_0} = \frac{1}{137}$$

$c = 2.988 \times 10^{10}$ cm/sec. Velocity of light

$e = 4.805 \times 10^{-10}$ esu $= 1.602 \times 10^{-20}$ emu. Electronic charge

$g \approx 2.5 \times 10^{-49}$ g cm$^5$/sec$^2$. Fermi constant in beta decay, p. 75

$h = 6.624 \times 10^{-27}$ erg-sec.. Planck constant

$\hbar = 1.054 \times 10^{-27}$ erg-sec.

$k = 1.38 \times 10^{-16}$ erg/degree. Boltzmann constant

$N_0 = 6.023 \times 10^{23}$ /mole. Avogadro's number

$r_e = 2.82 \times 10^{-13}$ cm. Classical electron radius

$\lambda_c = \hbar/mc = 3.86 \times 10^{-11}$ cm. Compton wavelength/$2\pi$

Deuteron:

n-p scattering $\sigma = 20.3$ barns

Triplet state: (p. 115)

$BE = 2.23$ Mev $= 3.58 \times 10^{-6}$ erg

$U_3 = -21.0$ Mev, Well depth

$r_0 = 2.82 \times 10^{-13}$ cm

n-p triplet scattering $\sigma_3 = 4.4$ barns

$a_3 = 0.59 \times 10^{-12}$ cm

Wave fcn phase at edge of well: $108^\circ$

Singlet state: (p. 120)

$U_1 = -11.5$ Mev, Well depth

n-p singlet scattering $\sigma_1 = 68$ barns

$a_1 = -2.32 \times 10^{-12}$ cm

Quadrupole moment $Q_D = 0.00273 \times 10^{-24}$ cm$^2$ (p. 15)

Magnetic Moments:

$\mu_{Bohr} = 0.9273 \times 10^{-20}$ erg/gauss

$\mu_{nuclear} = 5.04 \times 10^{-24}$ erg/gauss $= 1/1837\,\mu_{Bohr}$

$\mu_{electron} = -1.002\ \mu_{Bohr}$

$\mu_{proton} = 2.7896\ \mu_{nuclear}$

$\mu_{neutron} = -1.9103\ \mu_{nuclear}$

$\mu_{deuteron} = 0.85647\ \mu_{nuclear}$

Masses: (see p.2; mass formula, p.7)

$M_1 = 1$ amu $= 1.6603 \times 10^{-24}$ g $= 931$ Mev

$M_p = 1.00759$ amu

$M_n = 1.00898$ amu

$m = 1/1837.561\ M_H = 9.1066 \times 10^{-28}$ g $= 0.51$ Mev. Mass of electron

$M_{\pi^\pm} = 140.8$ Mev $= 276 \pm 6$ electron masses

$M_{\pi^0} = 138$ Mev $= 270$ electron masses

$M_{\mu^\pm} = 107$ Mev $= 210 \pm 4$ electron masses

Nuclear Spins:

| | |
|---|---|
| electron | ½ |
| neutron | ½ |
| proton | ½ |
| deuteron | 1 |
| tritium, $H^3$ | ½ |
| $Li^6$ | 1 |
| $Li^7$ | 3/2 |
| $Be^9$ | 3/2 |
| $B^{10}$ | 3 |
| $B^{11}$ | 3/2 |

Miscellaneous Constants, in Alphabetical Order:

Curie = $3.71 \times 10^{10}$ disintegrations/sec
Dose (see p. 18)
e.v. = $1.6 \times 10^{-12}$ erg
e.v. = 8100 $cm^{-1}$ for photon
e.v. = $11,600^\circ$ Abs. (setting 1 e.v. = kT)

k, propagation constant for electron $= \sqrt{\frac{2m}{\hbar^2}(KE \text{ in ergs})} = 0.51 \times 10^8 \sqrt{(KE \text{ in e.v.})}$

k, propagation constant for proton $= \sqrt{\frac{2M_p}{\hbar^2}(KE \text{ in ergs})} = 2.12 \times 10^{13} \sqrt{(KE \text{ in e.v.})}$

$\ell_R$ = radiation length (p. 47)
$\ell_R$ = 330 m in NTP Air
= 9.7 cm in aluminum
= .517 cm in lead
$\ell_P = \frac{9}{7} \ell_R$ = mean free path for pair production (p. 49)

$L_{air}$ = 57 g/$cm^2$. Geometrical collision length for air nuclei (p. 220)
Mev = $1.601 \times 10^{-6}$ erg
$R_{nucleus} = 1.5 \times 10^{-13} A^{1/3}$ (p. 6)
r, roentgen = that X-ray dose which, passing through STP air, leaves 83 ergs/g, or liberates 1 esu of positive ions per $cm^3$
Rydberg: $R_y$ = 13.52 e.v. = Rhc
R = 109,737 $cm^{-1}$, Rydberg constant for infinite mass
T (half-life, gamma decay, p. 96)
Velocity of thermal neutron = $2.74 \times 10^5$ cm/sec (thermal = $300^\circ$ Abs.) $= \sqrt{\frac{3kT}{m}}$
Velocity of thermal electron = $1.117 \times 10^8$ cm/sec
$x_0$ = 1030 g/$cm^2$, depth of standard atmosphere at sea level
Year = $3.16 \times 10^7$ seconds
$\lambda_{\text{thermal neutron}} = 1.81 \times 10^{-8}$ cm (thermal = $300^\circ$ Abs.)

$\lambda$ = (12340/electron volts) in Angstroms, for photon
$\rho_{air} = 1.225 \times 10^{-3}$ g/$cm^2$
$\sigma_{\text{Thompson}} = 0.66 \times 10^{-24}$ $cm^2$
$\tau$ (<u>mean</u> life; radioactive families, p. 17; gamma decay, p. 96)

# INDEX